Piezoelectric Materials and Devices - Practice and Applications

Piezoelectric Materials and Devices - Practice and Applications

Editor

Hitesh Dave

scitus
academics

Piezoelectric Materials and Devices - Practice and Applications
Edited by **Hitesh Dave**

ISBN: 978-1-68117-269-9
Library of Congress Control Number: 2016937437

SCITUS Academics LLC,
www.scitusacademics.com
Box No. 4766, 616 Corporate Way,
Suite 2, Valley Cottage,
NY 10989

Notice

Printed in United States of America on Acid Free Paper

Preface

Piezoelectric materials are materials which, when mechanically stressed, can generate electricity. They also work in the opposite direction, generating a strain by applying an electric field. The piezoelectric effect describes the relationship between electrical stress in solids and mechanical stress. The piezoelectric effect can be described in physics as the connection between electrostatics and mechanics. The piezoelectric effect was discovered in 1880 by the Jacques and Pierre Curie brothers. One of the first applications of the piezoelectric effect was an ultrasonic underwater detector developed during the First World War. The piezoelectric effect only occurs in non-conductive materials.

Focusing on the piezoelectric materials such as piezoelectric stacks in level sensors, pressure sensors, machining processes, acoustic wave velocity measurement, and non-destructive aeronautical structure testing; this book should be a valuable tool not only for researchers, but also for professional engineers, students and other experts in a variety of disciplines, both academic and industrial, who seek to better understand what has been done in the field recently and what kind of open problems exist.

Table of Contents

Chapter 1

The Application of Piezoelectric Materials in Machining Processes

Saeed Assarzadeh[1] and Majid Ghoreishi[1]

[1] Department of Mechanical Engineering, K. N. Toosi University of Technology, Tehran, Iran

1. INTRODUCTION

In 1880, Jacques and Pierre Curie discovered that pressure generates electrical charges in a number of crystals such as Quartz and Tourmaline, calling this phenomenon the "piezoelectric effect". Later, they noticed that electrical fields can also deform piezoelectric materials, showing an inverse effect comparing to their first observation. Nowadays, the effect has gained considerable practical attractions in miscellaneous industrial applications including machining processes as a broad range within manufacturing methods in our competitively global technologies.

This chapter focuses mainly on the usages of piezoelectric materials in two differently common practiced machining operations adopted by both manufacturers and research scholars.

The first study pays attention to the role of piezoelectric transducers as the core system in producing ultrasonic waves in ultrasonic machines. Ultrasonic machining (USM) process, as one of the popular nontraditional machining processes, is capable of chip removal from every brittle material, whether conductive or nonconductive, susceptible to failure under mechanical loads in conventional machining processes of whatsoever hardness, such as ceramics, glass, porcelains, etc. The viability and effectiveness of piezoelectric transformers with high rate of electro-mechanical conversion compared to their old magnetostrictive counterparts are described and analyzed. The USM process capabilities and applications are also succinctly introduced.

The second is the application of quartz as a piezoelectric material in dynamometers used to measure forces and torques during conventional machining processes, like turning, milling, drilling, and so on. The basic principles and features of how a piezoelectric-based dynamometer works are discussed along with the need to measure forces and torques through dynamometry.

2. ULTRASONIC MACHINING (USM) OF MATERIALS

Ultrasonic machining is an economically viable operation by which a hole or a cavity can be pierced in hard and brittle materials, whether electric conductive or not, using an axially oscillating tool. The tool oscillates with small amplitude of 10-15 μm at high frequencies of 18-40 KHz to avoid unnecessary noise and being above the upper frequency limit of the human ear, justifying the tem "ultrasonic" [1,2].

During tool oscillation, abrasive slurry (B4C and SiC) is continuously fed into the working gap between the oscillating tool and the stationary WP. The abrasive particles are, therefore, hammered by the tool into the WP surface, and consequently abrading the WP into a conjugate image of the tool form. Moreover, the tool imposes a static pressure ranging from 1N to some kilograms depending on the size of the tool tip, see Fig. 1. This static pressure is necessary to sustain the tool feed during machining. Owing to the fact that the tool oscillates and moves axially, USM is not limited to the production of circular holes. The tool can be made to the shape required, and hence extremely complicated shapes can be produced in hard materials. Beside machining domain, US techniques are applied in nondestructive testing (NDT), welding, and surface cleaning, as well as diagnostic and medical applications.

2.1. Elements of Process

The USM equipment shown in Fig. 2 has a table capable of orthogonal displacement in X and Y directions, and a tool spindle and carrying the oscillating system moving in direction Z perpendicular to the X-Y plane. The machine is equipped with a HF generator of a rating power of 600 W, and a two-channel recording facility to monitor important machining variables (tool displacement Z and oscillation amplitude ξ). A centrifugal pump is used to supplement the abrasive slurry into the working zone. Fig. 3 shows schematically the main elements of the equipment, which consist of the oscillating system, the tool feeding mechanism, and the slurry system.

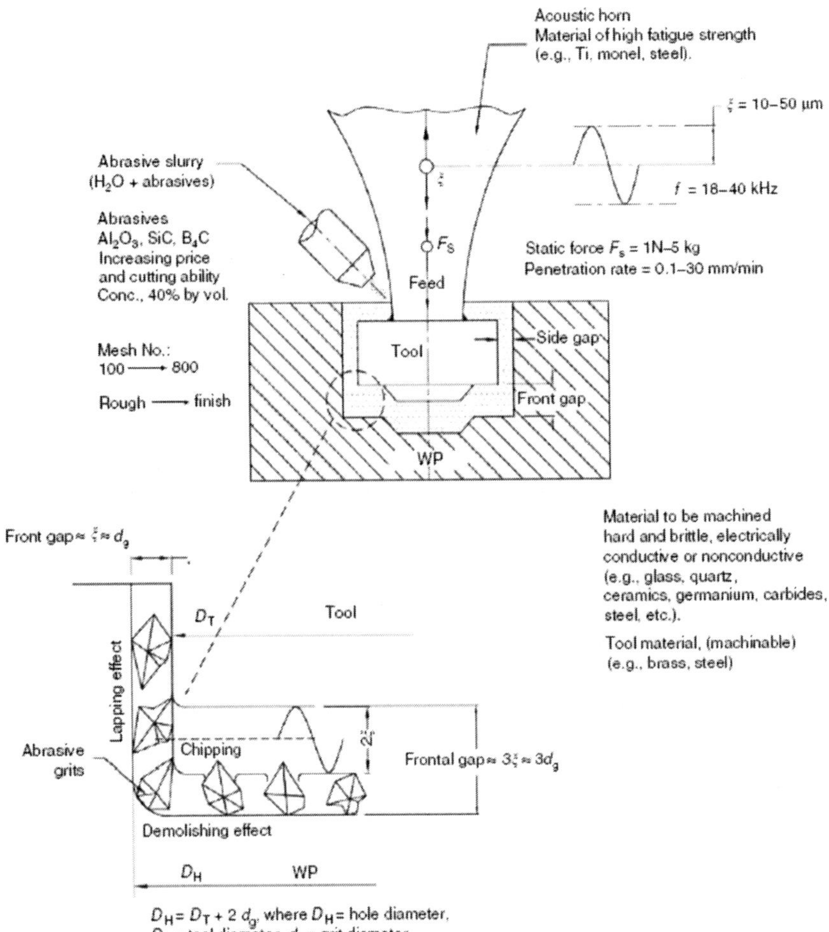

Figure 1. Characteristics of the USM process [1]

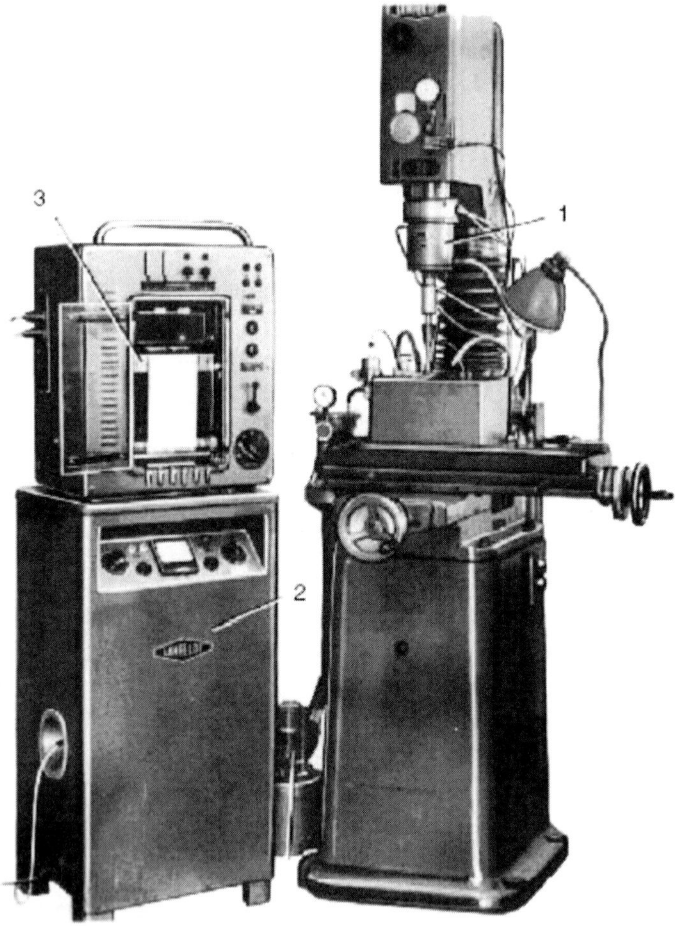

Figure 2. USM equipment [1]

2.1.1. Oscillating system and magnetostriction effect

The core element of each US machine, the oscillating system, includes a transducer in the acoustic head, a primary horn, and a secondary acoustic horn (see Fig. 4).

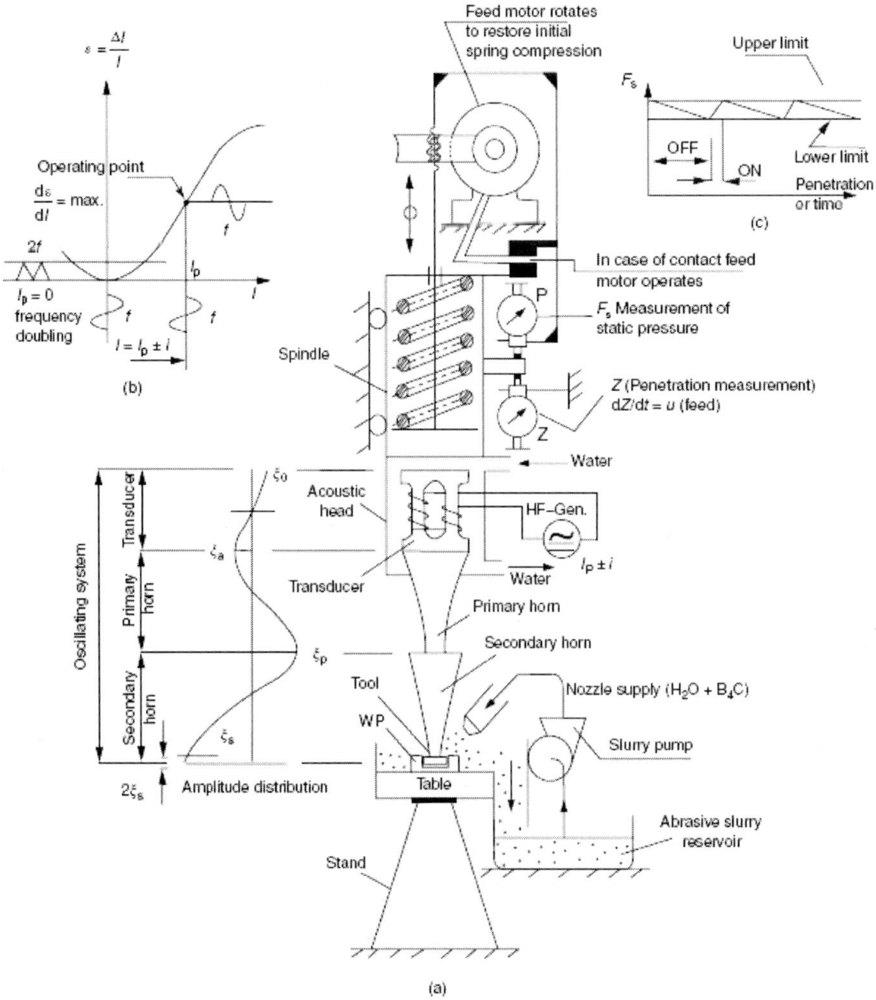

Figure 3. The schematic of complete vertical USM equipment [1]

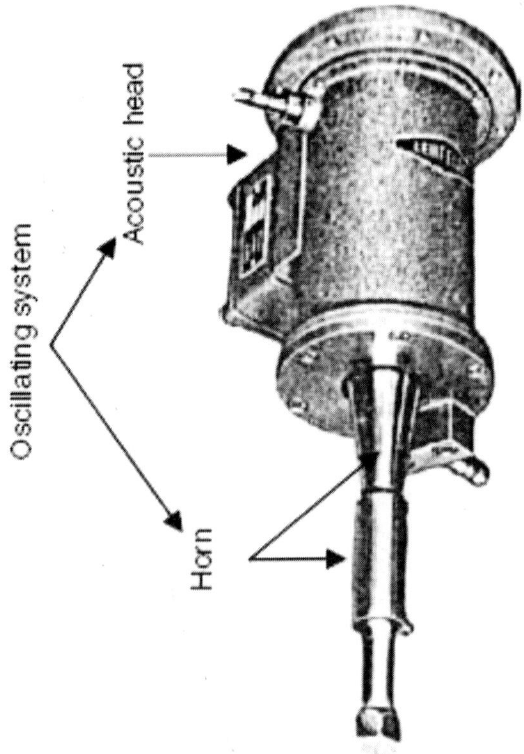

Figure 4. The oscillating system of USM equipment [1]

2.1.2. Acoustic transducer

This transforms electrical energy to mechanical energy in the form of oscillations. Magnetostrictive transducers are generally employed in USM, but piezoelectric ones may also be used.

The magnetostriction effect was first discovered by Joule in 1874. According to this effect, in the presence of an applied magnetic field, ferromagnetic metals and alloys change in length. The deformation can be positive or negative, depending on the ferromagnetic material. An electric signal of US-frequency f_r is fed in to a coil that is wrapped around a stack made of magnetostrictive material (iron-nickel alloy). This stack is made of laminates to minimize eddy current and hysteresis losses; moreover, it must be cooled to dissipate the generated heat (Fig. 3a). The alternating magnetic field produced by the HF-ac generator causes the stack to expand and contract at the same frequency.

To achieve the maximum magnetostriction effect, the HF-ac current i must be superimposed on an appropriate dc premagnetizing current I_p that must be exactly adjusted to attain an optimum or working point. This point corresponds to the inflection point ($d^2\varepsilon/dI^2=0$) of the magnetostriction curve, (Fig 3b). Without the application of premagnetizing direct current I_p, it is evident that the magnetostriction effect occurs in the same direction for a given ferromagnetic material irrespective of the field polarity, and hence the deformation will vary at twice the frequency $2f_r$ of the oscillating current providing the magnetic field (Fig. 3b). Therefore, the premagnetizing direct current I_p has the following functions:

- When precisely adjusted, it provides the maximum magnetostriction effect (maximum oscillating amplitude)
- It prevents the frequency doubling phenomenon

If the frequency of the ac signal, and hence that of the magnetic field, is tuned to be the same as the natural frequency of the transducer (and the whole oscillating system), so that it will be at mechanical resonance, then the resulting oscillation amplitude becomes quite large and the exciting power attains its maximum value.

2.1.3. Transducer length

The resonance condition is realized if the transducer length is *l*, which is equal to half of the wave length, λ (or positive integer number *n* of it).

Therefore,

$$l = \frac{n}{2}\lambda = \frac{\lambda}{2}, \quad if \;\; n = 1$$

and

$$\lambda = \frac{c}{f_r} = \frac{1}{f_r}\sqrt{\frac{E}{\rho}}$$

where

c = acoustic speed in magnetostrictive materials (m/s)

fr = resonant frequency (1/s)

E, ρ = Young's modulus (MPa) and density (kg/m^3) of the magnetostrictive material

Hence,

$$l = \frac{c}{2f_r} = \frac{1}{2f_r}\sqrt{\frac{E}{\rho}}$$

2.1.4. Piezoelectric transducers

A main drawback of magnetostriction transducers is the high power loss (η = 55%). The power loss is converted into heat, which necessitates the cooling of the transducer. In contrast, piezoelectric transducers are more efficient (η = 90%), even at higher frequencies (f = 25-40 KHz). Piezoelectric transducers utilize crystals like quartz and lead titanate-zirconate that undergoe dimensional changes proportional to the voltage applied. Similar to magnetostrictors, the length of crystal should be equal to half the wavelength of the sound in the crystal to produce resonant condition. At a frequency of 40 KHz, the resonant length *l* of the quartz crystal (E = 5.2×10^4 MPa, ρ = 2.6×10^3 kg/m^3) is equal to 57 mm. Sometimes a polycrystalline ceramic like barium titanate is used.

Piezoelectric transformers were first introduced into modern ultrasonic machines in the late 1960s. In 1970, Tyrrell [3] has described such a system. In essence, piezoelectric transducers are composed of small particles bound together by sintering; undergoing polarization by heating above the Curie point and placing it in an electric field such that orientation is preserved on cooling. A disc of the piezoelectric material which has a very high electromechanical conversion rating is sandwiched between two thick metal plates to form the ultrasonic horn. When a current of fixed frequency is fed to the horn the whole system is found to vibrate at some resonant frequency along the longitudinal axis; acoustically the motion is equal to one half a wavelength.

2.2. The Advantages and Disadvantages of Usm Process [4, 5]

Advantages

Some of the special priorities can be mentioned as follow:

- Intricate and complex shapes and cavities in both electric and nonelectric materials can be readily machined ultrasonically

- As the tool exhibits no rotational movement, the process is not limited to produce circular holes
- High dimensional accuracy and surface quality
- Especially, in the sector of electrically nonconductive materials, the USM process is not in competition with other nontraditional machining processes regarding accuracy and removal rates
- Since there is no temperature rise of the WP, no changes in physical properties or microstructure whatsoever can be expected
- However, the USM process has some disadvantages listing below.

Disadvantages:

- The USM is not capable of machining holes and cavities with a lateral extension of more than 25-30 mm with a limited depth of cut
- The tool suffers excessive frontal and side wear when machining conductive materials such as steels and carbides. The side wear destroys the accuracy of holes and cavities, leading to a considerable conicity error.
- Every job needs a special high-cost tool, which adds to the machining cost
- High rate of power consumption
- In case of blind holes, the designer should not allow sharp corners, because these cannot be produced by the USM.

2.3. The Applications of USM

It should be understood that the USM is generally applied to machining shallow cavities and forms in hard and brittle materials having a surface area not more than 10 cm^2. Some typical applications of USM are as follow:

- Manufacturing forming dies in hardened steel and sintered carbides
- Manufacturing wire drawing dies, cutting nozzles for jet machining applications in sapphire, and sintered carbides
- Slicing hard brittle materials such as glass, ceramics, and carbides
- Coining and engraving applications

- Boring, sinking, blanking,, and trepanning
- Thread cutting in ceramics by rotating the tool or the WP

Figure 5 illustrates some of the products produced ultrasonically.

a. Engraving a medal made of agate

b. Piercing and blanking of glass

c. Producing a fragile graphite electrode for EDM

d. Sinking a shearing die in hardened steel or WC

e. Production of outside contour and holes of master cutters made of zirconium oxide (ZrO_2) of a textile machine

f. Drilling fine holes Φ = 0.4 mm in glass

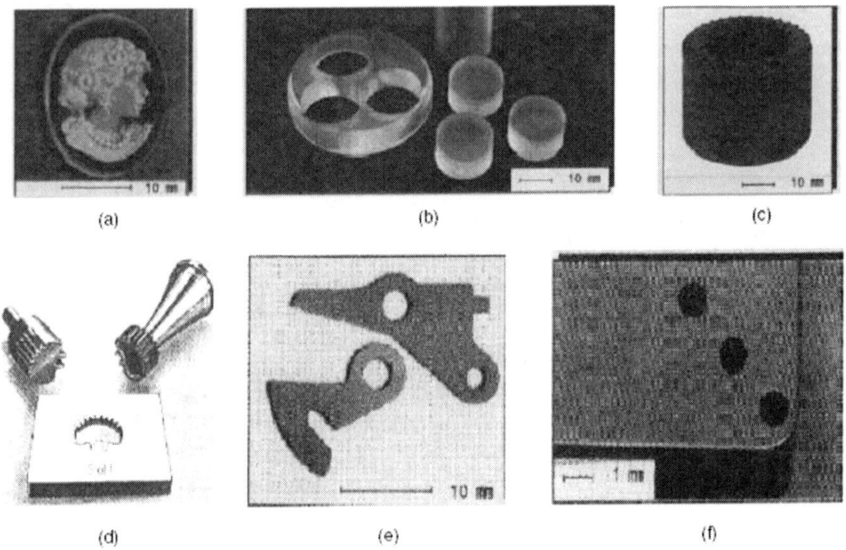

Figure 5. Some typical products by USM [1]

3. DYNAMOMETRY IN CONVENTIONAL MACHINING PROCESSES

During machining, the cutting tool exerts a force on the WP as it removes the machining allowance in the form of chips. Empirical values for estimating the cutting forces are no longer sufficient to reliably establish the optimum machining conditions. Depth of cut, feed rate, cutting speed, WP materials, tool material and geometry, and cutting fluid are just a few of the machining parameters governing the amplitude and direction of the cutting force.

The optimization of a machining process necessitates accurate measurement of the cutting force by a special device called a machine tool dynamometer, capable of measuring the components of the cutting force in a given coordinate system. It is a useful and powerful tool employed in a variety of applications in engineering research and manufacturing. A few examples of these applications are [1]:

- Investigating into the machinability of materials
- Comparing similar materials from different sources
- Comparing and selecting cutting tools
- Determining optimum machining conditions
- Analyzing causes of tool failure
- Investigating the most suitable cutting fluids
- Determining the conditions that yield the best surface quality
- Establishing the effect og fluctuating cutting forces on tool wear and tool life

The machine tool dynamometer is not standard equipment or a device that can be used on every machine. Rather, it is equipment especially designed to fulfill some desired requirements that adapt a specific machine type operating at a specific range of machining conditions.

3.1. Piezoelectric (Quartz) Dynamometers

Of the numerous piezoelectric materials, quartz is by far the most suitable one for force measurement, because it is stable material with constant properties. In its crystalline form, quartz is anisotropic, in that its material properties are not identical in all directions. Depending on the position in which they are cut out of the crystal, disks are obtained that are:

1. Sensitive only to pressure (longitudinal effect), Fig. 6a, which measure the main force component F_z (brown)

2. Sensitive only to shear in one particular direction (shear effect), Fig. 6b, which measures components F_x (blue) and F_y (green), perpendicular to F_z, as well as the torque M_z (red). Figure 6c illustrates the generalized multi-components with reference to a Cartesian coordinate system.

The piezoelectric force measuring principle differs fundamentally from their old traditional counterparts, the strain and displacement based dynamometers, in that it is an active system. When a force acts on a quartz element, a proportional electric charge appears on the loaded surfaces, meaning that it is not necessary to measure the actual deformation.

In piezoelectric dynamometers, the deflection is not more than a few micrometers at full load, whereas with conventional systems, several tenths of a millimeter may be needed. Thus, piezoelectric dynamometers are very stiff systems and their resonant frequency is high, so that even rapid events can be measured satisfactorily. Moreover, the individual components of the cutting force can be measured directly, eliminating any interference between measuring channels. Quartz dynamometers require no zero adjustment or balancing of the bridge circuit. It is just a matter of pressing a button, being ready for duty. The outstanding features of quartz dynamometers are [1, 6]:

- High rigidity, hence high resonant frequency
- Minimal deflections (few micrometers at full load)
- Wide measuring range
- Linear characteristics, free of hysteresis
- Lowest cross talk (typically under 1%)
- Simple in operation and without need for bridge balancing
- Compact design
- Unlimited life expectancy

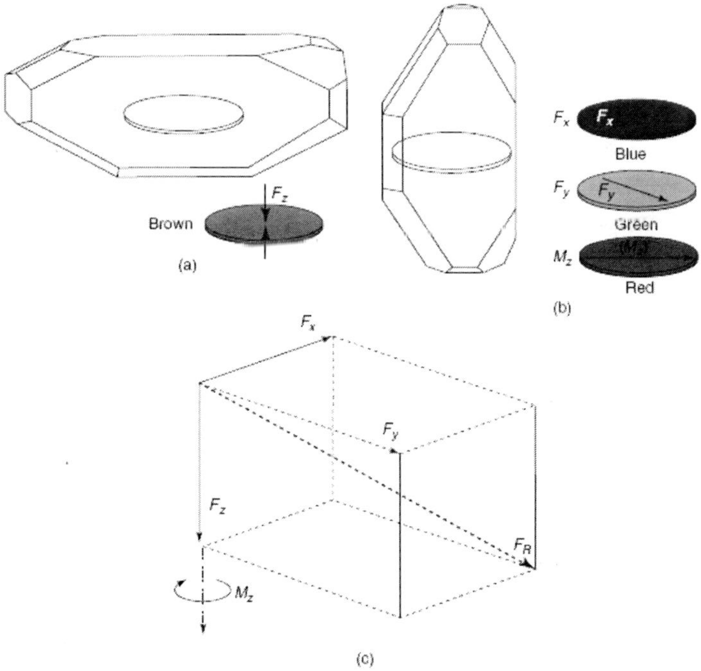

Figure 6. Disks of quartz crystals. (a) Pressure-sensitive, (b) shear-sensitive, (c) multi-components in reference to a Cartesian coordinate system [6]

3.2. Typical Piezoelectric Dynamometers

Piezoelectric dynamometers are efficiently used on the majority of machine tools. Three application examples are described below with their corresponding setup.

1. *Two-component piezoelectric drilling dynamometers.* Figure 7 illustrates a two-component drilling dynamometer in which shear-sensitive discs are organized in a circle with their shear-sensitive axes oriented to respond to the torque M_z (red), whereas pressure-sensitive disks are arranged and oriented to measure the trust load F_z (brown). A high preload is necessary because the shear forces must be transmitted by friction to measure the torque.

The two-component dynamometer, shown in Fig. 7, is suited for operations including drilling, thread cutting, countersinking, reaming, and so on. Torques and forces acting when machining holes from less than 1 mm to over 20 mm diameter can be measured satisfactorily by this dynamometer. A record of M_z and F_z is illustrated in Fig. 7, from which it is clearly seen that F_z rises steeply at the beginning (entry of tool chisel), followed by the gradual rise of the M_z component, as the latter is more affected by the force acting on the two drill lips.

1. *Three-component piezoelectric turning dynamometer.* This model includes several shear-sensitive quartz, with their shear-sensitive axes oriented to measure F_x (blue ring) and F_y (green ring), respectively. Their shear sensitive-axes are inclined to each other at an angle of 90^0, and both are contained in a housing to form a two-component force measurement element for F_x and F_y(Figure 8). Pressure-sensitive quartz disks are contained in a single housing to form a single-component force-measuring element for F_z (brown ring). Another alternative is illustrated in the construction shown in Figure 8, where three separate elements for measuring F_x, F_y, and F_z are sandwiched under high preload between a base plate and a top plate. The dynamometer is mounted on the lathe slide in place of a cross-slide. A record of the three components is shown also in the same figure, from which it is clear that F_x=F_y, meaning that the cut is preferred at an approach angle $\chi=45^0$.

2. *Three-component piezoelectric milling or grinding dynamometer.* Whole quartz rings may be employed. Two-shear-sensitive quartz pairs, for F_x (blue) and F_y (green), and a pressure-sensitive pair for F_z (brown), can be assembled in a common housing to form a three-component force-measuring element (Figure 9). The pressure-sensitive quartz are arranged in the middle so that they lie in the neutral axis under bending. During milling and grinding, the application point of the force varies a great deal. Consequently, dynamometers having four piece three-component force-measuring elements are employed. All the x, y, and z channels respectively are paralleled electrically. This makes the measurement independent of the momentary force application point. For bigger work, two dynamometers paralleled electrically and mechanically may be employed together. This system measures correctly independent of the point of force application. A typical output of the three-component milling dynamometer is shown in Fig. 9. The milling process has been performed under the following conditions:

- Status: Up milling
- Cutter diameter = 63 mm, helix β = 30^0, n = 90 rpm, Z = 12 teeth
- Feed u = 53 mm/min
- Depth of cut t = 3.5 mm

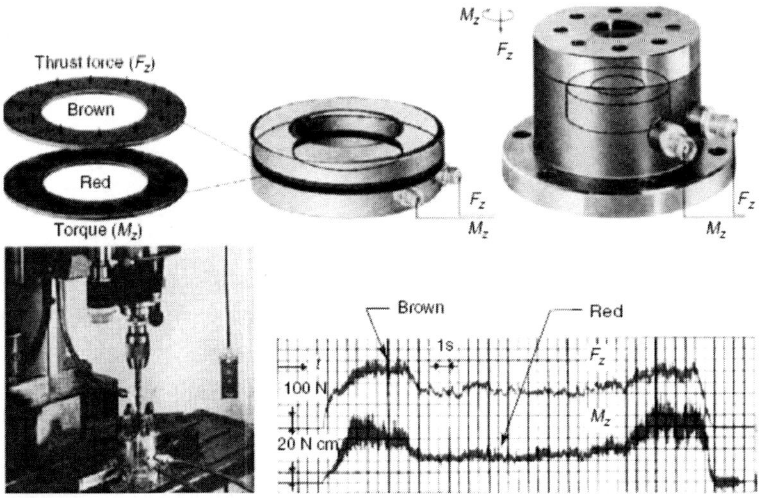

Figure 7. Two-component piezo drilling dynamometer [6]

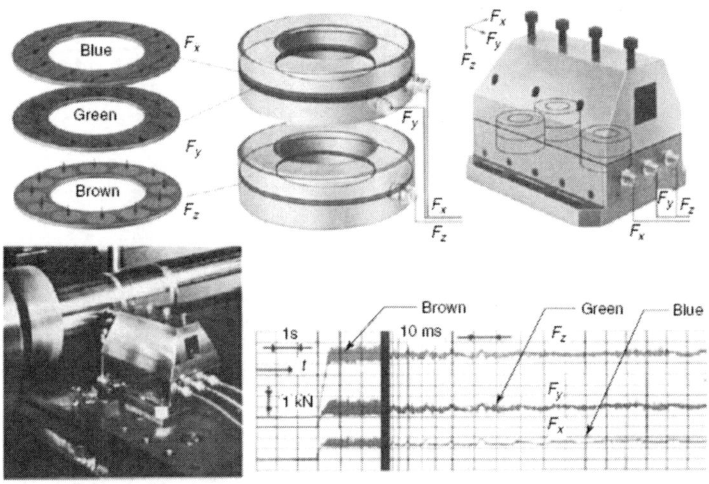

Figure 8. Three-component piezo turning dynamometer [6]

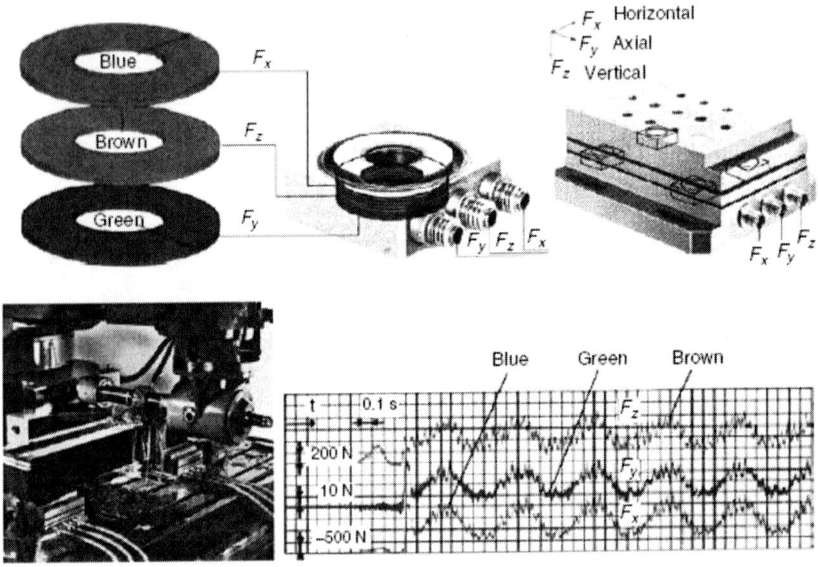

Figure 9. Three-component piezo milling dynamometer [6]

The severe periodic fluctuation in the measured forces is attributed to an eccentric motion of the cutter shaft. Superimposed are vibrations due to gearing of the machine. It is perfectly clear from the record that the setup shown is far from ideal. The force measure, therefore, sheds light on the machine tool behavior as well, and not just on the actual cutting operation.

4. CONCLUSION

In this chapter, attempt has been made to cover the basic details of utilizing piezoelectric materials in two different fields of machining processes. The first focused on the effectiveness of a piezoelectric transducer in producing ultrasonic waves as a cutting tool in ultrasonic machining processes. Compared to their old magnetostriction counterparts, the piezo ones demonstrate higher electro-mechanical efficiency. Besides, they are more compact as well as being simpler in design and operation. The second emphasizes the unique suitability of piezoelectric dynamometers in measuring various components of forces and torques generated during different kinds of conventional machining processes. The importance of measuring forces and torques in an on-line manner during every traditional machining operation for the purposes of modeling and optimization makes the use of a precise piezoelectric dynamometer an inseparable part in manufacturing industries as well as academic domains.

REFERENCES

1. Youssef Helmi A., El-Hofy, H., Machining technology: Machine tools and operations, CRC Press, 2008978-1-42004-339-6
2. J. A Mcgeough, Advanced methods of machining, Chapman and Hall, London and New York, 19880-41231-970-5
3. W. R Tyrrell, Rotary ultrasonic machining, Soc. Manufacturing Eng., USA, Paper MR70-5161970110
4. El-H Hofy, Advanced machining processes: Nontraditional and hybrid machining processes, McGraw Hill Book Com., 20050-07145-334-2
5. P. C Pandey, H. S Shan, Modern machining processes, Tata McGraw Hill Book Com., 20040-07096-553-6
6. http://www.Kistler.com

Chapter 2

Revisiting the Characterization of the Losses in Piezoelectric Materials from Impedance Spectroscopy at Resonance

Amador M. González [1,*], Álvaro García [2], César Benavente-Peces [1] and Lorena Pardo [2]

[1]ETS Ingeniería y Sistemas de Telecomunicación, Campus Sur UPM, Madrid 28031, Spain

[2]Instituto de Ciencia de Materiales de Madrid-CSIC, Madrid 28049, Spain

ABSTRACT

Electronic devices using the piezoelectric effect contain piezoelectric materials: often crystals, but in many cases poled ferroelectric ceramics (piezoceramics), polymers or composites. On the one hand, these materials exhibit non-negligible losses, not only dielectric, but also mechanical and piezoelectric. In this work, we made simulations of the effect of the three types of losses in piezoelectric materials on the impedance spectrum at the resonance. We analyze independently each type of loss and show the differences among them. On the other hand, electrical and electronic engineers include piezoelectric sensors in electrical circuits to build devices and need electrical models of the sensor element. Frequently, material scientists and engineers use different languages, and the characteristic material coefficients do not have a straightforward translation to those specific electrical circuit components. To connect both fields of study, we propose the use of accurate methods of characterization from impedance measurements at electromechanical resonance that lead to determination of all types of losses, as an alternative to current standards. We introduce a simplified equivalent circuit model with electrical parameters that account for piezoceramic losses needed for the modeling and design of industrial applications.

Keywords: piezoelectric materials; ceramics; mechanical losses; dielectric losses; piezoelectric losses; electromechanical resonances; material characterization; impedance spectroscopy

1. INTRODUCTION

1.1. Losses in Piezoelectric Materials

New electronic devices need better components to get better performance and improved functionalities. In the case of devices using the piezoelectric effect, the heart of the device consists of a piezoelectric element: a crystal [1]; but in most cases, poled ferroelectric ceramics (piezoceramics) [2], polymers [3] or composite materials [4]. The crystal structure of piezoelectrics is non-centrosymmetric, but randomly-oriented polycrystals are centrosymmetric. An induced macroscopic non-centrosymmetry is needed for a polycrystal to be piezoelectric. This induced symmetry can be provided by an external electric field (poling) or mechanical action (stretching).

Among the devices commonly using piezoelectric materials, we can find numerous sensors and actuators, which are classically used in telecommunications, medicine or industrial quality control, but new applications in energy storage or energy harvesting are also being developed nowadays.

A piezoelectric material or device can be driven by an electric or mechanical stimulus and responds with both electrical and mechanical reactions.

System losses are defined as the rate of energy provided to the system that cannot be transformed into work. Usually, we call this loss of energy dissipation. This definition needs to be applied, and frequently rewritten, for every process involving energy conversion.

Friction, dielectric dissipation, damping, *etc.*, are related to losses in piezoelectric materials.

A piezoelectric material has losses originating from the dielectric response to an electrical field, the mechanical response to applied stress, or its piezoelectric motion, or its strain response to the electric field, or conversely, the charge or voltage generation as a response to the applied stress. Losses in piezoelectric materials result in sample heating or noise production. These effects are detrimental in many applications, and this is why the understanding of loss mechanisms and knowledge of the actual value of the loss in the material becomes a key issue for the design of devices. Furthermore, the control of the loss mechanism is needed to optimize the efficiency of the electromechanical transduction and, consequently, the device performance. For this reason, many authors have treated this topic in ferro-piezoelectric ceramics and related materials [5,6,7,8,9,10,11,12,13,14,15]. To date, there is not a full agreement concerning the origin of the piezoelectric losses. Most authors consider that features in piezoceramic materials that are responsive to both the elastic and electric fields (such as certain point defects in crystals, non-180° ferroelectric-ferroelastic domains, second phase in the piezoelectric polymer materials, *etc.*) are a source of piezoelectric relaxation through the coupling of the elastic and dielectric losses.

Here, to study losses in piezoelectric materials we will take a harmonic stimulus and response. The use of a single harmonic stimulus allows us to use complex numbers for the physical constants. In this case, we separate the in-phase response as the real part of the constant and the in-quadrature response as the imaginary part.

Along this work, we will use the notation recommended in the IEEE Standard on Piezoelectricity [16].

1.2. Imaginary Part versus Phase Angle

We will take Hooke's law for elastic materials under a single harmonic stimulus as an example. In this case, S will be the unitary deformation, T the strength applied over the specimen, s the proportionality factor among them, the compliance of the material and ϑ the phase angle, which will allow us to characterize the time delay between the stimulus and the response (Figure 1). Then, for a material with mechanical losses, we obtain Equation (1):

$$S\cos(\omega t + \theta) = s\,T\cos(\omega t) \tag{1}$$

where we can expand the left part of the equation as Scosθcosωt+Ssinθsinωt. In order to solve this equation, we will use complex numbers to represent magnitudes and proportionality factors, then $Se^{i(\omega t+\theta)} = se^{i\theta}Te^{i\omega t}$, where we can write every quantity in the form (*real + i imaginary*):

$$\begin{aligned} Se^{i(\omega t+\theta)} &= S\cos(\omega t + \theta) + i\,S\sin(\omega t + \theta) \\ se^{i\theta} &= s\cos\theta + i\,s\sin\theta \\ Te^{i\omega t} &= T\cos\omega t + i\,T\sin\omega t \end{aligned} \tag{2}$$

Then, taking the real parts, we found:

$$S\cos\theta\cos\omega t - S\sin\theta\sin\omega t = sT\cos\theta\cos\omega t - sT\sin\theta\sin\omega t \tag{3}$$

This can be expressed alternatively as:

$$= |s|\,e^{i\theta} = s\cos\theta + \;i\;s\sin\theta = s' - i\;s'' \tag{4}$$

As can be seen, we write an explicit minus sign between the real and imaginary parts. This notation allow us to define *s''* as positive. This convention is the same used for dielectric losses.

To understand this convention, we must rewrite the unitary deformation as:

$$S\cos(\omega t + \theta) = S\cos\omega(t - \tau) \tag{5}$$

In this way, we use the new variable, τ, as the time delay. This time delay must be positive, due to the causality principle, and then, the phase delay becomes $\vartheta = -\omega\tau \le 0$. Moreover, $\theta \geq -\frac{\pi}{2}$, because for lower values than this, the term we find $s\cos\vartheta < 0$, and the response should be opposite the stimulus. All of this implies:

$$-\frac{\pi}{2} \leqslant \theta \leqslant 0, \; and \; s'' = s\sin\theta \leqslant 0 \tag{6}$$

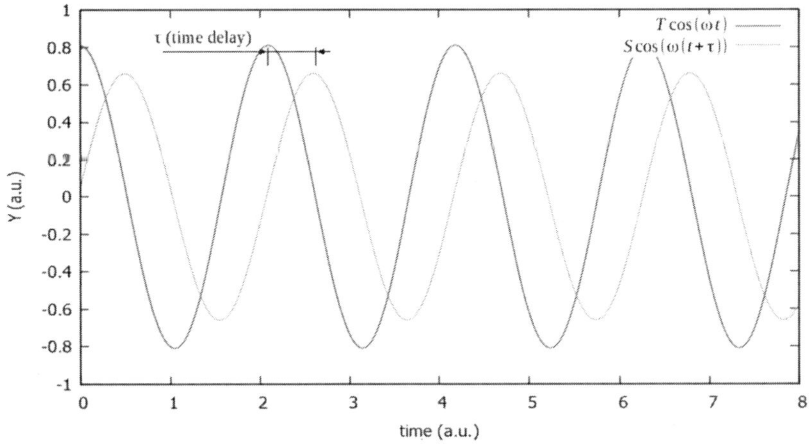

Figure 1. Single harmonic signal and delayed or lossy response.

1.3. Dielectric and Mechanical Losses

For a wide range of piezoelectric applications, under adiabatic and linear response conditions, we can use the well-known linear equations of the piezoelectricity:

$$\begin{aligned} S &= s\,T + d\,E \\ D &= \epsilon E + dT \end{aligned} \tag{7}$$

where *S*, *T* and *s* were explained above, *E* and *D* are the electric field and electric displacement, respectively, ϵis the dielectric permittivity and *d* is the charge piezoelectric constant. All of these parameters must be described by complex numbers, *z*, and hence, we can define their corresponding loss tangent as:

$$\tan\delta_z = \frac{Im\,(z)}{Re\,(z)} \tag{8}$$

According to Holland [5], the imaginary parts of the elastic compliance and dielectric permittivity represent the mechanical and electrical losses. Both of them must be explained from the point of view of a delay in the response and may be used to calculate the energy lost as friction in the mechanical equation or as the Joule effect in the electrical one. Then, the three characteristic constants must be written as:

$$\begin{aligned} \hat{s} &= s' - i\, s'' \\ \hat{\epsilon} &= \epsilon' - i\, \epsilon'' \\ \hat{d} &= s' - i\, d'' \end{aligned} \tag{9}$$

or, as an alternative:

$$\begin{aligned} \hat{s} &= s'(1 - i\, \tan\delta_s) \\ \hat{\epsilon} &= \epsilon'(1 - i\, \tan\delta_\epsilon) \\ \hat{d} &= d'(1 - i\, \tan\delta_d) \end{aligned} \tag{10}$$

where the three imaginary parts, as well as the losses' tangent **are positive numbers.**

1.4. The Matrix of Coefficients

All of these quantities are tensors (second order for the elastic magnitudes, S_{ij} and T_{ij}, and the dielectric permittivity, ϵTik, third order for the piezoelectric constant, d_{ijk}, and forth order for the elastic compliance, sEijkl), and we must rewrite the linear equations as:

$$\begin{aligned} S_{ij} &= s^E_{ijkl} T_{kl} + d_{kij} E_k \\ D_i &= d_{ikl} T_{kl} + \epsilon^T_{ik} E_k \end{aligned} \tag{11}$$

In order to get the simplest representation of the mechanical constants and magnitudes, we will make the usual index contraction (i,j = 1,2,3; p,q = 1,...6); thus, $s^E_{ijkl} = s^E_{pq},\ d_{ijk} = d_{pk},$ and then, the final form for the piezoelectric equations will be:

$$\begin{aligned} S_p &= s^E_{pq} T_q + d_{pk} E_k \\ D_i &= d_{iq} T_q + \epsilon^T_{ik} E_k \end{aligned} \tag{12}$$

where $S_{ij} = S_p,\ (i = j \rightarrow p = 1,2,3)$ corresponds to extensional stresses along the primary axes and $2S_{ij} = S_p\,(i \neq j \rightarrow p = 4,5,6)$ corresponds to shear stresses.

For every crystalline class, due to crystal symmetries, we have different matrices of coefficients. In this work, we will use only those corresponding to electrically-poled piezoceramics (6-mm group of symmetry) [17], which can be used for other kind of materials (such as composites or ferroelectric polymers).

In this case, Direction 3 is assumed as that corresponding to the polar axis, and Directions 1 and 2 are equivalents. Then, for the non-zero coefficients, we find the following relations: $s_{11}=s_{22}$; $s_{13}=s_{23}=s_{31}=s_{32}$; $s_{44}=s_{55}$; $s_{66}=2(s_{11}-s_{12})$; $d_{31}=d_{32}$; $d_{15}=d_{24}$; $\epsilon_{11}=\epsilon_{22}$.

Additionally, the whole matrix has only 10 independent elements (five elastic, three piezoelectric and two dielectric coefficients):

$$\begin{pmatrix} S_1 \\ S_2 \\ S_3 \\ S_4 \\ S_5 \\ S_6 \\ D_1 \\ D_2 \\ D_3 \end{pmatrix} = \begin{pmatrix} s_{11} & s_{12} & s_{13} & \cdot & \cdot & \cdot & \cdot & \cdot & d_{31} \\ s_{12} & s_{11} & s_{13} & \cdot & \cdot & \cdot & \cdot & \cdot & d_{31} \\ s_{13} & s_{13} & s_{33} & \cdot & \cdot & \cdot & \cdot & \cdot & d_{33} \\ \cdot & \cdot & \cdot & s_{44} & \cdot & \cdot & \cdot & d_{15} & \cdot \\ \cdot & \cdot & \cdot & \cdot & s_{44} & \cdot & d_{15} & \cdot & \cdot \\ \cdot & \cdot & \cdot & \cdot & \cdot & s_{66} & \cdot & \cdot & \cdot \\ \cdot & \cdot & \cdot & \cdot & d_{15} & \cdot & \epsilon_{11} & \cdot & \cdot \\ \cdot & \cdot & \cdot & d_{15} & \cdot & \cdot & \cdot & \epsilon_{11} & \cdot \\ d_{31} & d_{31} & d_{33} & \cdot & \cdot & \cdot & \cdot & \cdot & \epsilon_{33} \end{pmatrix} \begin{pmatrix} S_1 \\ S_2 \\ S_3 \\ S_4 \\ S_5 \\ S_6 \\ D_1 \\ D_2 \\ D_3 \end{pmatrix} \quad (13)$$

1.5. Piezoelectric Losses

We have no reason to assume that the piezoelectric coefficients in the matrix are not complex quantities if both elastic and dielectric coefficients are complex. In fact, Holland [5] probes the necessity of using an imaginary part for the piezoelectric constants and found the limits imposed by thermodynamical considerations.

The simplest explanation is that the imaginary part corresponds to losses during the energy conversion. From the point of view of the piezoelectric coefficient *d*:

$$d = \frac{\partial S}{\partial E} = \frac{\partial D}{\partial T} \tag{14}$$

We can read these losses as a delay between stimulus and response (as we did above) or as a result of crossed interactions between the two forms of energy (frictional losses under electric fields and the increment of the electric resistance under mechanical stresses).

The ratio between energies gets us the efficiency of the mechanical conversion and is called the electromechanical coupling factor, defined as:

$$k^2 = \frac{W^2_{piezo}}{W_{elastic} W_{electric}} \tag{15}$$

where Wpiezo, Welastic and Welectric are their corresponding energies. This factor is strongly dependent on the conditions of the transduction (quasistatic, dynamic) and, of course, the coefficients involved. For dynamical conversion (as those due to a harmonic stimulus), the coupling factor can be found by means of the relation:

$$k^2_{ip} = \frac{d^2_{ip}}{s^E_{pq} \epsilon^T_{ij}} \tag{16}$$

We make use of complex quantities to include information about losses. It must be remarked that this works correctly only for single harmonic stimulus and response, as used with resonant methods of characterization, and only for the linear response range. In the rest of the cases, the representation of losses is more difficult, such that we must take into account non-linear effects, intermodulation, coupling between modes, *etc.*, but the fundamental physics of the material losses can be studied under the simplest conditions.

Although the three characteristic constants of the material (d_{ip}, s^E_{pq} and ϵ^T_{ij}) are independent, there are limits, described by Holland [5], for the values of their imaginary parts. As an example, for the length vibration of a bar poled in this direction, the corresponding constants will be d_{33}, s^E_{33}, ϵ^T_{33}, and the condition:

$$\left(d''_{33}\right)^2 \leqslant s^{E}_{11}{}'' \cdot \epsilon^{T}_{33}{}'' \tag{17}$$

Even more, the constants s33 and ϵ33 are different before and after the resonance and are related by the equations:

$$\begin{aligned} s^{D}_{33} &= s^{E}_{33}\left(1 - k^{2}_{33}\right) \\ \epsilon^{S}_{ii} &= \epsilon^{T}_{ii}\left(1 - k^{2}_{33}\right) \end{aligned} \tag{18}$$

where superscripts *E* and *T* correspond to "free" conditions (frequencies below that the resonance) and *D* and *T*correspond to "clamped" conditions (above the resonance). That means that for the relation between the three constants, in this case, the clamped compliance includes not only the mechanical response, but the interaction with the dielectric and piezoelectric one. We find the same behavior for the permittivity that includes the mechanical and piezoelectric responses. Their respective losses will be, of course, related.

Noticeably, the fact that piezoelectric coefficients of piezo materials have a complex form indicates the presence of extrinsic contributions to the piezoelectricity in these materials, regardless of the actual mechanisms involved [18].

Piezoelectric losses are the most controversial [13,19] and difficult to measure independently, because when the material deforms under the action of the piezoelectric effect, it also must endure mechanical loss. The hysteresis is also indicative of the loss. Alternatively to the resonance method, a number of methods were proposed for the measurement of the phase angle between strain and field [14,18]. These alternative measurements [14] yield piezoelectric loss values in the linear range that are in good agreement with iterative methods at the resonance [20,21].

2. MATERIAL CHARACTERIZATION FROM ELECTROMECHANICAL RESONANCES

One of the most common ways to obtain an easy and accurate characterization is the resonant method. It consists of getting the complex impedance spectrum of a resonator, including at least one electromechanical resonance. We consider the material sample as a resonant cavity (a propagation media and its boundaries). Knowing the geometrical factors of the resonant cavity and a mathematical model for the impedance function that takes into account this resonance, we can obtain the elastic, piezoelectric and dielectric coefficients involved in this particular resonance. For reasons explained later, in some cases, the description in terms of admittance is easier than the impedance representation. We must, of course, use the complex form for both, that is: $Z = R + i\,X$ and $Y = G + i\,B$.

The most general function describing the impedance of a sample shape as a thin plate in the neighborhood of a resonance is:

$$Z = \frac{1}{i\omega\epsilon\frac{S}{\ell}}\left(1 - k^2 F(\omega, s)\right) \tag{19}$$

where S is the area of the electrodes, ℓ the distance between them, ω the angular frequency and ϵ the dielectric permittivity. The dielectric capacitance of the sample will be $C=\epsilon S/\ell$, then we can observe that this relation corresponds to the characteristic impedance of a capacitor, $Z=(i\omega C)-1$, modified by the resonant term, $k^2F(\omega,s)$. The electromechanical coupling factor, k^2, was previously defined, and the function $F(\omega,s)$depends on the shape of the sample and the speed of the sound across the material (that implies a dependence on the elastic coefficient) and is responsible for the resonance.

2.1. Shapes and Modes

As we treat the sample as a resonant cavity, shape and dimension are the most important parameters to take into account for the study of the standing waves inside and getting information from them of the material.

The set of linear equations of the piezoelectricity is written under the assumption of the existence of an external force and an electric field (by means of a voltage in electrodes). Nevertheless, depending on the contour conditions, we can rewrite such linear equations as:

$$\begin{pmatrix} T \\ D \end{pmatrix} = \begin{pmatrix} c^E & e \\ e & \epsilon^S \end{pmatrix} \begin{pmatrix} S \\ E \end{pmatrix}, \begin{pmatrix} S \\ D \end{pmatrix} = \begin{pmatrix} s^E & d \\ d & \epsilon^T \end{pmatrix} \begin{pmatrix} T \\ E \end{pmatrix}$$
$$\begin{pmatrix} S \\ E \end{pmatrix} = \begin{pmatrix} s^D & g \\ -g & (\epsilon^T)^{-1} \end{pmatrix} \begin{pmatrix} T \\ D \end{pmatrix}, \begin{pmatrix} T \\ E \end{pmatrix} = \begin{pmatrix} c^D & h \\ -h & (\epsilon^S)^{-1} \end{pmatrix} \begin{pmatrix} S \\ D \end{pmatrix} \tag{20}$$

Then, we find five new constants: one mechanical, *c*, the stiffness, one dielectric, *β*, that we prefer to write as ϵ^{-1}, and three piezoelectric constants, *g* (called voltage piezoelectric constant), *h* and *e*.

2.2. The Iterative Method

To get a complete characterization of a piezoceramic, we must get a set of 10 different parameters, the independent elements of the characteristic matrix. Since these materials exhibit non-negligible losses, it is needed to use alternative characterization methods to that proposed by the current standards [16], which do not account for all losses. Numerous authors have developed such alternative modes, and the interested reader can find reference to them in [22]. New methods keep on being published on this topic; among others, those based on the iterative modification of finite elements simulations to match measurements in the best possible way are noticeable [23,24].

For determining these elements in a given resonant mode, we propose here Alemany's method [20,25] and the available free-ware to use it [26]. This fully-automatic iterative method, based on Smits´ one [27], solves numerically the analytical expression for the wave equation of a given mode of resonance to obtain the parameters involved in it. Typically, the directly calculated parameters from a resonance mode are one elastic, one dielectric and one piezoelectric parameter (except for the radial mode of disks, which provides two elastic constants). The selection, based on the standards for measurements [16], of three resonator shapes and four modes of resonance, together with the smart combination of the parameters directly calculated from these analytical expressions with the remaining parameters, allow us to get the whole set of complex material parameters [28]. This method of analysis of the complex impedance curves has been also applied to the determination of the parameters from overtone resonances in the radial [29] and thickness

[30] modes of thin disks to account for the frequency dependence of these parameters. Besides, the principles for the application of this method to the determination of the properties of self-standing films were developed and applied to lead zirconate titanate (PZT) cantilevers [31].

Using Alemany's iterative method, we get the complex values of every parameter and, then, information about losses and the relations between them. This procedure is not limited to piezoceramics, and we used it for the characterization of ferroelectric polymers (Figure 2) and composites.

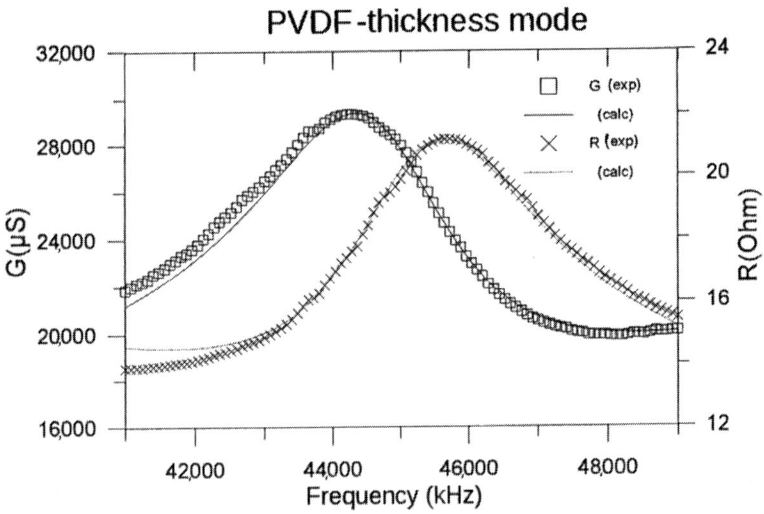

Figure 2. Example of the characterization of a piezoelectric polymer (PVDF: polyvinylidene fluoride).

The main limit of the resonance method consists of the coupling between modes. This problem cannot be solved as suggested by the standard IEEE-176 [16], because there is no a simple superposition of modes, but a coupling with energy exchange between them.

The impedance spectra of the shear plates used in standard characterization methods from resonance, in-plane poled and excited in thickness, always show additional peaks, satellite resonances, around that of the main resonance. These correspond to natural modes of vibration of plates, such as contour modes, and are unavoidable, since they are excited simultaneously to the shear mode. When using an alternative shear geometry [32], thickness poled and in-plane excited, it is possible, by tuning the aspect ratio of the sample (Length and width to thickness = *L*,*w*/*t*), through the reduction of the thickness, to obtain uncoupled shear modes in a periodical way, even below the standard ratio of *L*,*w*/*t* = 10. Accurate values of impedance around resonance and anti-resonance frequencies are required to determine accurately the material parameters, including losses, which can only be obtained from uncoupled modes. For the optimum sample aspect ratios obtained in the first four periods, the dispersion in the so-measured parameters are 0.4% for the k_{15} coupling factor and for the directly-calculated parameters, sE55, e15 and ϵS11, less than 2%, in their real part, and 10%, 12% and 13% in their imaginary part, respectively. This is a unique result to the authors' knowledge, given the difficulties encountered to measure even the real part of this coefficient by other methods [33]. An additional advantage of the use of the thickness poled shear plate for the full characterization of the piezoceramics is that, since it can be obtained from the disk after measuring it, in principle, this will allow the characterization from two resonators, a thin disk, thickness poled and a long rod or bar, length poled [34]. Finally, the higher consistency with the parameters is obtained from the thickness poled disk, since for both resonators, the spatial distribution of the polarization is identical. Due to this enhanced consistency, a finite element modelling, based on the full matrix characterization using thickness poled shear plates, was successfully tested for both the impedance spectra and displacement patterns. Good agreement with the experimental, electrical and laser interferometry measurements, respectively, was achieved in a range between 325 kHz and 1.2 MHz.

2.3. Plotting Data

There are no standards for graphing the piezoelectric characterization in resonant mode. Most of the time, researchers represent as-measured data from the impedance analyzer, *i.e.*, the absolute value, $|Z|$, and the argument, θZ, of the impedance. Using this plot, we can get a good idea about the electric anti-resonance, which is clearly shown by the maximum electrical impedance.

For some vibrational modes, as thickness extensional modes, where the direction of the electrical excitation and movement are parallel, due to the boundary conditions, the mechanical resonance corresponds to the maximum of the impedance, *i.e.*, the electric anti-resonance. However, for modes where the mechanical movement is perpendicular to the direction of the electrical excitation, e.g., length expansion of thin bars thickness poled or radial modes of thickness poled thin discs, the mechanical resonance corresponds to the minimum of the electrical impedance, *i.e.*, the electrical resonance. This is due to the fact that the electric open/short circuit conditions are interchanged with the mechanical ones.

We use the frequency of the mechanical resonance to determine the complex mechanical compliance, *s*, or stiffness, *c*:

$$s_{ij}^{E} = \frac{1}{4\rho l^2 f_s^2}\left(1 - i\frac{\Delta f_s}{f_s}\right), \; or \; c_{ij} = 4\rho t^2 f_p^2 \left(1 - i\frac{\Delta f_p}{f_p}\right) \quad (21)$$

These frequencies do not correspond to the maximum or minimum of the *Z*-plot (usually called: fn, fm), which give us wrong values for this purpose in lossy materials, and we must use instead the maxima of the real parts of admittance and impedance (fs, fp) [16].

There is another pair of resonance-anti-resonance frequencies; those correspond to the zeros for the values of the susceptance or reactance: fr, fa, respectively. In Figure 3, the three possible anti-resonance frequencies are shown in admittance-*GB* and impedance-*RX* plots.

For lossless resonators, these frequencies are close, but the difference is very important for lossy and/or weak resonators.

The authors currently use a dual plot that includes the real parts of the impedance (conductance, *G*) and the admittance (resistance, *R*) that includes both peaks (Figure 4). Alemany's [20,25] method uses information from anti-resonance to increase the accuracy in the characterization, including all losses.

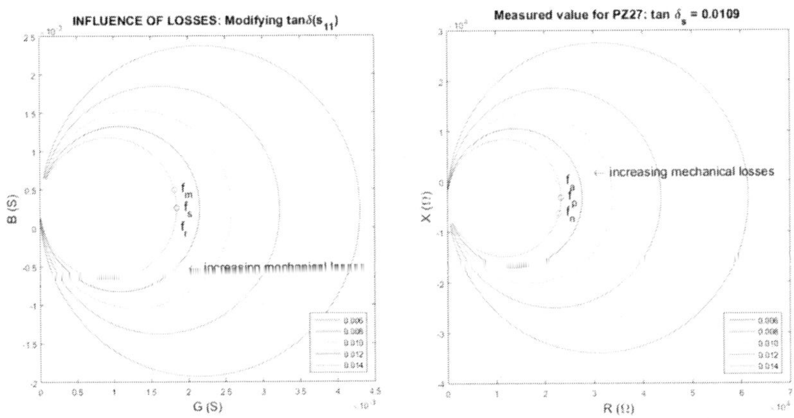

(a) (b)

Figure 3. (**a**) *G-B* and (**b**) *R-X* plots showing the "position" of the three possible resonance frequencies (in the *G-B* plot): f_r, zero reactance; f_s, maximum conductance; f_m, maximum absolute admittance; and the three possible anti-resonance frequencies (at *R-X* plot): f_a, zero susceptance; f_p, maximum resistance; f_n, maximum absolute impedance.

The display of the Java version for the Alemany's method software is shown in Figure 4. The experimental data and the reproduced spectrum for the first overtone of the radial mode of a PZ27 (Ferroperm Piezoceramics A/S) thin disc of t = 1 mm and D = 20 mm are plotted [25]. The reproduction is carried out by introducing the calculated material parameters back into the analytical expression used for the iterative numerical solution of the impedance measurement:

$$Y = i\omega \frac{S}{t} \epsilon_{33}^{T} \left(1 - \frac{k_p^2}{\dfrac{1}{2 - \Im(\xi)} - \dfrac{1}{1 + \sigma^p}} \right) \qquad (22)$$

where ω is the driven frequency, S the area of electrodes, t the thickness, ϵ_{33}^{T} the dielectric permittivity at constant stress, k_p the planar coupling factor, $\sigma = -s_{12}^{E}/s_{11}^{E}$ Poisson's planar ratio and $\Im(\xi)$ the so-called Onoe' s function, defined by:

$$\mathfrak{J}(\xi) = \frac{\xi J_0(\xi)}{J_1(\xi)}, \text{ with } \xi = \omega a\sqrt{\frac{c_{11}^p}{\rho}} \tag{23}$$

where J_0 and J_1 are Bessel functions of the first kind, and of zeroth and first order, respectively, a is the radius of the disc, ρ the density and $c_{11}^p = s_{11}^E / \left[\left(s_{11}^E\right)^2 - \left(s_{12}^E\right)^2 \right]$ the stiffness of the planar mode.

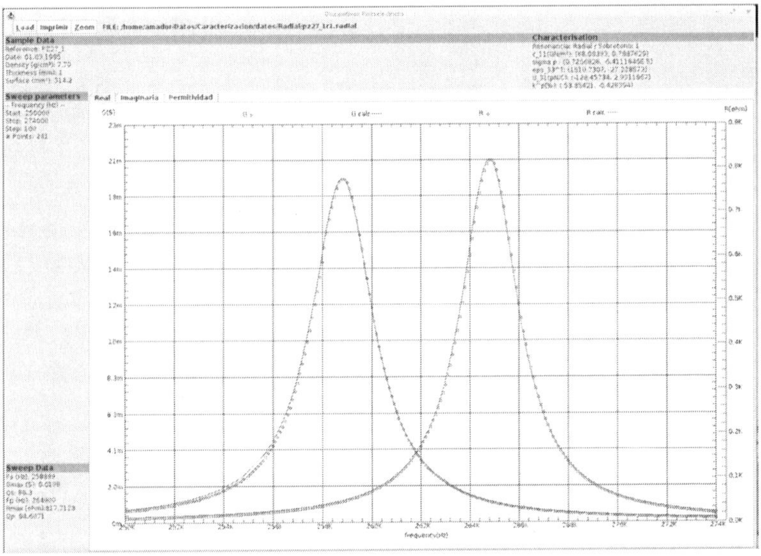

Figure 4. Presentation of the results obtained using Alemany's method for measurement at the first overtone of the radial resonance mode of a PZ27 (Ferroperm Piezoceramics A/S) thin disc.

As that is a monomodal resonance, there is a perfect agreement between the measured and the reproduced spectra.

3. STUDYING LOSSES

3.1. What Losses Look Like?

For a piezoelectric resonator, as we mentioned above, it can be expressed as the loss tangent: tanδ, but also using its inverse, the *Q*-factor, defined as:

$$\tan\delta = \frac{1}{Q} = \frac{\Delta f}{f_0} \tag{24}$$

where f_0 is the resonant frequency and Δf is the FWHM (full with at half maximum). For a pure mechanical resonator, the loss tangent of the resonator corresponds to those of the elastic constant, and then, this is equivalent to:

$$\tan\delta = \frac{s''}{s'} = -\frac{c''}{c'} \tag{25}$$

with s', s'' as the real and imaginary parts of the compliance and c', c'' the real and imaginary parts of the stiffness.

This loss tangent can be assigned to both resonance and anti-resonance peaks. As mentioned above, the lowest frequency, the so-called "free" elastic stiffness, c^E_{pq}, is that corresponding to constant E conditions, and the higher frequency, the "clamped" stiffness, c^E_{pq}, corresponds to constant D.

In the following, we will study the influence of the different loss factors on the behavior of the sample at the electromechanical resonance, *i.e.*, in the frequency range between the resonance and the anti-resonance frequencies, of two piezoceramic resonator geometries of the same material, Ferroperm A/S PZ27, a Navy type II piezoceramic (Figure 5). One geometry is a thickness poled rectangular bar (L = 31.3 mm, w = 1.46 mm and t = 0.65 mm) at the length extensional resonance mode; the other one is a thickness poled thin disc (t = 1 mm and D = 20 mm) at the thickness mode. Both of them are electrically excited in thickness. Experimental measurements on these samples and the relevant complex material parameters (Table 1) calculated from them using Alemany's [20,29] method were used as a starting point of the simulations.

The spectra were reproduced from these material parameters using MATLAB® R2015a software and the corresponding analytical solution of the wave equation for each considered resonance mode [20,29]. Then, the losses were both increased and decreased from that experimental reference, modifying the imaginary part of one specific constant each time (elastic, dielectric and piezoelectric). The following Figures show the so-obtained parameterized spectra.

Table 1. Experimental characterization parameters obtained by Alemany's method for two samples of PZ27 piezoceramic.

Shape	*Rod*
mode	*length extensional*
Thickness (m)	0.65×10^{-3}
length (m)	31.3×10^{-3}
width (m)	1.46×10^{-3}
density (kg m^{-3})	7700
ϵ_{33}^{T}	$(1682 - i\,25)\,\epsilon_0$
s_{11}^{E} (m^2 N^{-1})	$(16.011 - i\,0.175)\,10^{-12}$
d_{31} $\left(\mathrm{C\,N^{-1}}\right)$	$(-159.22 + i\,3.0)\,10^{-12}$
k_{31}	$0.3259 + i\,0.0103$
Shape	*Disc*
mode	*thickness extensional*
Thickness (m)	1.00×10^{-3}
radius (m)	20.0×10^{-3}
density $\left(\mathrm{kg\,m^{-3}}\right)$	7700
ϵ_{33}^{S}	$(818.6 - i\,23.9)\,\epsilon_0$
c_{33}^{D} $\left(\mathrm{N\,m^{-2}}\right)$	$(136.9 + i\,0.i)\,10^9$
h_{33} $(V\,m^{-1})$	$(1.995 + i\,0.036)\,10^9$
k_t	$0.4592 + i\,0.0004$

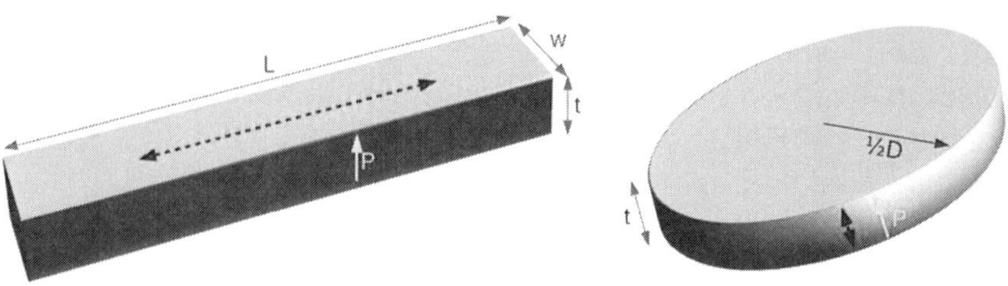

Figure 5. The two resonator geometries used in the simulation of the effect of the losses on the resonance.

3.2. Influence of the Mechanical Losses

The effect produced by the change of the elastic constant loss tangent is shown in Figure 6. In this case, we show the resonance-anti-resonance peaks of the length extensional mode of a bar, and then, we use the elastic compliances E11 and simulate the effect of varying losses around the measured one $(\tan\delta\left(s_{11}^{E}\right) = 0.0109)$. Both peaks change simultaneously and get wider with the increase of the losses, as expected. This is the classical result: low-Q (high mechanical loss) resonators presents wide peaks.

As we mentioned above, for the radial mode of resonance of thin disks, it is necessary take into account two mechanical constants, c_{11}^{P} and σ^{P}, which represent the contour stiffness constant and the planar Poisson modulus, and those are related to the most common compliances s_{11}^{E}, s_{12}^{E} by means of the relations:

$$c_{11}^{P} = \frac{s_{11}^{E}}{\left(s_{11}^{E}\right)^{2} - \left(s_{12}^{E}\right)^{2}} \qquad \sigma^{P} = -s_{12}^{E}/s_{11}^{E} \tag{26}$$

In this case, the imaginary part of σ^{P} seems to be null within the experimental errors in all materials that we studied, thus meaning that $\tan\delta\left(s_{11}^{E}\right) = \tan\delta\left(s_{12}^{E}\right)$, and then:

$$\tan\delta = -\frac{s_{11}^{E\,\prime\prime}}{s_{11}^{E\,\prime}} = \frac{\left(c_{11}^{P}\right)^{\prime\prime}}{\left(c_{11}^{P}\right)^{\prime}} \tag{27}$$

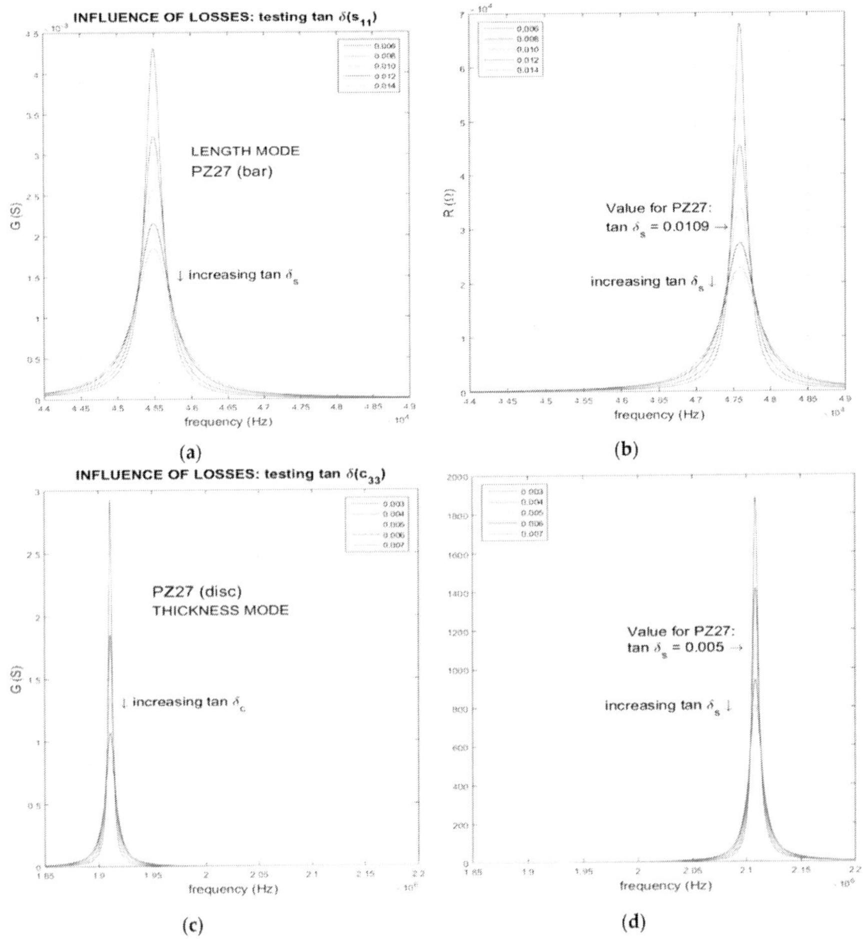

Figure 6. Influence of mechanical losses on the (**a**) and (**c**) resonance G-peaks; and (**b**) and (**d**) anti-resonance R-peaks. The increment of the losses of the elastic parameter (compliance) affects both peaks in the same way.

3.3. Influence of the Dielectric Losses

The influence of modifying the loss factor of the dielectric permittivity is shown in Figure 7. Note the increasing of the width in the peak corresponding to the mechanical anti-resonance with the increase of the dielectric losses, while the mechanical resonance peak remains unchanged. For the thickness poled bar vibrating along its longest dimension, the mechanical anti-resonance corresponds to the electric one; thus, the *R*-peak changes.

For the thin disc, thickness poled, in the thickness resonance, the mechanical anti-resonance corresponds to the electric resonance, in this case the *G*-peak exhibits variation in its width.

Then, the effect of the dielectric losses is observed only in the mechanical anti-resonance getting a lower value of the *Q*-factor with the increase of the dielectric losses.

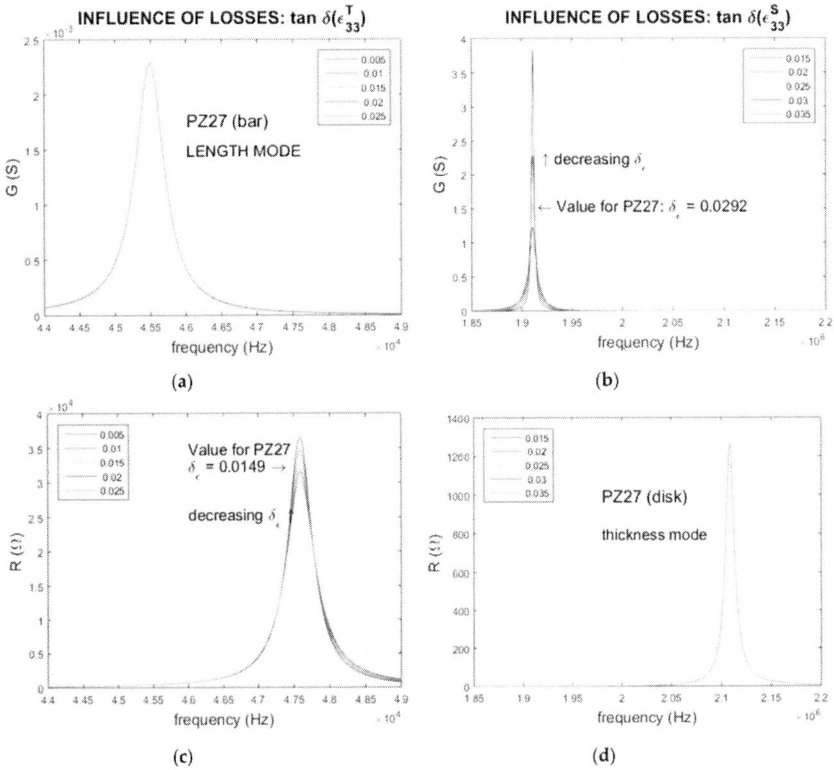

Figure 7. Influence of the dielectric losses on the (**a**) resonance and (**c**) anti-resonance peaks for length extensional mode of a thickness poled bar; dielectric losses modify only (see the *R*-peak) the elastic anti-resonance peak (also electric anti-resonance). Dielectric losses influence on the (**b**) resonance and (**d**) anti-resonance for thickness extensional mode of a thin disc, thickness poled; dielectric losses modify the elastic anti-resonance, which in this case corresponds to the *G*-peak (electric resonance).

3.4. Influence of the Piezoelectric Losses (Piezoelectric Modulus)

The influence of the piezoelectric losses, while keeping both mechanical and dielectric losses constant, is shown in Figure 8.

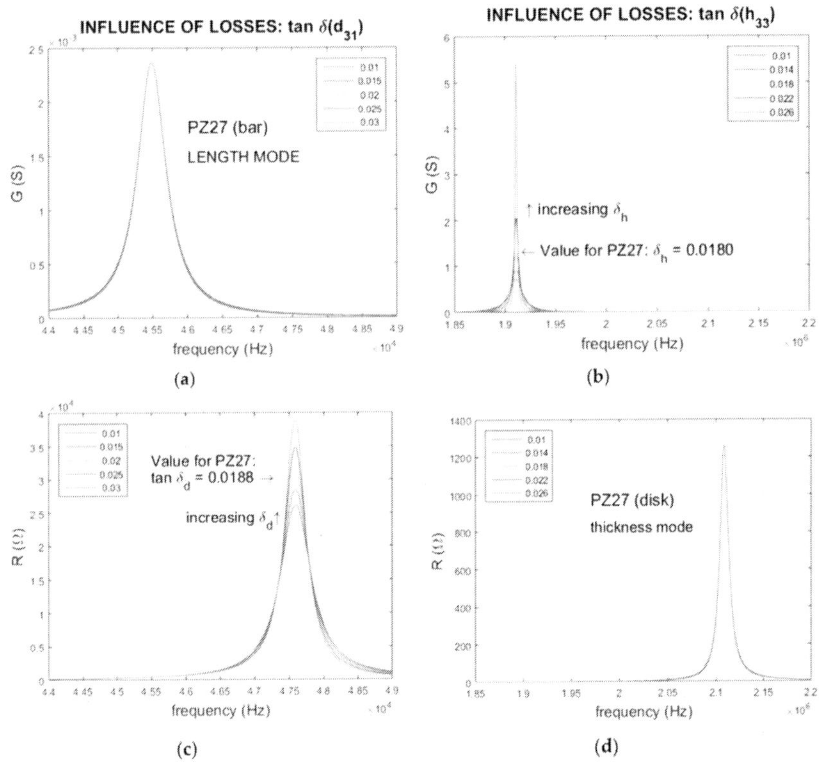

Figure 8. Influence of piezoelectric losses on the (**a**) resonance and (**c**) anti-resonance peaks for length extensional mode of a thickness poled bar; loss ofd_{31} modifies the mechanical anti-resonance peak (*R*-peak). Influence of piezoelectric losses on the (**b**) resonance and (**d**) anti-resonance peaks for thickness mode of a thin disc, the losses of the h_{33} factor modify the mechanical anti-resonance peak (*G*-plot).

Here, the simulation of small variations around the measured value for the imaginary part of the piezoelectric constant involved in every resonant mode (d31 for the length extensional mode of the bar and h33 for the thickness mode of the thin disk) is carried out. It is noticeable that their influence is different from that observed in the two previous simulations. As for the dielectric losses, for both studied resonances, the change of piezoelectric losses only affects the mechanical anti-resonance. However, increasing losses in this case produce narrower peaks (getting a higher *Q*-factor that corresponding to the resonance).

This anisotropy of loss factors (resonance and anti-resonance) is observed in materials with high piezoelectric losses (also reported by Uchino [10]), but could be masked by the opposite effect of the dielectric losses; thus, an accurate method is necessary to get the correct characterization.

4. LOSSES IN DEVICES

Piezoelectric materials are developed to be integrated in a device or system. It is possible to characterize devices instead of materials [35], but it is not so easy to use this information to design a device.

On the one hand, electrical engineers developed toolboxes (such as SPICE: Simulation Program with Integrated Circuit Emphasis) to make the design and implementation of a system easier by means of equivalent circuits, but if we want to include a piezoelectric resonator in their databanks we need to speak a common language, *i.e.*, we must give the equivalent circuit of our element.

On the other hand, from early times of the study of piezoelectric materials, the strategies to determine the electric components of the equivalent circuit of a piezoelectric resonator from electrical measurements around an electromechanical resonance were a matter of concern [36].

In the beginning, standards proposed the use of the simplest resonant circuit (Figure 9), but do not provide a proper resonance peak shape and describe only the fundamental resonance or one of its overtones. Besides, the mechanical losses are represented by the resistance in the "motional" LCR branch; the dielectric losses must be accounted for by additional series and parallel resistances, but piezoelectric losses require complex components.

More accurate models have been developed and tested for electromechanical transducers [37,38,39], but typically, if one can get from them a higher accuracy, this is at the expense of the increasing complexity of the model and the use of non-linear or more sophisticated elements as transformers.

As a general consideration, a model should not be more complex than the object that it represents.

In this work, we introduce a simplified alternative equivalent circuit model [40], given by the use of a transmission line as one of its components (Figure 10) due to the analogies between the distributed elements of the transmission line and the media through which the acoustic waves propagate [41].

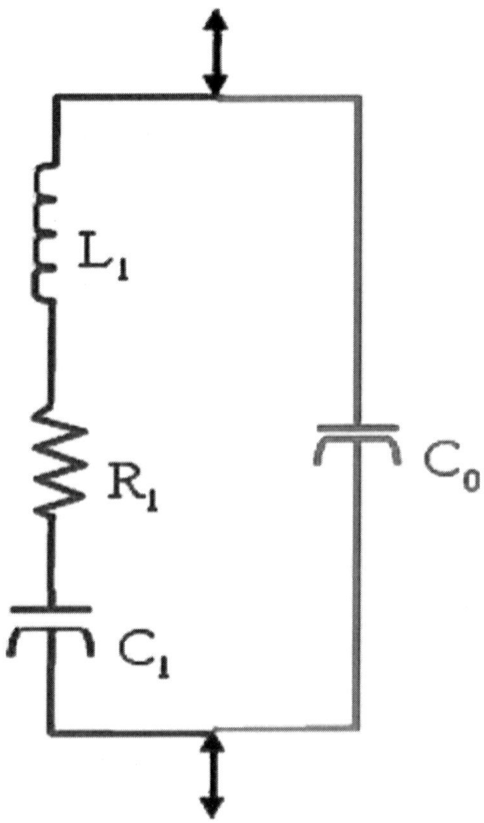

Figure 9. Elementary model of a piezoelectric resonator based on passive elements.

Using a model like Püttmer's [42], mechanical and electrical losses can be easily introduced to model a low-*Q*piezoelectric resonator. Figure 11 shows the description of this model using ORCAD/PSPICE. The figure depicts the different circuit elements modeling the electrical behavior of the piezoelectric material once integrated into the full device [43]. This example corresponds to a simple device, an ultrasonic emitter and receiver, using a single piezoelectric element. Figure 12 draws the impedance (modulus and phase) as a function of the frequency resulting from the device simulation. The use of the lossy transmission line is appropriate for simulating both the multiple resonances and losses. It is worth noting that the results obtained by using the proposed simulation model reproduce the overtones of the resonances. Dependent generators in the model allow simulating mode coupling. This simple model can be extended to account separately for the piezoelectric losses from an accurate material characterization. Work is in progress to this aim.

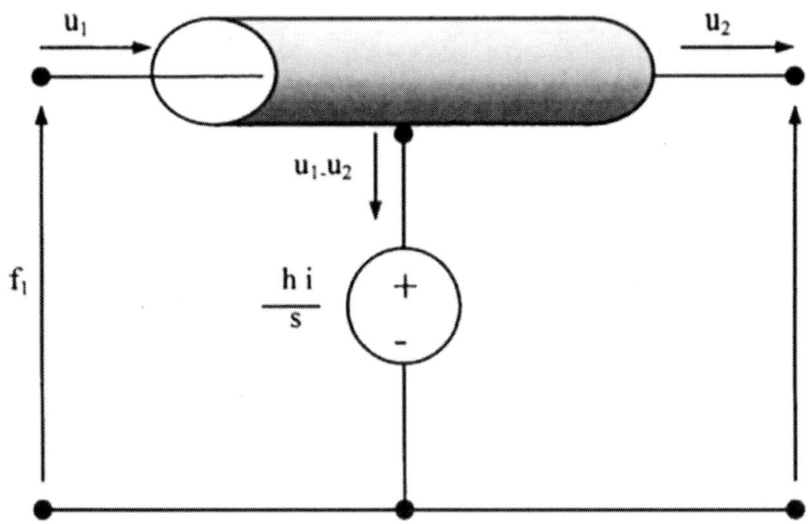

Figure 10. Analogous circuit based on a transmission line for the thickness-mode transducer.

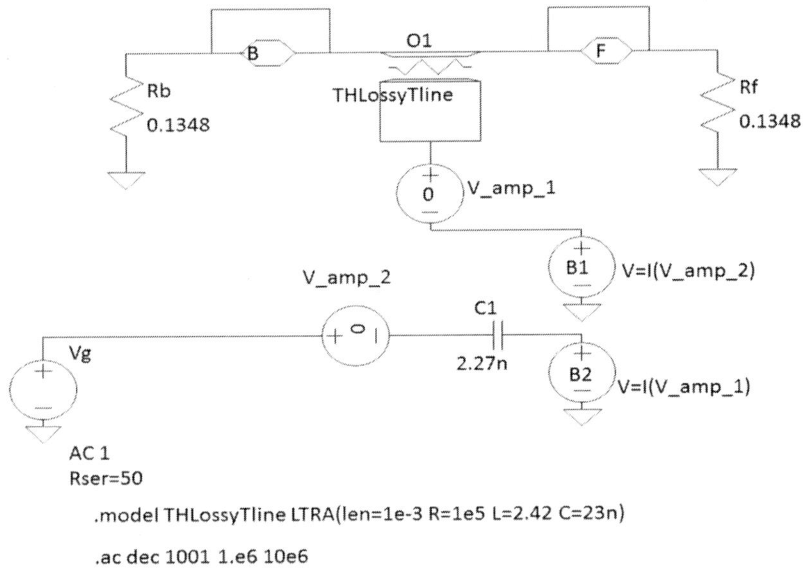

Figure 11. PSPICE schematic, which describes the equivalent circuit of a given device (under simulation) using a single piezoelectric element, represented by the lossy transmission line.

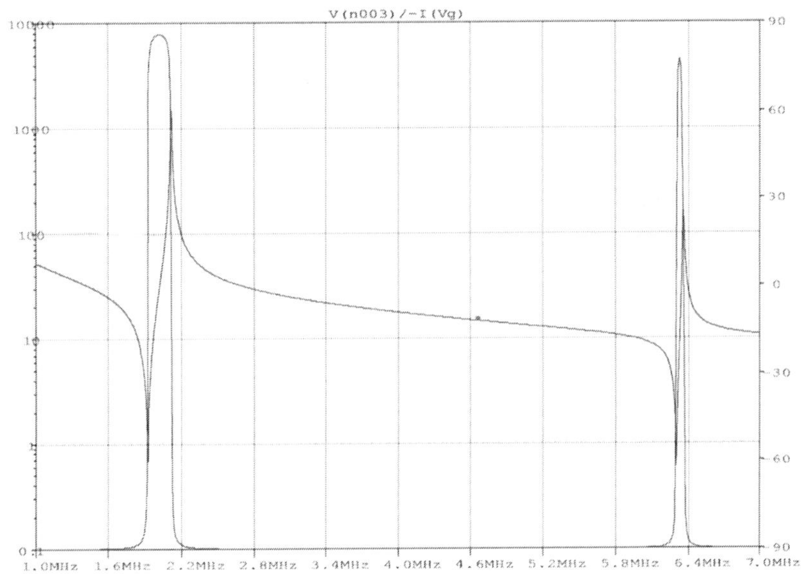

Figure 12. Plot of the modelled impedance of a device using a single piezoelectric element.

5. CONCLUSIONS

Fundamentals of the losses in piezoelectric materials were reviewed. The analysis of the influence of each type of loss (dielectric, elastic and piezoelectric) on the resonance and anti-resonance peaks for resonators with the acoustic wave parallel and perpendicular to the electrical excitation was carried on. Analysis reveals that each kind of loss has a characteristic effect. To connect the field of material researchers with electrical and electronic engineers we proposed, on the one hand, the use of accurate methods of characterization from impedance measurements at resonance and, on the other hand, simplified equivalent circuit models using transmission lines.

ACKNOWLEDGMENTS

Amador M. González acknowledges the Spanish MEC Project MAT2013-40722-R. Lorena Pardo acknowledges the Spanish CSIC Project #201060E069.

AUTHOR CONTRIBUTIONS

Amador M. González conceived and coordinated the manuscript. Amador M. González, Lorena Pardo and Álvaro García worked on the impedance spectroscopy measurements and material characterization from resonance. Amador M. González and César Benavente-Peces worked on the modeling using equivalent circuits.

REFERENCES

1. Cady, W.G. *Piezoelectricity: An Introduction to the Theory and Applications of Electromechanical Phenomena in Crystals*; Mc Graw Hill Book Co.: New York, NY, USA, 1946.
2. Jaffe, B.; Cook, W.R.; Jaffe, H. *Piezoelectric Ceramics*; Academic Press: New York, NY, USA, 1971.
3. Fukada, E. History and recent progress in piezoelectric polymers. *IEEE Trans. Ultrason. Ferroelectr. Freq. Control* **2000**, *47*, 1277–1290.

4. Newnham, R.E.; Skinner, D.P.; Cross, L.E. Connectivity and piezoelectric-pyroelectric composites. *Mat. Res. Bull.* **1978**, *13*, 525–536.

5. Holland, R. Representation of dielectric, elastic, and piezoelectric losses by complex coefficients. *IEEE Trans. Sonics Ultrason.* **1967**, *14*, 18–20.

6. Martin, G.E. Dielectric, Elastic and Piezoelectric Losses in Piezoelectric Materials. In Proceedings of the 1974 Ultrasonics Symposium, University of Wisconsin-Milwakee, Milwaukee, WI, USA, 11–14 November 1974; pp. 613–617.

7. Arlt, G.; Dederichs, H. Complex elastic, dielectric and piezoelectric constants by domain wall damping in ferroelectric ceramics. *Ferroelectrics* **1980**, *29*, 47–50.

8. Damjanovic, D.; Gururaja, T.R.; Jang, S.J.; Cross, L.E. Temperature behavior of the complex piezoelectric d_{31}, coefficient in modified lead titanate ceramics. *Mater. Lett.* **1986**, *4*, 414.

9. Sherrit, S; Mukherjee, B.K. The use of complex material constants to model the dynamic response of piezoelectric materials. In Proceedings of the Ultrasonics Symposium, Sendai, Japan, 5–8 October 1998; Volume 1, pp. 633–640.

10. Uchino, K.; Zhuang, Y.; Ural, S.O. Loss determination methodology for a piezoelectric ceramic: New phenomenological theory and experimental proposals. *J. Adv. Dielectr.* **2011**, *1*, 17–31.

11. Mezheritsky, A.V. Efficiency of excitation of piezoceramic transducers at antiresonance frequency. *IEEE Trans. Ultrason. Ferroelectr. Freq. Control* **2002**, *49*, 484–494.

12. Lu, X.; Hanagud, S.V. Extended Irreversible Thermodynamics Modeling for Self-Heating and Dissipation in Piezoelectric Ceramics. *IEEE Trans. Ultrason. Ferroelectr. Freq. Control* **2004**, *51*, 1582–1592.

13. Hagiwara, M.; Hoshina, T.; Takeda, H.; Tsurumi, T. Identicalness between Piezoelectric Loss and Dielectric Loss in Converse Effect of Piezoelectric Ceramic Resonators. *Jpn. J. Appl. Phys.* **2012**, *51*, 09LD10.

14. Cain, M.G.; Stewart, M. Losses in Piezoelectrics via Complex Resonance Analysis. In *Characterisation of Ferroelectric Bulk Materials and Thin Films*; Cain, M.G., Ed.; Springer Series in Measurement Science and Technology: New York, NY, USA, 2014; Volume 2.

15. Liu, G.; Zhang, S.; Jiang, W.; Cao, W. Losses in ferroelectric materials. *Mater. Sci. Eng. R Rep.* **2015**, *89*, 1–48.

16. *IEEE Standard on Piezoelectricity*; ANSI/IEEE Std. 176–1987; IEEE Society: Washinton D.C., WA, USA, 1987.

17. Nye, J.F. *Physical Properties of Crystals*; Oxford University Press Inc.: New York, NY, USA, 1957.

18. Vicente, J.M.; Jimenez, B. Frequency dependence of the piezoelectric d_{31} coefficient as a function of ceramic tetragonality. *Ferroelectrics* **1992**, *134*, 157–162.

19. Jimenez, B.; Vicente, J.M. Influence of mobile 90° domains on the complex elastic modulus of PZT ceramics.*J. Phys. D: Appl. Phys.* **2000**, *33*, 1525–1535.

20. Alemany, C.; Pardo, L.; Jimenez, B.; Carmona, F.; Mendiola, J.; Gonzalez, A. Automatic iterative evaluation of complex material constants in piezoelectric ceramics. *J. Phys. D Appl. Phys.* **1994**, *27*, 148.

21. Piezoelectric Resonance Analysis Program. Available online: http://www.tasitechnical.com/products.html(accessed on 30 October 2015).

22. Pardo, L. Piezoelectric materials for power ultrasonic transducers. In *Power Ultrasonics*; Gallego-Juárez, J.A., Graff, K., Eds.; Woodhead Publishing Ltd.: Cambridge, UK, 2014.

23. Rupitsch, S.J. Complete Characterization of Piezoceramic Materials by Means of Two Block-Shaped Test Samples. *IEEE Trans. Ultrason. Ferroelectr. Freq. Control* **2015**, *62*, 1403–1413.

24. Kiyono, C.Y.; Peréz, N.; Silva, E.C.N. Determination of full piezoelectric complex parameters using gradient-based optimization algorithm. *Smart Mater. Structs* **2016**.

25. Alemany, C.; Gonzalez, A.; Pardo, L.; Jimenez, B.; Carmona, F.; Mendiola, J. Automatic determination of complex constants of piezoelectric lossy materials in the radial mode. *J. Phys. D Appl. Phys.* **1995**, *28*, 945–956.

26. Experimental Techniques (ICMM-CSIC). Available online: http://icmm.csic.es/gf2/medidas.htm (accessed on 30 October 2015).

27. Smits, J.G. Iterative method for accurate determination of the real and imaginary parts of the materials coefficients of piezoelectric ceramics. *IEEE Trans. Sonics Ultrason.* **1976**, *23*, 393–402.

28. Algueró, M.; Alemany, C.; Pardo, L.; Gonzalez, A.M. Method for obtaining the full set of linear electric, mechanical and electromechanical coefficients and all related losses of a piezoelectric ceramic. *J. Am. Ceram. Soc.* **2004**, *87*, 209–215. Gonzalez, A.M.; Alemany, C. Determination of the frequency dependence of characteristic constants in lossy piezoelectric materials. *J. Phys. D: Appl. Phys.* **1996**, *29*, 2476–2482.

29. Gonzalez, A.M.; Alemany, C. Estudio de los sobretonos del modo de vibración en espesor de resonadores piezoeléctricos con pérdidas. *Bol. Soc. Esp. Ceram. Vid.* **1995**, *34*, 368–371.

30. Pardo, L.; Jiménez, R.; García, A.; Brebøl, K.; Leighton, G.; Huang, Z. Impedance measurements for determination of the elastic and piezoelectric coefficients of films. *Adv. Appl. Ceram.: Struct. Funct. Bioceram.* **2010**, *109*, 156–161.

31. Pardo, L.; García, A.; de Espinosa, F.M.; Brebøl, K. Shear Resonance Mode Decoupling to Determine the Characteristic Matrix of Piezoceramics For 3-D Modelling. *IEEE Trans. Ultrason. Ferroelectr. Freq. Control* **2011**, *58*, 646–657.

32. Guennou, M.; Dammak, H.; Djémi, P.; Moch, P.; Pham-Thi, M. Electromechanical properties of single domain PZN–12%PT measured by three different methods. *Solid State Sci.* **2010**, *12*, 298–301.

33. Sherrit, S.; Masys, T.J.; Wiederick, H.D.; Mukherjee, B.K. Determination of the reduced matrix of the piezoelectric, dielectric, and elastic material constants for a piezoelectric

material with C∞ symmetry. *IEEE Trans. Ultrason. Ferroelectr. Freq. Control* **2011**, *58*, 1714–1720.

34. Gonzalez, A.M.; de Frutos, J.; Duro, M.C. Procedure for the Characterization of Piezoelectric Samples in Non-standard Resonant Modes. *J. Eur. Ceram. Soc.* **1999**, *19*, 1285–1288.

35. Martin, G.E. Determination of Equivalent-Circuit Constants of Piezoelectric Resonators of Moderately Low Q by Absolute-Admittance Measurements. *J. Acoust. Soc. Am.* **1954**, *26*, 413–420.

36. Tilmans, H.A.C. Equivalent circuit representation of electromechanical transducers: I. Lumped -parameter systems. *J. Micromech. Microeng.* **1996**, *6*, 157–176.

37. Tilmans, H.A.C. Equivalent circuit representation of electromechanical transducers: II. Distributed-parameter systems. *J. Micromech. Microeng.* **1997**, *7*, 285–309.

38. Sherrit, S.; Leary, S.P.; Dolgin, B.P.; Bar-Cohen, Y. Comparison of the Mason and KLM equivalent circuits for piezoelectric resonators in the thickness mode. In Proceedings of the Ultrasonics Symposium, Caesars Tahoe, NV, USA, 17–20 October 1999; pp. 921–926.

39. Guisado, A.; Torres, J.L.; González, A.M. Study of Equivalent Circuits of Piezoceramics to Use in Simulations with Pspice. *Ferroelectrics* **2003**, *293*, 307–319.

40. Ballato, A. Modeling piezoelectric and piezomagnetic devices and structures via equivalent networks. *IEEE Trans. Ultrason. Ferroelectr. Freq. Control* **2001**, *48*, 405–414.

41. Püttmer, A.; Hauptmann, P.; Lucklum, R.; Krause, O.; Henning, B. SPICE Model for Lossy Piezoceramic Transducers. *IEEE Trans. Ultrason. Ferroelectr. Freq. Control* **1997**, *44*, 60–66.

42. Alcocer, E.M. Modelado y Simulación de Sensores y Actuadores Para Ultrasonidos. Bachelor's Thesis, Tecnical University of Madrid-EUIT Telecomunicación, Madrid, Spain, 2003. Available online: http://marte.biblioteca.upm.es/ (accessed on 21 January 2016).

Chapter 3

Nonlinear Bending of Piezoelectric Cylindrical Shell Reinforced with BNNTs under Electro-Thermo-Mechanical Loadings

Jinhua Yang*, **Pengjun Zhang**

School of Civil Engineering and Architecture, Changsha University of Science & Technology, Changsha, China

ABSTRACT

Under combined electro-thermo-mechanical loadings, the nonlinear bending of piezoelectric cylindrical shell reinforced with boron nitride nanotubes (BNNTs) is investigated in this paper. By employing nonlinear strains based on Donnell shell theory and utilizing piezoelectric theory including thermal effects, the constitutive relations of the piezoelectric shell reinforced with BNNTs are established. Then the governing equations of the structure are derived through variational principle and resolved by applying the finite difference method. In numerical examples, the effects of geometric nonlinear, voltage, temperature, as well as volume fraction on the deflection and bending moment of axisymmetrical piezoelectric cylindrical shell reinforced with BNNTs are discussed in detail.

Keywords: Nonlinear Bending, Piezoelectric, Cylindrical Shell, BNNT, Electro-Thermo-Mechanical Loadings

1. INTRODUCTION

BNNTs are similar to CNTs in structure and their extraordinary mechanical properties, but are different in that BNNTs possess higher temperature resistance to oxidation and stronger piezoelectric characteristics. Also, unlike CNTs, BNNTs have stable semiconducting behavior with a large band gaps regardless of radius and chirality of the structure. This property of BNNTs makes them promising candidate materials in a large variety of nanosized electronic and photonic devices. Therefore, BNNTs seem to be more suitable as reinforcement in composite structures due to their high resistance to oxidation at elevated temperatures [1], outstanding mechanical properties [2] and high thermal conductivity [3]. With the development of science and technology, a new sort of smart nanocomposites, with piezoelectric material as matrix and BNNTs as the reinforcement, has attracted increasing interests in both research and engineering communities. It is noted that the investigations on this new smart nanocomposites are limited in number and most discuss the linear problem. Therefore, it is necessary to do more extensive researches on the nonlinear behavior for this structure.

At present, most researches are limited to discussing the behavior of piezoelectric structure without reinforce- ment of BNNTs. Yao et al. [4] presented static behaviors of piezoelectric cantilever actuator under large electric field. Shen [5] studied the nonlinear bending for a simply supported, shear deformable cross-ply laminated plate with piezoelectric actuators subjected to a transverse uniform or sinusoidal load combined with electrical loads and in thermal environments. Shegokar et al. [6] deals with the stochastic nonlinear bending response of functionally graded materials beam with surface bonded piezoelectric layers subjected to thermo-electro-mechanical loadings. Narita et al. [7] illustrated an analytical and experimental study of nonlinear bending response and domain wall motion in piezoelectric laminated actuators under electric fields. Beldica et al. [8] analyzed the bending of nonlinear viscoelastic beams with small or large deformations. Yan et al. [9] investigated the time- dependent behavior of a simply supported, angle-ply piezoelectric laminate in cylindrical bending with viscoelastic interfaces. Narita et al. [10] discussed the static electromechanical displacement and polarization switching properties of piezoelectric laminated actuators under three point bending. Using a variational formulation, Li et al. [11] developed a size-dependent functionally graded piezoelectric beam model. Based on the local Petrov- Galerkin approach, Sladek et al. [12] proposed a meshless method for plate bending analysis with functionally graded piezoelectric material properties. Employing Euler-Bernoulli beam theory and the physical neutral surface concept, Fu et al. [13] presented the thermo-piezoelectric buckling, nonlinear free vibration and dynamic stability for the piezoelectric functionally graded beams. None of the above mentioned works have considered the behavior of piezoelectric structure reinforced with BNNTs.

Recently, some researches about the static buckling of piezoelectric shell reinforced with BNNTs have emerged. Buckling of BNNTs in a PVDF elastic medium subjected to combined electro-thermo-mechanical loadings was investigated by Salehi-Khojin and Jalili [14] who showed that applying direct and reverse voltages to BNNT changed buckling loads for any axial and circumferential wave-numbers. Mosallaie Barzoki et al. [15] studied torsional linear buckling of a PVDF cylindrical shell reinforced by BNNTs with an elastic core under the same loading condition as [14] were investigated, indicating that buckling strength increased substantially as harder foam cores were employed. Using virtual displacement method based on nonlocal cylindrical piezoelasticity continuum shell theory, Arani et al. [16] discussed the axial buckling of double-walled Boron Nitride nanotubes embedded in an elastic medium under combined electro-thermo-mechanical loadings. Up to now, to the best of authors' knowledge, the research on nonlinear bending of piezoelectric shell reinforced with BNNTs has not been reported in the open literature.

Motivated by the considerations, we aim to study the nonlinear bending of axisymmetrical piezoelectric shell reinforced with BNNTs under combined electro-thermo-mechanical loadings. Using the difference method in space, the whole problem is resolved by the iteration method. Numerical results in nonlinear deflection and bending moment of axisymmetrical piezoelectric cylindrical shell reinforced with BNNTs are presented for different values of voltage, temperature, volume fraction and so on.

2. BASIC EQUATIONS

Consider that a piezoelectric cylindrical shell reinforced with BNNTs has midsurface radius R, thickness h, length L and mass density ρ_0

(see Figure 1). The shell is referred to the coordinate system (x, y, z) in which x and y are the axial and circumferential directions of the shell and z is in the direction of the inward normal to the middle surface. The origin of the coordinate system is located at the end of the shell on the middle plane. The shell is subjected to transverse static load q, applied voltage V and a uniform temperature rise ΔT.

2.1. Strain Displacement Relationships

Supposing that $\bar{u}, \bar{v}, \bar{w}$ denote the axial, circumferential and radial displacement of an arbitrary point on the shell, and the corresponding displacement components of middle surface are u, v and w, then the displacement components of piezoelectric cylindrical shell can be written as

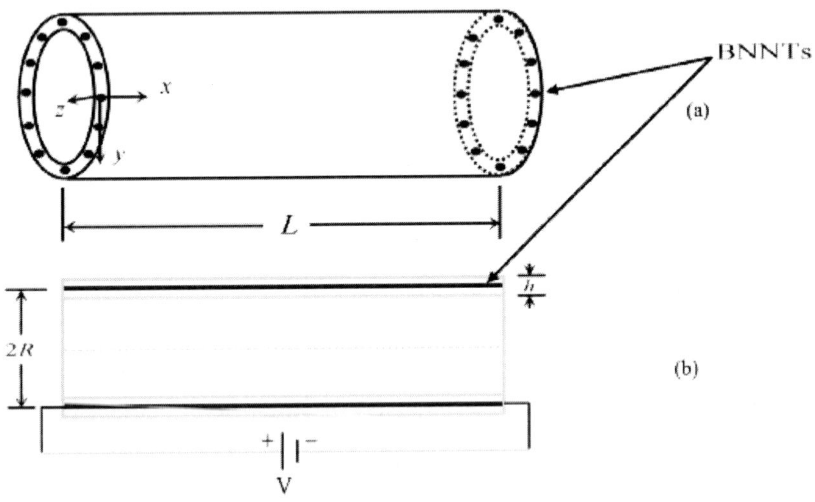

Figure 1. (a) Geometry of piezoelectric cylindrical shell reinforced with BNNTs; (b) section of cylindrical shell.

$$\bar{u}(x,y,z) = u(x,y) - zw_{,x}(x,y);\ \bar{v}(x,y,z) = v(x,y) - zw_{,y}(x,y)$$
$$\bar{w}(x,y,z) = w(x,y) \tag{1}$$

where the inferior mark (,) denotes the partial derivative to variables coordinate.

Based on classical shell theory with von Kármán-Donnell type kinematic relations, the nonlinear strain-dis- placement relations can be expressed as

$$\bar{\varepsilon}_x = \varepsilon_x + z\kappa_x \qquad \bar{\varepsilon}_y = \varepsilon_y + z\kappa_y \qquad \bar{\varepsilon}_{xy} = \varepsilon_{xy} + z\kappa_{xy} \tag{2}$$

where $\varepsilon_x, \varepsilon_y, \varepsilon_{xy}$ are the strain components on the middle surface and $\kappa_x, \kappa_y, \kappa_{xy}$ are the change values of curvatures on the middle surface, and

$$\varepsilon_x = u_{,x} + \frac{1}{2}w_{,x}^2, \quad \varepsilon_y = v_{,y} - \frac{w}{R} + \frac{1}{2}w_{,y}^2, \quad \varepsilon_{xy} = u_{,y} + v_{,x} + w_{,x}w_{,y}$$
$$\kappa_x = -w_{,xx}, \quad \kappa_y = -w_{,yy}, \quad \kappa_{xy} = -2w_{,xy} \tag{3}$$

2.2. Constitutive Equations

The constitutive relationship of a piezoelectric structure under combined mechanical, thermal and electrical loadings can be expressed as follows [15]

$$\begin{Bmatrix} \sigma_x \\ \sigma_y \\ \sigma_z \\ \sigma_{yz} \\ \sigma_{xz} \\ \sigma_{xy} \end{Bmatrix} = \begin{bmatrix} C_{11} & C_{12} & C_{13} & 0 & 0 & 0 \\ C_{12} & C_{22} & C_{23} & 0 & 0 & 0 \\ C_{13} & C_{23} & C_{33} & 0 & 0 & 0 \\ 0 & 0 & 0 & C_{44} & 0 & 0 \\ 0 & 0 & 0 & 0 & C_{55} & 0 \\ 0 & 0 & 0 & 0 & 0 & C_{66} \end{bmatrix} \cdot \left(\begin{Bmatrix} \bar{\varepsilon}_x \\ \bar{\varepsilon}_y \\ \bar{\varepsilon}_z \\ \bar{\varepsilon}_{yz} \\ \bar{\varepsilon}_{xz} \\ \bar{\varepsilon}_{xy} \end{Bmatrix} - \begin{Bmatrix} \alpha_x \\ \alpha_y \\ \alpha_z \\ 0 \\ 0 \\ 0 \end{Bmatrix} \Delta T \right) - \begin{bmatrix} e_{11} & 0 & 0 \\ e_{12} & 0 & 0 \\ e_{13} & 0 & 0 \\ 0 & e_{24} & 0 \\ 0 & 0 & e_{35} \\ 0 & 0 & 0 \end{bmatrix} \begin{Bmatrix} E_x \\ E_y \\ E_z \end{Bmatrix} \tag{4}$$

$$\begin{Bmatrix} D_x \\ D_y \\ D_z \end{Bmatrix} = \begin{bmatrix} e_{11} & e_{12} & e_{13} & 0 & 0 & 0 \\ 0 & 0 & 0 & e_{24} & 0 & 0 \\ 0 & 0 & 0 & 0 & e_{35} & 0 \end{bmatrix} \left(\begin{Bmatrix} \bar{\varepsilon}_x \\ \bar{\varepsilon}_y \\ \bar{\varepsilon}_z \\ \bar{\varepsilon}_{yz} \\ \bar{\varepsilon}_{xz} \\ \bar{\varepsilon}_{xy} \end{Bmatrix} - \begin{Bmatrix} \alpha_x \\ \alpha_y \\ \alpha_z \\ 0 \\ 0 \\ 0 \end{Bmatrix} \Delta T \right) - \begin{bmatrix} \varepsilon_{11}^* & 0 & 0 \\ 0 & \varepsilon_{22}^* & 0 \\ 0 & 0 & \varepsilon_{33}^* \end{bmatrix} \begin{Bmatrix} E_x \\ E_y \\ E_z \end{Bmatrix} \tag{5}$$

where $\alpha_\kappa (\kappa = x, y, z)$, ΔT and E_κ represent respectively, thermal expansion coefficient, temperature rise and electric field, and $C_{ij}, e_{ij}, \varepsilon_{ii}^* (i, j = 1, \cdots, 6)$ are elastic constants, piezoelectric constant, dielectric constants, respectively. The material constants of the structure can be calculated using "XY (or YX) rectangle model" [17]. The closed-form formula used in "X model" (or "Y model") expressing the mechanical, the thermal and the electrical properties of the material are as follows [15]:

$$
\begin{aligned}
C_{11} &= \frac{C_{11}^{r} C_{11}^{m}}{V_f C_{11}^{m} + \left(1-V_f\right) C_{11}^{r}} \\
C_{12} &= C_{11}\left[\frac{V_f C_{12}^{r}}{C_{11}^{r}} + \frac{\left(1-V_f\right) C_{12}^{m}}{C_{11}^{m}}\right] \\
C_{13} &= C_{11}\left[\frac{V_f C_{13}^{r}}{C_{11}^{r}} + \frac{\left(1-V_f\right) C_{13}^{m}}{C_{11}^{m}}\right] \\
C_{22} &= V_f C_{22}^{r} + \left(1-V_f\right) C_{22}^{m} + \frac{C_{12}^{2}}{C_{11}} - \frac{V_f \left(C_{12}^{r}\right)^2}{C_{11}^{r}} - \frac{\left(1-V_f\right)\left(C_{12}^{m}\right)^2}{C_{11}^{m}} \\
C_{44} &= V_f C_{44}^{r} + \left(1-V_f\right) C_{44}^{m} \\
C_{55} &= \frac{A}{B^2 + AC} \\
C_{66} &= \frac{C_{66}^{r} C_{66}^{m}}{V_f C_{66}^{m} + \left(1-V_f\right) C_{66}^{r}} \\
e_{31} &= C_{11}\left[\frac{V_f e_{31}^{r}}{C_{11}^{r}} + \frac{\left(1-V_f\right) e_{31}^{m}}{C_{11}^{m}}\right] \\
e_{32} &= V_f e_{32}^{r} + \left(1-V_f\right) e_{32}^{m} + \frac{C_{12} e_{31}}{C_{11}} - \frac{V_f C_{12}^{r} e_{31}^{r}}{C_{11}^{r}} - \frac{\left(1-V_f\right) C_{12}^{m} e_{31}^{m}}{C_{11}^{m}} \\
e_{33} &= V_f e_{33}^{r} + \left(1-V_f\right) e_{33}^{m} + \frac{C_{13} e_{31}}{C_{11}} - \frac{V_f C_{13}^{r} e_{31}^{r}}{C_{11}^{r}} - \frac{\left(1-V_f\right) C_{13}^{m} e_{31}^{m}}{C_{11}^{m}} \\
e_{24} &= V_f e_{24}^{r} + \left(1-V_f\right) e_{24}^{m} \\
e_{15} &= \frac{B}{B^2 + AC} \\
\varepsilon_{11}^{*} &= \frac{C}{B^2 + AC} \\
\varepsilon_{22}^{*} &= V_f \varepsilon_{22}^{*r} + \left(1-V_f\right) \varepsilon_{22}^{*m} \\
\varepsilon_{33}^{*} &= V_f \varepsilon_{33}^{*r} + \left(1-V_f\right) \varepsilon_{33}^{*m} - \frac{e_{31}^{2}}{C_{11}} + \frac{V_f \left(e_{31}^{r}\right)^2}{C_{11}^{r}} + \frac{\left(1-V_f\right)\left(e_{31}^{m}\right)^2}{C_{11}^{m}}
\end{aligned}
\tag{6}
$$

where

$$A = \frac{V_f C_{55}^r}{\left(e_{15}^r\right)^2 + C_{55}^r \varepsilon_{11}^{*r}} + \frac{\left(1 - V_f\right) C_{55}^m}{\left(e_{15}^m\right)^2 + C_{55}^m \varepsilon_{11}^{*m}}$$

$$B = \frac{V_f C_{15}^r}{\left(e_{15}^r\right)^2 + C_{55}^r \varepsilon_{11}^{*r}} + \frac{\left(1 - V_f\right) C_{15}^m}{\left(e_{15}^m\right)^2 + C_{55}^m \varepsilon_{11}^{*m}}$$

$$C = \frac{V_f \varepsilon_{11}^{*r}}{\left(e_{15}^r\right)^2 + C_{55}^r \varepsilon_{11}^{*r}} + \frac{\left(1 - V_f\right) \varepsilon_{11}^{*m}}{\left(e_{15}^m\right)^2 + C_{55}^m \varepsilon_{11}^{*m}} \tag{7}$$

Superscripts r and m refer to the reinforced and matrix components of the composite, respectively. V_f is the vol% of the reinforced BNNTs in matrix.

2.3. Governing Equations

For the piezoelectric cylindrical shell reinforced with BNNTs, the total potential energy Π can be written as

$$\Pi = -U + W \tag{8}$$

where U represents the strain energy and W represents the work done by the transverse load.

The expression of the strain energy is

$$U = \frac{1}{2}\iiint_V \sigma_i \bar{\varepsilon}_i \mathrm{d}V - \frac{1}{2}\iiint_V E_i D_i \mathrm{d}V \tag{9}$$

Considering Equations (4) and (5), as well as the zigzag structure for BNNTs employed here, and the longitudinal arrangement of strips in matrix, makes $E_y = E_z = 0$. Hence, Equation (9) becomes:

$$U=\frac{1}{2}\iiint_V\{\bar{\varepsilon}_x-\alpha_x\Delta T,\bar{\varepsilon}_y-\alpha_y\Delta T,\bar{\varepsilon}_{xy},-E_x\}\begin{bmatrix}C_{11}&C_{12}&0&e_{11}\\C_{12}&C_{22}&0&e_{12}\\0&0&C_{66}&0\\e_{11}&e_{12}&0&-\varepsilon_{11}^{\bullet}\end{bmatrix}\begin{Bmatrix}\bar{\varepsilon}_x-\alpha_x\Delta T\\\bar{\varepsilon}_y-\alpha_y\Delta T\\\bar{\varepsilon}_{xy}\\-E_x\end{Bmatrix} \quad (10)$$

Letting V is the voltage applied on both ends of shell, then

$$E_x=V/L \quad (11)$$

The work done by the transverse load $q(x,y)$ is

$$W=\iint_A q(x,y)\,w\mathrm{d}x\mathrm{d}y \quad (12)$$

Applying the variational principle $(\delta\Pi=0)$, the nonlinear governing equations of piezoelectric cylindrical shell reinforced with BNNTs can be derived as

$$\begin{aligned}&N_{x,x}+N_{xy,y}=0\\&N_{xy,x}+N_{y,y}=0\\&M_{x,xx}+2M_{xy,xy}+M_{y,yy}+N_y/R+N_xw_{,xx}+2N_{xy}w_{,xy}+N_yw_{,yy}+q=0\end{aligned} \quad (13)$$

where

$$\begin{Bmatrix}N_x\\N_y\\N_{xy}\end{Bmatrix}=\begin{bmatrix}A_{11}&A_{12}&0\\A_{12}&A_{22}&0\\0&0&A_{66}\end{bmatrix}\begin{Bmatrix}\varepsilon_x\\\varepsilon_y\\\varepsilon_{xy}\end{Bmatrix}-\begin{Bmatrix}N_x^T\\N_y^T\\N_{xy}^T\end{Bmatrix}-\begin{Bmatrix}N_x^P\\N_y^P\\N_{xy}^P\end{Bmatrix}, \quad (14)$$

in which

$$\begin{Bmatrix} N_x^T & M_x^T \\ N_y^T & M_y^T \\ N_{xy}^T & M_{xy}^T \end{Bmatrix} = \int_{-h/2}^{h/2} \begin{bmatrix} C_{11} & C_{12} & 0 \\ C_{11} & C_{22} & 0 \\ 0 & 0 & C_{66} \end{bmatrix} \begin{Bmatrix} \alpha_x \\ \alpha_y \\ 0 \end{Bmatrix} (1, z) \Delta T \mathrm{d}z \quad , (15)$$

In the above equations, the A_{ij}, D_{ij} are the tensile and bending rigidity and they can be defined as

$$\left(A_{ij}, D_{ij}\right) = \int_{-h/2}^{h/2} C_{ij} \left(1, z^2\right) \mathrm{d}z \qquad (i, j = 1,2,6) \quad (16)$$

Under the axisymmetrical circumstances, the circumferential displacement $v = 0$ and u, w is only the function of coordinate x. Hence, the second equation of Equation (12) is automatic balance and it can be omitted. Then by Equations (3), (14) and introducing the following dimensionless parameters,

$$\xi = \frac{x}{L}, \quad W = \frac{w}{h}, \quad U = \frac{u}{L}, \; Q = \frac{qL^4}{A_{22}h^3}, \; M_\xi = \frac{L^2 M_x}{A_{22}h^3}, \; H = \frac{L}{h}, \quad K = \frac{L^2}{Rh} \quad (17)$$

the nonlinear governing equations of axisymmetrical piezoelectric shell reinforced with BNNTs under electro- thermo-mechanical loadings can be reduced as

$$S_{11A}U_{,\xi\xi} - S_{12A}\frac{K}{H^2}W_{,\xi} + S_{11A}\frac{1}{H^2}W_{,\xi}W_{,\xi\xi} = 0$$

$$-\frac{1}{12}S_{11D}W_{,\xi\xi\xi\xi} + \left(S_{11A}H^2U_{,\xi} + \frac{S_{11A}}{2}W_{,\xi}^2 - S_{12A}KW - H^2\left(S_{11A}\alpha_x + S_{12A}\alpha_y\right)\Delta T - H^2 e_{11}E_x \right) W_{i,\xi\xi}$$

$$+S_{12A}KH^2U_{,\xi} + \frac{S_{12A}K}{2}W_{,\xi}^2 - S_{22A}K^2W - H^2K\left(S_{12A}\alpha_x + S_{22A}\alpha_y\right)\Delta T - H^2Ke_{12}E_x + S_{22A}Q = 0 \quad (18)$$

where, $S_{ijA} = A_{ij}/h, S_{ijD} = D_{ij}/\left(h^3/12\right)$.

Supposing the both ends of the shell are clamped, then the dimensionless boundary conditions are respectively as follows:

$$\begin{aligned} &N_\xi(0) = 0, \qquad W(0) = 0, \qquad W_{,\xi}(0) = 0 \\ &N_\xi(1) = 0, \qquad W(1) = 0, \qquad W_{,\xi}(1) = 0 \end{aligned} \quad (19)$$

where

$$N_{\xi} = S_{11A}\left(U_{,\xi} + \frac{1}{2H^2}W_{,\xi}^2\right) - S_{12A}\frac{K}{H^2}W - \left(S_{11A}\alpha_x + S_{12A}\alpha_y\right)\Delta T - e_{11}E_x \tag{20}$$

3. SOLUTION METHODOLOGY

For seeking the solution of differential Equation (18) with boundary condition (19), the dimensionless displacement functions W and U are dispersed in time-space domain to obtain their approximate solution.

Difference method is adopted in space domain. For the disposal of linear item, taking $W_{,\xi\xi}$ as example, we have

$$W_{,\xi\xi} = \frac{1}{(\Delta\xi)^2}\left[W(i+1,j) - 2W(i,j) + W(i-1,j)\right] \tag{21}$$

Referring to difference scheme, the difference expressions of the other linear items in governing equation can be easily achieved.

Then the nonlinear items of governing equations are linearized and can be written as follows [18],

$$(x \cdot y)_j = (x)_j (y)_{jp} \tag{22}$$

in which $(y)_{jp}$ is the value of the former iterative. For the primary iteration, secondary extrapolation method is introduced to obtain the value of $(y)_{jp}$, that is

$$(y)_{jp} = \Delta_1 (y)_{j-1} + \Delta_2 (y)_{j-2} + \Delta_3 (y)_{j-3} \tag{23}$$

As for different iterations, the coefficients Δ_1, Δ_2 and Δ_3 are decided as follows:

$$\begin{aligned} &j=1:\Delta_1=1, \quad \Delta_2=0, \quad \Delta_3=0 \\ &j=2:\Delta_1=2, \quad \Delta_2=-1, \quad \Delta_3=0 \\ &j\geq 3:\Delta_1=3, \quad \Delta_2=-3, \quad \Delta_3=1 \end{aligned} \tag{24}$$

After the equations and conditions are linearized and disposed by using the finite difference method, the nonlinear partial differential equations are transformed into linear algebraical equations expressed by difference schemes. These algebraic equations are solved by using the iteration method. For every step, the iterative lasts until the difference of the present value and the former is smaller than 0.01%, then continues the calculation of the next step.

4. NUMERICAL RESULTS AND DISCUSSION

The nonlinear bending of piezoelectric shell reinforced with BNNTs under electro-thermo-mechanical loadings is investigated in the following calculations. The geometrical parameter of the shell is $L/R=5/3, R/h=30$. The material used for matrix is PVDF and the reinforced material is BNNT. The material constants are listed in Table 1.

In the following figures (Figures 2-5), the vertical ordinate W and M_ξ are the dimensionless deflection and bending moment of each point on shell along x.

The effects of geometric nonlinear on the bending of piezoelectric shell reinforced with BNNTs are presented in Figure 2(a) and Figure 2(b). The volume fraction $V_f=0.6$, the voltage $V=0$ and the temperature rise $\Delta T=0$.

From the two figures, it can be noticed that the dimensionless deflection and bending moment of the shell in linear case is greater than that in nonlinear case, and this phenomenon becomes more evident when the transverse load Q increases. As we know, the linearity case is based on the limited deformation consumption, and the higher order item in the geometric relations is neglected while it is in consideration for the nonlinear case. So in some sense it can be concluded that the linear lowly predicts the stiffness of the structure. In order to reflect the property of the piezoelectric shell reinforced with BNNTs accurately, the consideration of the nonlinear effect is very necessary.

Figure 3 shows the effect of positive and negative voltage on the nonlinear bending of piezoelectric shell reinforced with BNNTs. The volume fraction $V_f = 0.6$, the temperature rise $\Delta T = 0$ and the mechanical load is taken as $Q = 150$. From the figure, it is observed that applying negative voltage to BNNT decreases the deflection and bending moment. This is due to the fact that applying negative voltage creates polarization in the BNNT in the longitudinal direction, and leads to its contraction. This makes the structure of BNNT more compact and strong, and correspondingly increases the structure's stiffness. Therefore, the deflection and bending moment of the structure decrease. Figure 3 also depicts the results of deflection and bending moment when applying positive voltage. As expected, the deflection and bending moment increase compared to normal situation, and the results can be explained using the similar concept as mentioned above.

The effects of temperature on the nonlinear bending of the shell are presented in Figure 4. The volume fraction $V_f = 0.6$, the voltage $V = 0$ and the mechanical load is taken as $Q = 150$ in the figure. As can be seen, the deflection as well as the bending moment increases when the temperature increases.

The effect of volume fraction on the nonlinear bending of piezoelectric shell reinforced with BNNTs is discussed in Figure 5. The voltage is $V = 0$, the temperature rise is $\Delta T = 0$ and the mechanical load is $Q = 150$ in this figure. It can be noticed that the deflection and the bending moment decrease when the volume fraction of BNNT in matrix (V_f) increases. That is due to the fact that the increase of volume fraction would increase the stiffness of the structure, and thus the deflection decreases.

Table 1. Mechanical, electrical and thermal properties of PVDF and BNNT.

$$C_{11} = 238.24(\text{GPa}),\quad C_{22} = 23.6(\text{GPa}),\ C_{12} = 3.98(\text{GPa}),\ C_{66} = 6.43(\text{GPa}),\ e_{11} = -0.135(\text{C/m}^2)$$

$$e_{12} = -0.145(\text{C/m}^2),\quad \varepsilon_{11}^{*} = 1.1.68\times10^{-8}(\text{F/m}),\quad \alpha_x = 7.1\times10^{-5}(1/\text{K}),\quad \alpha_y = 7.1\times10^{-5}(1/\text{K})$$

$$E = 1.8(\text{TPa}),\quad \nu = 0.34,\quad e_{11} = 0.95(\text{C/m}^2),\quad \alpha_x = 1.2\times10^{-6}(1/\text{K}),\quad \alpha_y = 0.6\times10^{-6}(1/\text{K})$$

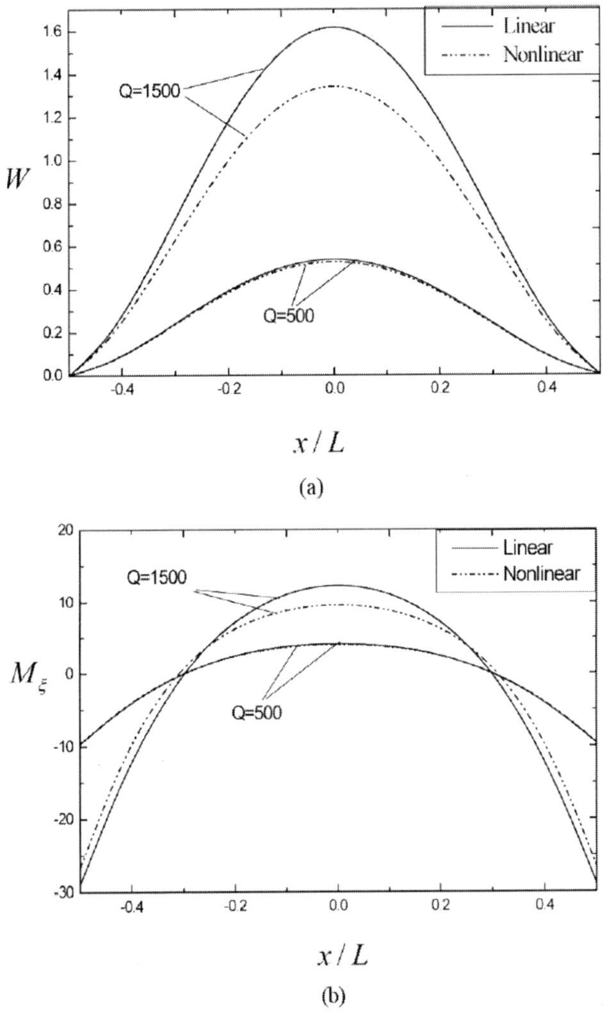

Figure 2. Effect of geometric nonlinear on bending of piezoelectric shell reinforced with BNNTs; (a) Deflection of each point along x; (b) Bending moment of each point along x.

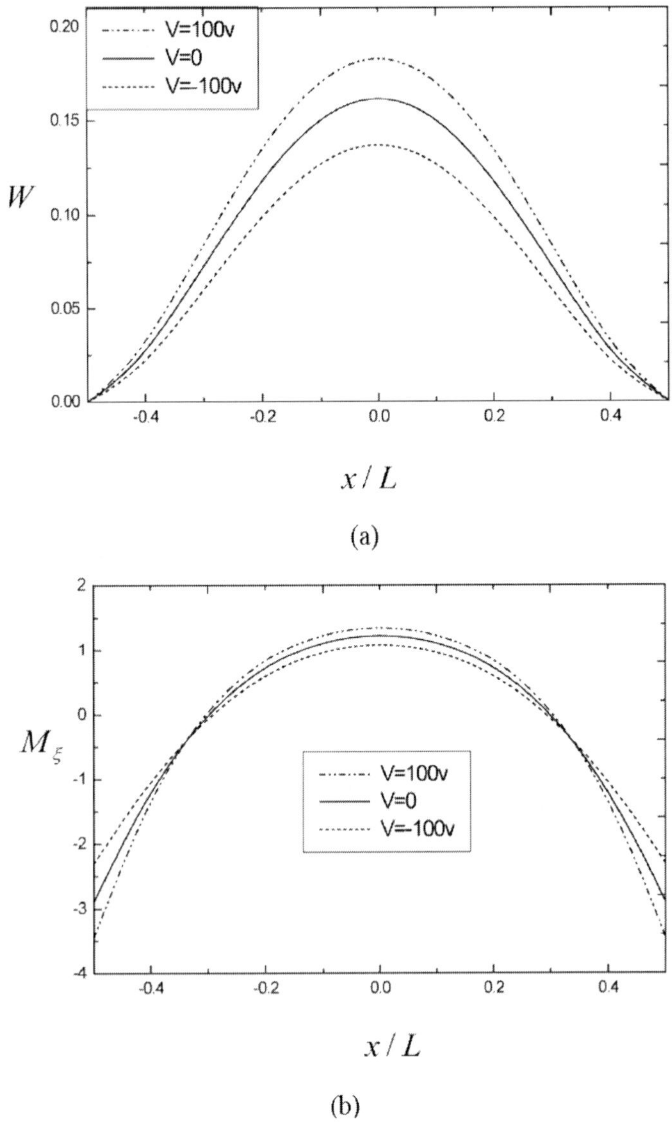

Figure 3. Effect of voltage on nonlinear bending of piezoelectric shell reinforced with BNNTs; (a) Deflection of each point along x; (b) Bending moment of each point along x.

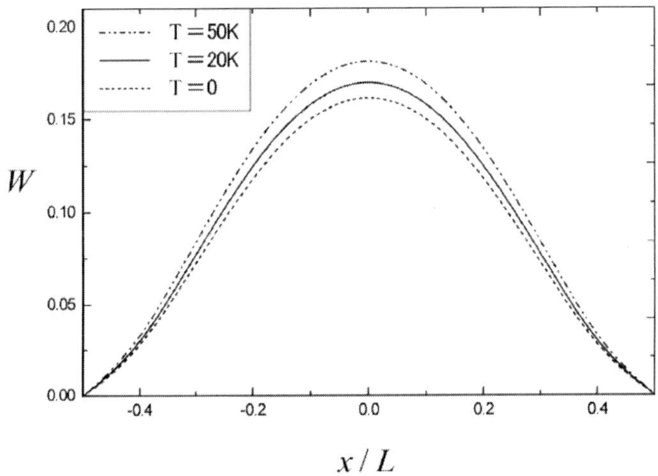

(a)

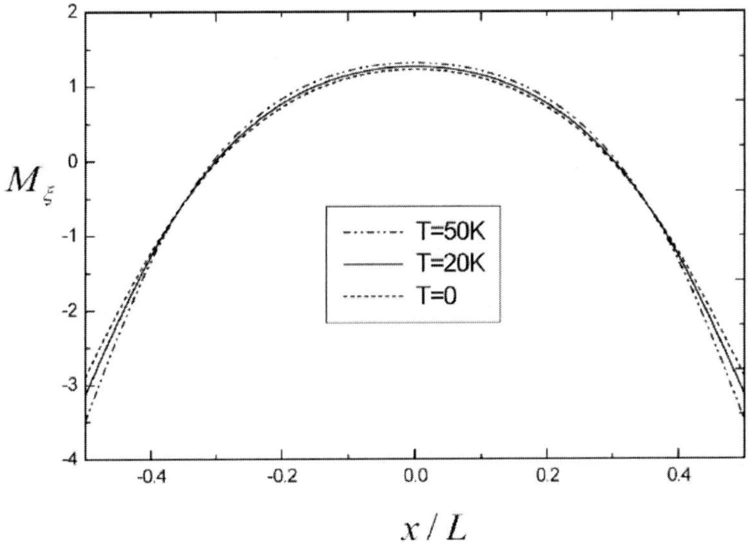

(b)

Figure 4. Effect of temperature on nonlinear bending of piezoelectric shell reinforced with BNNT; (a) Deflection of each point along x; (b) Bending moment of each point along x.

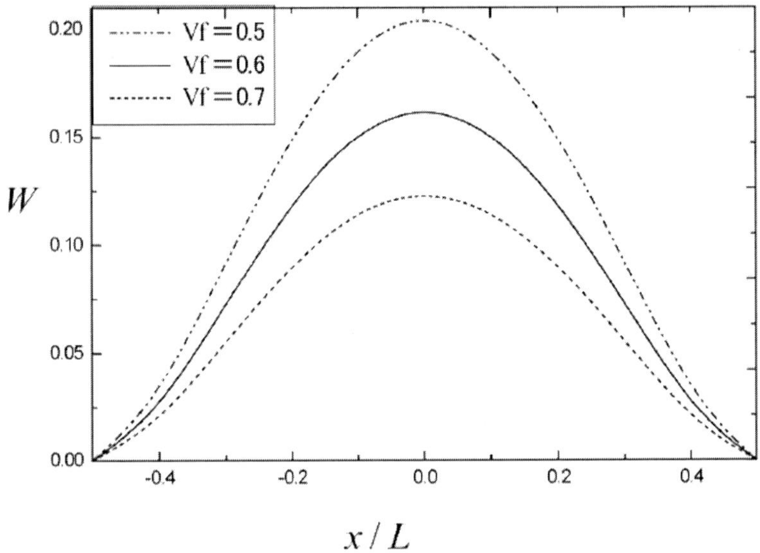

(a)

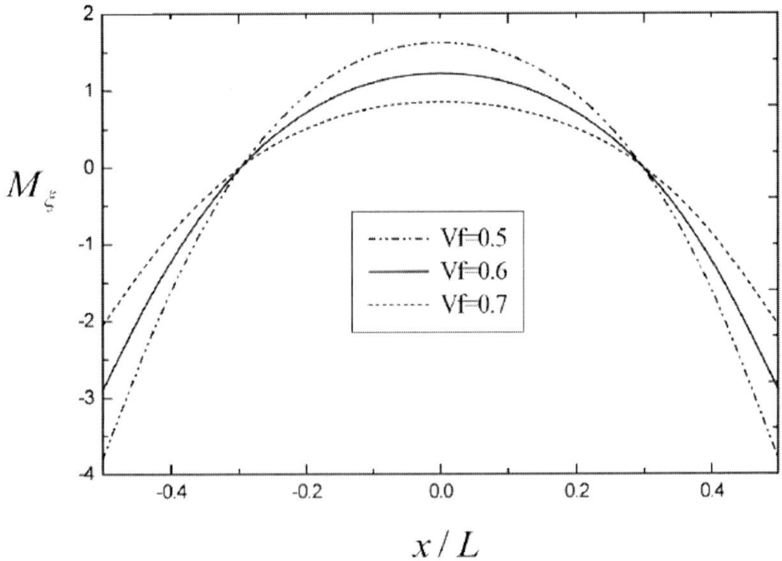

(b)

Figure 5. Effect of volume fraction on nonlinear bending of piezoelectric shell reinforced with BNNT. (a) Deflection of each point along x; (b) Bending moment of each point along x.

5. CONCLUSIONS

In present study, the governing equations of nonlinear bending are presented for piezoelectric cylindrical shell reinforced with BNNTs under combined electro-thermo-mechanical loadings. Results indicate that some parameters, including geometric nonlinear, voltage, temperature, volume fraction and so on, have significant influence on the deflection and bending moment of the shell. The following conclusions may be drawn from the present work:

1) The deflection and bending moment of the shell in linear case is greater than that in nonlinear case, and the nonlinear effect enhances when the transverse load Q increases.
2) Applying positive and negative voltage to BNNT leads to increase and decrease of the deflection and bending moment.
3) The deflection as well as the bending moment increases with the increase of temperature, and decreases when the volume fraction of BNNT in matrix increases.

ACKNOWLEDGEMENTS

The work described in this paper was supported by National Natural Science Foundation of China (No: 11102028, 11202038, 11172051) and Research Foundation of Education Bureau of Hunan Province, China (No: 11B005).

CITE THIS PAPER

JinhuaYang,PengjunZhang, (2015) Nonlinear Bending of Piezoelectric Cylindrical Shell Reinforced with BNNTs under Electro-Thermo-Mechanical Loadings. *Materials Sciences and Applications*,**06**,743-752.

REFERENCES

1. Chen, Y., Zou, J., Campbell, S.J. and Le Caer, G. (2004) Boron Nitride Nanotubes: Pronounced Resistance to Oxidation. Applied Physics Letters, 84, 2430-2432.

2. Suryavanshi, A.P., Yu, M.F., Wen, J.G., Tang, C.C. and Bando, Y. (2004) Elastic Modulus and Resonance Behavior of Boron Nitride Nanotubes. Applied Physics Letters, 84, 2527-2529.

3. Terrones, M., Romo-Herrera, J.M., Cruz-Silva, E., López-Urías, F., Muñoz-Sandoval, E., Velázquez-Salazar, J.J., Terrones, H., Bando, Y. and Golberg, D. (2007) Pure and Doped Boron Nitride Nanotubes. Materials Today, 10, 30-38.

4. Yao, L.Q., Zhang, J.G., Lu, L. and Lai, M.O. (2004) Nonlinear Static Characteristics of Piezoelectric Bending Actuators under Strong Applied Electric Field. Sensors and Actuators A: Physical, 115, 168-175.

5. Shen, H.-S. (2004) Nonlinear Bending Analysis of Unsymmetric Cross-Ply Laminated Plates with Piezoelectric Actuators in Thermal Environments. Composite Structures, 63, 167-177.

6. Shegokar, N.L. and Lal, A. (2013) Stochastic Nonlinear Bending Response of Piezoelectric Functionally Graded Beam Subjected to Thermoelectromechanical Loadings with Random Material Properties. Composite Structures, 100, 17-33.

7. Narita, F., Shindo, Y. and Mikami, M. (2005) Analytical and Experimental Study of Nonlinear Bending Response and Domain Wall Motion in Piezoelectric Laminated Actuators under Electric Fields. Acta Materialia, 53, 4523-4529.

8. Beldica, C.E. and Hilton, H.H. (2001) Nonlinear Viscoelastic Beam Bending with Piezoelectric Control-Analytical and Computational Simulations. Composite Structures, 51, 195-203.

9. Yan, W., Wang, J. and Chen, W.Q. (2014) Cylindrical Bending Responses of Angle-Ply Piezoelectric Laminates with Viscoelastic Interfaces. Applied Mathematical Modelling, 38, 6018-6030.

10. Narita, F., Shindo, Y. and Hayashi, K. (2005) Bending and Polarization Switching of Piezoelectric Laminated Actuators under Electromechanical Loading. Computers & Structures, 83, 1164-1170.

11. Li, Y.S., Feng, W.J. and Cai, Z.Y. (2014) Bending and Free Vibration of Functionally Graded Piezoelectric Beam Based on Modified Strain Gradient Theory. Composite Structures, 115, 41-50.

12. Sladek, J., Sladek, V., Stanak, P., Zhang, C.Z. and Wünsche, M. (2013) Analysis of the Bending of Circular Piezoelectric Plates with Functionally Graded Material Properties by a MLPG Method. Engineering Structures, 47, 81-89.

13. Fu, Y.M., Wang, J.Z. and Mao, Y.Q. (2012) Nonlinear Analysis of Buckling, Free Vibration and Dynamic Stability for the Piezoelectric Functionally Graded Beams in Thermal Environment. Applied Mathematical Modelling, 36, 4324-4340.

14. Salehi-Khojin, A. and Jalili, N. (2008) Buckling of Boron Nitride Nanotube Reinforced Piezoelectric Polymeric Composites Subject to Combined Electro-Thermo-Mechanical Loadings. Composites Science and Technology, 68, 1489-1501.

15. Mosallaie Barzoki, A.A., Arani, A.G., Kolahchi, R. and Mozdianfard, M.R. (2012) Electro-Thermo-Mechanical Torsional Buckling of a Piezoelectric Polymeric Cylindrical Shell Reinforced by DWBNNTs with an Elastic Core. Applied Mathematical Modelling, 36, 2983-2995.

16. Arani, A.G., Amir, S., Shajari, A.R. and Mozdianfard, M.R. (2012) Electro-Thermo-Mechanical Buckling of DWBNNTs Embedded in Bundle of CNTs Using Nonlocal Piezoelasticity Cylindrical Shell Theory. Composites Part B: Engineering, 43, 195-203.

17. Tan, P. and Tong, L.Y. (2001) Micro-Electromechanics Models for Piezoelectric-Fiber-Reinforced Composite Materials. Composites Science and Technology, 61, 759-769.

18. Fu, Y.M. (1997) Nonlinear Dynamic Analysis of Structures. Jinan University Press, Guangzhou.

Chapter 4

Functional Piezocrystal Characterisation under Varying Conditions

Xiaochun Liao [1,2,†], Zhen Qiu [3,†], Tingyi Jiang [2,†], Muhammad R. Sadiq [1], Zhihong Huang [2], Christine E. M. Demore [1] and Sandy Cochran [4,*]

[1]*Institute for Medical Science and Technology, University of Dundee, 1 Wurzburg Loan, Dundee DD2 1FD, UK*

[2]*School of Science and Engineering, University of Dundee, Dundee DD1 4HN, UK*

[3]*Department of Electronics and Electrical Engineering, University of Strathclyde, Glasgow G1 1XW, UK*

[4]*School of Engineering, University of Glasgow, Glasgow G12 8QQ, UK*

ABSTRACT

Piezocrystals, especially the relaxor-based ferroelectric crystals, have been subject to intense investigation and development within the past three decades, motivated by the performance advantages offered by their ultrahigh piezoelectric coefficients and higher electromechanical coupling coefficients than piezoceramics. Structural anisotropy of piezocrystals also provides opportunities for devices to operate in novel vibration modes, such as the d_{36} face shear mode, with domain engineering and special crystal cuts. These piezocrystal characteristics contribute to their potential usage in a wide range of low- and high-power ultrasound applications. In such applications, conventional piezoelectric materials are presently subject to varying mechanical stress/pressure, temperature and electric field conditions. However, as observed previously, piezocrystal properties are significantly affected by a single such condition or a combination of conditions. Laboratory characterisation of the piezocrystal properties under these conditions is therefore essential to fully understand these materials and to allow electroacoustic transducer design in realistic scenarios. This will help to establish the extent to which these high performance piezocrystals can replace conventional piezoceramics in demanding applications. However, such characterisation requires specific experimental arrangements, examples of which are reported here, along with relevant results. The measurements include high frequency-resolution impedance spectroscopy with the piezocrystal material under mechanical stress 0–60 MPa, temperature 20–200 °C, high electric AC drive and DC bias. A laser Doppler vibrometer and infrared thermal camera are also integrated into the measurement system for vibration mode shape scanning and thermal conditioning with high AC drive. Three generations of piezocrystal have been tested: (I) binary, PMN-PT; (II) ternary, PIN-PMN-PT; and (III) doped ternary, Mn:PIN-PMN-PT. Utilising resonant mode analysis, variations in elastic, dielectric and piezoelectric constants and coupling coefficients have been analysed, and tests with thermal conditioning have been carried out to assess the stability of the piezocrystals under high power conditions.

Keywords: piezocrystal; piezoelectric characterisation; high power; high resolution; high stress field; high temperature field; high electric drive field; mode shape; thermal response

1. INTRODUCTION

Piezocrystals of the relaxor-PT type such as $(x)Pb(Mg_{1/3}Nb_{2/3})O_3$-$(1-x)PbTiO_3$ (PMN-PT), $(x)Pb(In_{1/2}Nb_{1/2})O_3$-$(1-x-y)$-$Pb(Mg_{1/3}Nb_{2/3})O_3$-$(y)PbTiO_3$ (PIN-PMN-PT) and Mn-doped PIN-PMN-PT (Mn:PIN-PMN-PT), respectively termed Generations I, II and III, have demonstrated electrical, mechanical and piezoelectric properties that can be translated into higher electroacoustic transducer performance than can be achieved with conventional piezoceramics [1] and PMN-PT has now established itself as the material of choice in biomedical ultrasound transducers [2,3,4,5]. However, this application has relatively undemanding requirements in terms of average power and operation at elevated temperature and pressure because of the output power and temperature safety limitations imposed by medical applications, and this matches well with the limited coercive field, E_c, mechanical quality factor, Q_m, and ferroelectric rhombohedral-tetragonal (F_{RT}) phase transition and Curie temperatures (T_{RT} and T_C) of PMN-PT. In contrast, there are numerous applications, for example in underwater sonar, the oil and gas industries, medical therapy and ultrasound-actuated tools, where high average power handling and elevated temperature and pressure may be encountered [6,7]. These applications have prompted research into PIN-PMN-PT and Mn:PIN-PMN-PT, the former now available commercially (e.g., CTG Advanced Materials, Bolingbroke, IL, USA) and the latter in small quantities for testing purposes (TRS Technologies, State College, PA, USA).

Almost all piezocrystal production is based on the Bridgman technique [8,9], in which individual boules are grown in Pt crucibles, typically of 4″ diameter for commercial supply. Unlike piezoceramic, the production of which by extrusion permits continuous and close to uniform processing, piezocrystal boule growth carries the possibility of boule-to-boule variation and, more fundamentally, involves inherent compositional gradients in the PT fraction. Continuous growth has been under development for some time [10] to reduce this problem but still awaits commercial adoption. All these lead to greater variations in material properties during production than are typical of piezoceramic. In turn, there is greater demand for routine measurement of the characteristics of piezocrystal. For the design of many electroacoustic transducers, the full elastoelectric matrix is required [11]. This can be measured with a standard process [12] based on analysis of the complex electrical impedance spectra of several material samples, each supporting a single, primary ultrasonic mode. Because of the possibility of variation between samples and the difficulty in achieving self-consistency in measurements from different samples, alternative techniques have been developed using ultrasound transit time measurement and resonant ultrasound spectroscopy [13,14] to obtain the elastic parameters, with the results combined with electrical measurements to obtain the full elastoelectric matrix. In this way, the material properties can be obtained from just one sample. This technique has also been extended to measurement at elevated temperatures [14] but not yet at pressures above ambient.

Whilst the full elastoelectric matrix is necessary for detailed transducer design using contemporary computer-modelling and virtual prototyping tools such as finite element analysis (FEA, e.g., PZFlex, Weidlinger Associates Inc., Mountain View, CA, USA), it is also possible to gain significant insight into piezocrystal properties and behaviour through characterisation of specific sample configurations [15]. This approach is appropriate when new materials are in development, both piezocrystals and other types such as Pb-free piezoceramics [16]. Although only limited subsets of material properties can be obtained, they are nevertheless valuable in controlling material production, in estimation of likely transducer properties, and sometimes for transducer design using constrained models [17]. Crucially, this approach also allows the materials to be characterised under practical conditions of elevated temperature, above ambient uniaxial pressure, and at high electrical excitation levels, the latter having the potential to cause self-heating and to elicit the effects of coercive field limits. In addition, it is possible to explore behaviour relating to specific material properties, such as acoustic modes relating to the unusually high d_{31}, d_{32}, d_{15} and d_{36} coefficients of relevant piezocrystal cuts [1], which can have many practical applications.

To evaluate the suitability of piezocrystals for practical applications, characterisation is currently undertaken in both industrial and academic research in relation to performance, consistency and stability. In industry, a datasheet is often supplied on delivery, including parameters such as permittivity, ε_{ij}, piezoelectric constants, d_{ij}, elastic compliance constants, s_{ij}, electromechanical coupling coefficients, k_{ij}, and frequency constant, N_{ij}. These parameters define the material's functionality in a particular vibration mode and always in an ambient environment. By examining the parameters in the datasheets, consistency across different piezocrystal samples can be checked within a batch or between different batches. It is rare to check the piezocrystal stability under varying conditions in industrial production unless a new application specifically encourages collaboration with academics, e.g., to examine a newly designed composition or mode [18]. Collaboration between industry and academic research then motivates more detailed and comprehensive characterisation [19,20] and the possibility of new standards such as the one under development, under the auspices of the IEEE, for relaxor-based piezocrystal for transducer and actuator applications [21].

In academic research, performance characterisation frequently includes measurement of the full elastoelectric matrix to gain understanding of a material under test [22,23,24]. This full matrix characterisation can be combined with investigation of compositional variation, domain engineering and special cuts to explore possible ultrahigh piezoelectric constants and novel vibration modes [8,24,25,26,27]. In addition to the full elastoelectric matrix, some key parameters are also measured, including k_{ij}, dielectric loss, tanδ_e, and mechanical quality factor, Q_m. For particular applications, a figure of merit (FOM) may also be calculated, such as $k_{ij}^2Q_m$ for active underwater transducers or d_hg_h/tanδ_e for passive underwater hydrophones where d_h and g_h are the piezoelectric constants under hydrostatic conditions [1].

Stable piezocrystal performance intrinsically relies on phase stability and can thus deteriorate significantly during phase transitions caused by external conditions such as temperature, pressure and electrical excitation. Stability is always explained in relation to E_c, T_C and T_{RT}, and can be quantified as a percentage variation resulting from applied conditions, e.g. $\Delta d_{ij}/d_{ij}$ [28,29]. Characterisation of piezocrystals under deliberately conditioned fields is used to mimic real applications; this is crucial to understand performance variation, to design piezocrystal transducers with FEA, and to apply piezocrystal devices in practice. Characterisation as a function of temperature is usually conducted in an oven or thermal chamber; in this study, we also report self-heating as a parameter in stability testing and characterisation. Such self-heating directly relates to the mechanisms of dielectric, piezoelectric and mechanical losses, the equations for which have been derived for the common length-extensional (LE), length-thickness-extensional (LTE) and thickness-shear (TS) modes in piezoceramics but not yet in piezocrystal [30,31]. Pressure can be applied hydrostatically or along specific axes. Whilst hydrostatic pressure may appear most relevant, many high power transducer designs for which piezocrystal is suitable, such as the Tonpilz type for underwater SONAR [32,33] and various ultrasonically-actuated devices [8], include a uniaxial prestress. The simple uniaxial configuration can thus generate valuable results [34] and is one focus of this paper. In practice, piezocrystal may be subject to combinations of temperature and pressure making the situation more complicated and performance stability worse, and there is thus a pressing need for experimental exploration of such situations.

This paper focuses on application-oriented characterisation of the three generations of piezocrystals, all grown conventionally with the Bridgman method. The relatively diverse measurement methods that have been employed are described in Section 2. These include electrical impedance spectroscopy with standard instrumentation, which allows high average-power excitation when configured appropriately. In Section 3, a functional characterisation system is introduced including environmental and excitation factors relating to potential passive and active applications respectively. Section 4 outlines measurement conditions and results. The use of standard measurement instrumentation is combined with environmental heating in the range from ambient temperature to approximately 200 °C and with the application of uniaxial pressure up to 60 MPa, consistent with full ocean depth in underwater SONAR systems. High-average power excitation is combined with thermal imaging and control techniques to allow specific piezocrystal temperatures to be set and examined. The use of laser Doppler vibrometry (LDV) is illustrated through the direct measurement of piezoelectric coefficients and comparison is made with the authors' use of the Berlincourt method. Ultrasonic transit-time testing can also be utilised to obtain elastic properties. In Section 5, the resulting material characteristics are related to fundamental properties, including crystallographic phase transitions and changes in elastic, dielectric and piezoelectric properties, and comparisons with other measurement techniques are discussed. Conclusions are drawn in Section 6 and measurement avenues open to further exploration are considered.

2. CHARACTERISATION METHODS FOR PIEZOELECTRIC MATERIALS

2.1. Material Properties and Measurement Methods

The translation of microstructural characteristics of piezocrystals such as crystal phase, domain configuration and polarization dynamics into macroscopic properties determines functional performance, with the macroscopic properties mathematically represented by material properties in tensor form. Using reduced notation, these are collated in matrices which can take multiple forms according to convenience. Here, we assume the full elastoelectric matrix contains elastic (stiffness), dielectric and piezoelectric properties, c_{ij}, ε_{ij} and d_{ij}. For functional evaluation, another two performance characteristics k_{ij} and Q_m are also often used. The measurement of all these characteristics can be made using: (1) the Berlincourt method; (2) electrical impedance spectroscopy; (3) LDV; and (4) ultrasonic methods. Not all four methods can characterise all the properties; compatibility between properties and the measurement methods is shown in Table 1.

Table 1. Compatibility between piezocrystal properties and measurement methods.

Properties	Measurement Methods			
	Berlincourt Method	Electrical Impedance Spectroscopy	Laser Doppler Vibrometry	Ultrasonic Methods
Dielectric Properties		√		
Piezoelectric Properties	√	√	√	
Elastic Properties		√		√
Electromechanical Coupling Coefficient		√	√	
Mechanical Quality Factor		√	√	

The *Berlincourt method* [35] measures the fundamental piezoelectric charge coefficient, d_{ij}, based on the direct piezoelectric effect, when electrical charge is generated from applied mechanical force. During quasi-static measurement, the mechanical force is applied at a frequency lower than any resonance in the sample under test. The Berlincourt method is thus insensitive to sample geometry, as long as electrodes are present to make electrical contact. It can obtain comparable d_{33} values from both a tall, narrow LE mode bar and from a thin thickness-extensional (TE) mode plate, if the two samples are both fully polarised. Variations of d_{33} values in practical measurement may be induced by incomplete polarisation, as previously reported for PMN-PT piezocrystal plates with different thicknesses [36]. Commercial d_{33} meters provide a fast, convenient way to measure d_{31} and d_{15} values with specially designed fixtures (e.g., ZJ-6B, CTG Advanced Materials, Bolingbroke, IL, USA).

Electrical impedance spectroscopy of a piezocrystal sample is achieved by applying an AC signal with frequency swept over a selected range. The current amplitude and phase are recorded and the complex impedance is obtained with reference to the driving voltage [37]. Based on the impedance spectrum, electrical and mechanical resonances and their corresponding frequencies can be identified, allowing evaluation of the piezoelectric and elastic properties strongly coupled to particular resonant modes. Many commercial impedance analysers are available, but the same technique can also be realised with less specialised instrumentation, as we describe later for high power characterisation. Methods of analysis of electrical impedance spectra in the IEEE standard [12] specify multiple resonant geometries with poling directions and recommended aspect ratios to ensure the target resonant behaviour is the main motion in the specified direction, well separated from any others in frequency. Based on the resonance frequencies of these samples, all the fundamental piezoelectric constants can be calculated. However, this requires multiple samples which can lead to inconsistency in piezocrystal characterisation.

LDV is a direct, non-invasive way to measure d_{ij} based on the inverse piezoelectric effect. In this method, the LDV measures the surface displacement when a non-resonant electrical field is applied across the electrodes of a given sample; d_{ij} can be calculated from the measured displacement, applied voltage and sample geometry. The vibration boundary conditions must be strictly respected by mounting the sample on a nodal plane and the measurement frequency should differ from any resonance in the sample to avoid the interference from the effect of Q_m. LDV can also be used with frequency sweeping across a resonance to evaluate both k_{ij} and Q_m.

A simple ultrasonic method based on the pulse-echo technique can be used to measure the elastic properties of solid materials. The full set of stiffness constants can be determined with longitudinal and shear waves applied in different orientations. It has been suggested that this method provides better accuracy in determining the elastic constants of piezoelectric material than the impedance spectroscopy method [38], but it has insufficient accuracy for the piezoelectric constants. A combined ultrasonic-resonance method has therefore been developed by Cao *et al.* [39], and this also reduces the number of samples required. Recently, the ultrasound method has been combined with inverse electrical impedance spectroscopy to determine the full elastoelectric property matrix from only one sample [13]. This technique guarantees self-consistency of the full property matrix by eliminating sample-to-sample variation.

2.2. Other Methods for Sample Size Reduction and Loss Mechanism Characterisation

Some other newly proposed characterisation methods for piezocrystals are in development. Drawing on previous work in geology and passive material characterization, resonant ultrasound spectroscopy (RUS) has been used to determine the elastic properties of a hard lead zirconate titanate (PZT) composition using one sample under different temperature conditions [14]. However, the single sample is usually a cube which RUS requires to be mounted on two body diagonal corners. Care must be taken to avoid damage during preloading, and the mounting method makes it difficult to test these samples under high stress conditions. Another method has been proposed based on loss mechanisms [30] using loss factor equations derived for the modes associated with k_{33}, k_{31}, k_{15} and k_t. Based on these, the loss factors can be easily characterised for dielectric, piezoelectric and elastic losses rather than to measure the hysteresis between stress, strain, electric displacement and electric field. This method may help understanding of heating effects in piezoelectric transducers, but it has not yet been applied to piezocrystals.

2.3. Requirements of Piezocrystal Samples

For many years, the IEEE standard method and its precursors [11] have been used to determine the full elastoelectric matrix. This requires a separate material sample with a specified geometry for each of several defined propagation modes, as shown in Figure 1. The number of samples differs according to the crystallographic symmetry of the material system to be characterised. The (001)-poled piezocrystal has 4 mm symmetry and therefore requires five samples, operating in the TE, TS, LE and LTE modes, with the latter also required 45°-rotated. Two of these samples raise particular issues. The TS sample must be poled in one direction then have the poling electrodes removed and replaced on a different pair of surfaces; none of the other samples requires this preparation. The LE sample is specified to have a width:thickness aspect ratio of 1:10. Because of the compositional gradient along the length of a piezocrystal boule, combined with a relatively short section with properties that are acceptable in terms of the performance of electroacoustic transducers, even a lateral cross-section as small as 1 mm square, from which a 10 mm long LE bar is cut, may result in significant compositional variation within the sample. In addition, long samples can elicit problems in poling and samples of small cross-section have high impedance. The specified aspect ratio for the LE bar is therefore sometimes adjusted, e.g., to 1:5 or in the authors' experience and as published [40,41] as low as 1:3, provided that unimodal LE operation is observed explicitly.

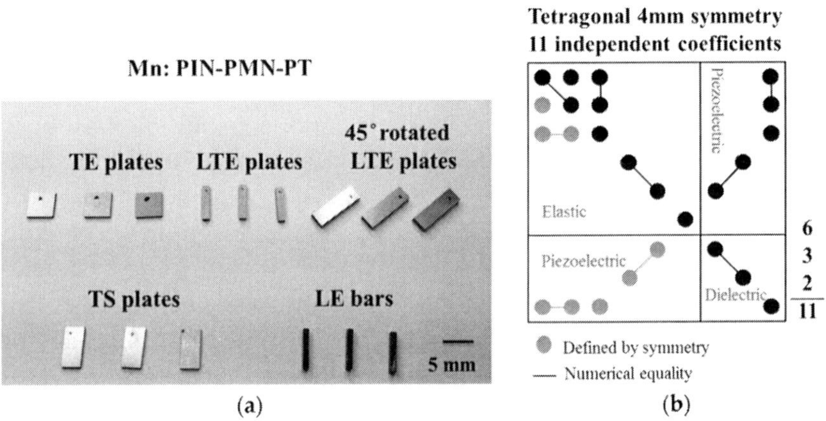

Figure 1. (**a**) Sample set of Mn:PIN-PMN-PT for full elastoelectric matrix; (**b**) Dependence of elastic, piezoelectric and dielectric constants for 4 mm symmetry in full elastoelectric matrix.

Given the variation in material properties along a crystal boule and potential variations in poling, the use of multiple samples to provide subsets of the elastoelectric matrix raises the possibility of inconsistency in the matrix and this has been found to be the case in practice [15]. Therefore, the alternative methods outlined previously are under development, with the aim to reduce the number of samples required [13,14]. However, such methods leave two issues unresolved. First, it would be very unusual to use a multimodal piezoelectric material configuration in a practical electroacoustic transducer, as the intention is usually to base such transducers on unimodal configurations to avoid performance degradation by cross-coupling of the modes. Thus, the single sample on which characterisation is based is likely to differ from the configurations used in practical transducers with consistency of poling requiring particular attention. Second, compositional variation along the length of the boule and potential variation through insufficiently controlled growth mean that measurement even of a single sample does not represent the whole boule. Nevertheless, single sample characterisation overcomes difficulties with self-consistency because of sample-to-sample variation and is thus advantageous compared with the standard method.

An alternative approach is to obtain incomplete but useful measurements from selected material configurations. For example, for the many electroacoustic transducers utilising the TE mode, very useful information can be gained by characterising TE mode samples and the same principle applies for other modes. Since most transducers intentionally exploit only one mode, a great deal of value can be gained from such characterisation in estimating transducer performance, checking sample-to-sample uniformity and elucidating behaviour under different conditions of temperature and pressure. This approach is the one taken here with measurements illustrated in the following sections based on TE, LTE and novel d_{36} face shear (FS) mode behaviours. In this context, the samples on which most results in this paper are based are TE mode with typical thickness of 0.5 mm and essentially unimodal behaviour achieved over the fundamental ultrasonic frequency range by meeting the constraint of a width:thickness ratio of 10:1. Thus, most samples have dimensions 5 mm × 5 mm × 0.5 mm. Furthermore, a representative range of properties is often used. Here, we focus on the dielectric coefficient, ε_{r33}^{S}, the stiffness, c_{33}^{D}, the piezoelectric coefficient, e_{33}, and the electromechanical coupling coefficient, k_t, as these provide useful information across a wide range of materials and applications.

3. PIEZOELECTRIC FUNCTIONAL CHARACTERISATION SYSTEM

The system on which the work reported here is based is shown schematically in Figure 2. Key components are an electrical impedance analyser (e.g., 4395A, Agilent/Keysight, Santa Clara, CA, USA) and a material testing system (MTS, SSID 5966, Instron Ltd., High Wycombe, UK) with which a bespoke, fan-assisted oven is integrated, allowing penetration of the roof and floor by the MTS applicator bars. These instruments are controlled with a PXI-based PC running software implemented in LabVIEW (National Instruments, Houston, TX, USA). This software is able to select a particular pressure profile with multiple levels in the range 0–60 MPa and simultaneously control the oven temperature in the range from ambient to 200 °C. Data can be acquired over multiple contiguous frequency ranges, thus extending the number of sampling points arbitrarily beyond the standard limit of 801 points from the impedance analyser, consequently enhancing the measurement resolution. In practice, the system is operated in a mode in which a given temperature is set, starting with the lowest temperature required, then the pressure is incremented, applied uniaxially with brass contact pads dry-coupled to the sample under test to minimise energy leakage from it. After the pressure has reached the specified maximum and the first set of measurements has been completed, the pressure is reduced to the minimum, the temperature is increased to the next set level, and the pressure cycle is repeated, as shown in Figure 3g. Dependency of material properties on previous conditions must thus be considered in the analysis of the results, which is based on calculation of material properties from resonant frequencies observed in the impedance spectra. This basic configuration focuses on the environmental factors of elevated temperature and pressure, targeting passive piezocrystal applications.

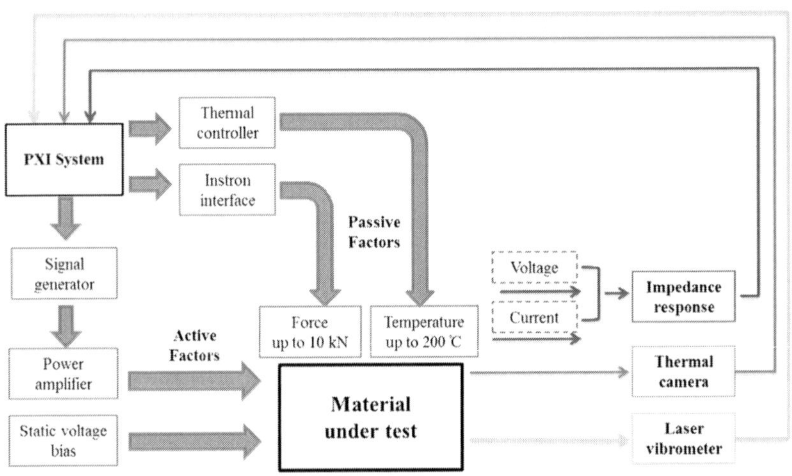

Figure 2. Piezoelectric functional characterisation system diagram.

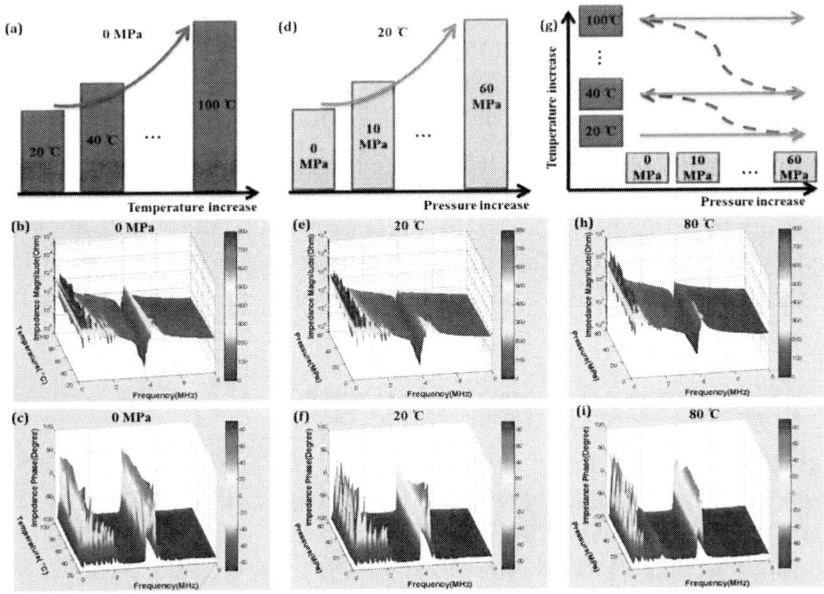

Figure 3. Characterisation of Mn:PIN-PMN-PT: (**a**) Configuration for variation of temperature at pressure of 0 MPa with measurement of (**b**) impedance magnitude and (**c**) impedance phase; (**d**) Characterisation configuration for variation in pressure at ambient temperature with measurement of (**e**) impedance magnitude and (**f**) impedance phase; (**g**) Characterisation configuration for variation in temperature and pressure with measurement of (**h**) impedance magnitude and (**i**) impedance phase.

For characterisation of piezocrystals under active application conditions, the basic system is complemented by other measurement capabilities. If necessary, and of particular relevance to PMN-PT because of E_c values in the range 1–2 kV·mm^{-1}, an electrical field bias can be applied with a high voltage power supply. For the TE samples defined previously, V = 1 kV is sufficient for many measurements. A scanning LDV can provide information about modal behaviour and as noted previously, can be used directly to determine the piezoelectric modulus through a few single-point measurements of displacement of the relevant surface of the sample under off-resonant excitation. It is also possible to measure electrical impedance at high average-power excitation levels by driving the sample under test with a power amplifier and determining voltage, current and their relative phase with appropriate probes. This technique has been taken further by observing heating of the sample under test with a thermal imaging camera (TIM-160, Micro-Epsilon Ltd., Ortenburg, Germany). This self-heating measurement provides information on both temporal thermal behaviour and impedance spectral response at a set temperature achieved by controlling the applied voltage and current. In summary, the material characterisation setup, described above, can be used experimentally to characterise piezocrystal material in both passive and active electroacoustic transduction conditions.

4. PIEZOCRYSTAL CHARACTERISATION

Based on the designed piezoelectric functional characterisation system, measurements were performed on three generations of piezocrystals. Most measurements were carried out on Generation II and III, 33% PIN-PMN-PT and Mn:PIN-PMN-PT samples, respectively, while we also include measurement results on PMN-29% PT samples from the literature [15] where they provide useful data for this paper. Effects of increased environmental temperature and pressure were investigated under passive conditions, *i.e.* with the piezocrystal samples un-driven. Electric field effects were determined with the piezocrystal driven at various voltages. In addition, LDV was used under ambient conditions to measure the piezoelectric constants d_{33}, d_{31}, d_{15} and d_{36}, with the results compared with the values obtained by the Berlincourt method and electrical impedance spectroscopy. Although only applied to Mn:PIN-PMN-PT, this is sufficient to demonstrate characterisation with LDV.

4.1. Environmental Temperature and Pressure Effects

4.1.1. Impedance Spectra with Elevated Temperature and Pressure

As an illustration of our passive characterization methods, Figure 3 presents electrical impedance spectra of a 5 mm × 5 mm × 0.5 mm TE mode plate of Mn:PIN-PMN-PT with varying temperature, *T*, and pressure, *P*. Although rarely found in the materials literature, this is a familiar data presentation format for transducer developers evaluating material and device performance. The frequency ranges for these graphs were chosen to centre the TE-mode resonance with lateral resonances also appearing at much lower frequencies.

Although these minor modes are not analysed further here, they provide clear evidence of behavioral changes with environmental conditions. As described previously, temperature and uniaxial compressive stress were applied simultaneously to the sample, following a pre-programmed testing protocol. Although impedance spectra were recorded for every condition and further analysed in later sections, only selected results are presented here: Figure 3a–c present results for 20 °C ≤ T ≤ 100 °C with P ambient; similarly, Figure 3d–f present results for 0 MPa ≤ P ≤ 60 MPa at ambient temperature; and Figure 3g–i show the combined effects of T and P, with T = 80 °C and 0 MPa ≤ P ≤ 60 MPa.

As expected, Mn:PIN-PMN-PT is relatively stable over the temperature range up to 100 °C which is most important for applications and which was used for the measurements shown in Figure 3, taking into account that T_{RT} > 120 °C. The resonance frequencies and magnitudes for the minor lateral resonances and the main TE resonance decrease as only T increases. Similarly, the application of only P impedes vibration in both the planar and thickness directions, manifesting as decreases in the resonance frequencies and magnitudes; the effect of pressure is clear even at levels as low as P = 5 MPa. Combined T and P have increased influence on piezoelectric behaviour, with a significant decrease and then a slight recovery for P > 40 MPa, as seen in Figure 3i.

4.1.2. Property Variation with Environmental Temperature

Using the IEEE resonance method, it is possible to extract specific material properties from the data presented in Figure 3. With the TE mode plates under test, ε_{r33}^S, k_t, e_{33} and c_{33}^D can be derived from:

$$\varepsilon_{r33}^{S} = t \cdot C / (A \cdot \varepsilon_0) \tag{1}$$

$$k_t = \sqrt{\frac{\pi f_s}{2 f_p} \tan\left(\frac{\pi}{2} \cdot \frac{f_p - f_s}{f_p}\right)} \tag{2}$$

$$c_{33}^{D} = 4\rho t^2 f_p{}^2 \tag{3}$$

$$e_{33} = k_t \sqrt{\varepsilon_{33}{}^{S} \cdot c_{33}^{D}} \tag{4}$$

where t is sample thickness, ρ is density and A is the lateral surface area of the sample plate. C is the capacitance measured at a frequency double the sample's mechanical resonance frequency; f_p is the equivalent circuit parallel resonance frequency measured at peak electrical impedance magnitude, and f_s is the equivalent circuit series resonance frequency measured at minimum electrical impedance magnitude.

Figure 4 presents the variations in the four selected piezoelectric properties with elevated environmental temperature for all three generations of piezocrystals. These measurements were conducted in the range 20 °C ≤ T≤ 180 °C. For PMN-PT, for which the expected range for use in applications is approximately 20–80 °C, $\varepsilon_{r33}{}^{S}$shows reasonable uniformity up to T~80 °C. A perturbation is then observed in the interval 90 °C < T < 120 °C, corresponding with the phase transition zone [42]. With temperature increasing towards 130 °C, $\varepsilon_{r33}{}^{S}$ experiences a significant increase which indicates the material is depoled and piezoelectric behaviour disappears. The same trends are observed for k_t, e_{33} and $c_{33}{}^{D}$. In comparison, PIN-PMN-PT maintains piezoelectric behavior and is stable for T ≤ 120 °C. A phase transition then occurs in the range 120 °C < T < 160 °C, about 30 °C higher than for PMN-PT. For T > 160 °C, the piezoelectric behaviour deteriorates significantly, as expected when approaching T_C, reported at around 180 °C [43]. Mn:PIN-PMN-PT is expected to be less temperature dependent than PIN-PMN-PT and PMN-PT because of the presence of Mn dopant [25]. In Figure 4, the phase transition of Mn:PIN-PMN-PT is shown in the range 140 °C < T < 170 °C, and T_C is not approached at or below 180 °C, as expected from the literature [27]. Table 2 lists the variations of the four properties in details, over the range of 20 °C ≤ T ≤ 80 °C before PMN-PT piezocrystal experiences the F_{RT} phase transition.

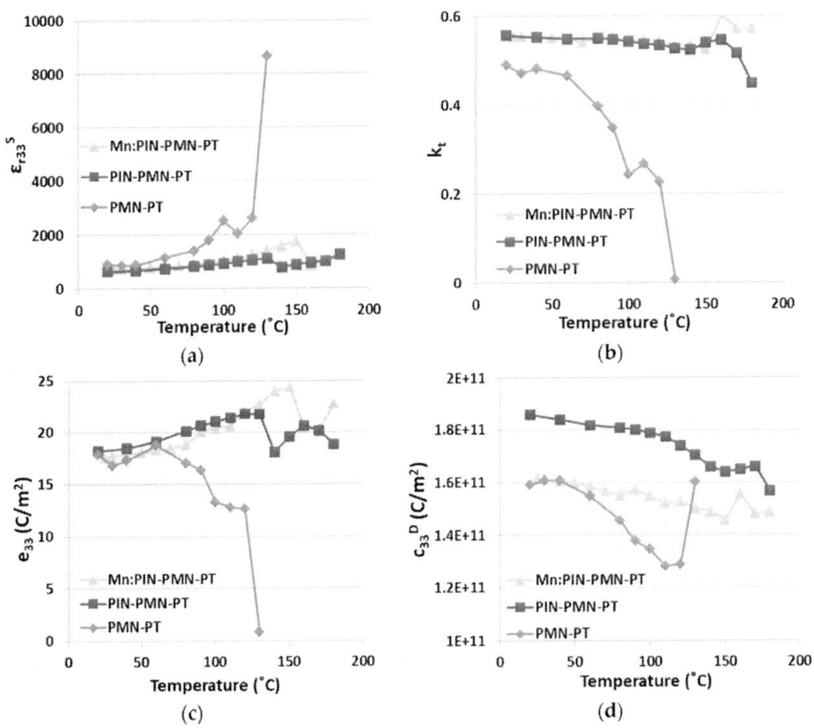

Figure 4. The effect of *T* on three generations of piezocrystals (**a**) ε_{r33}^{S}; (**b**) k_t; (**c**) e_{33}; and (**d**) c_{33}^{D}. (PMN-PT results adapted from [15]).

Table 2. Variation with *T* for three generations of piezocrystals, 20°C ≤ *T* ≤ 80°C (Reference *T* = 20°C).

(20 °C < *T* < 80 °C)	ε_{r33}^{S} (%)	k_t (%)	e_{33} (%)	c_{33}^{D} (%)
PMN-PT	+53.7	−19.2	+5.2, −5.9	+1.1, −8.5
PIN-PMN-PT	+29.0	−1.4	+10.4	−2.7
Mn:PIN-PMN-PT	+26.7	−1.1	+6.9	−4.1

+: maximum variation above the reference; −: minimum variation below the reference.

4.1.3. Property Variation with Uniaxial Pressure

Figure 5 presents the variations of ε_{r33}^{S}, k_t, e_{33} and c_{33}^{D} with applied uniaxial pressure for all three generations of piezocrystals. The results presented for PIN-PMN-PT and Mn:PIN-PMN-PT piezocrystal were obtained using the automatic system described in Section 3 and the results for PMN-PT are taken from our previous work for comparison. Considering the range 0 MPa ≤ *P* ≤ 20 MPa under which the PMN-PT sample was tested, *P* appears to affect PMN-PT material relatively little compared with the other two materials. This may be because different test fixtures were used to contact the samples in these two sets of experiments.

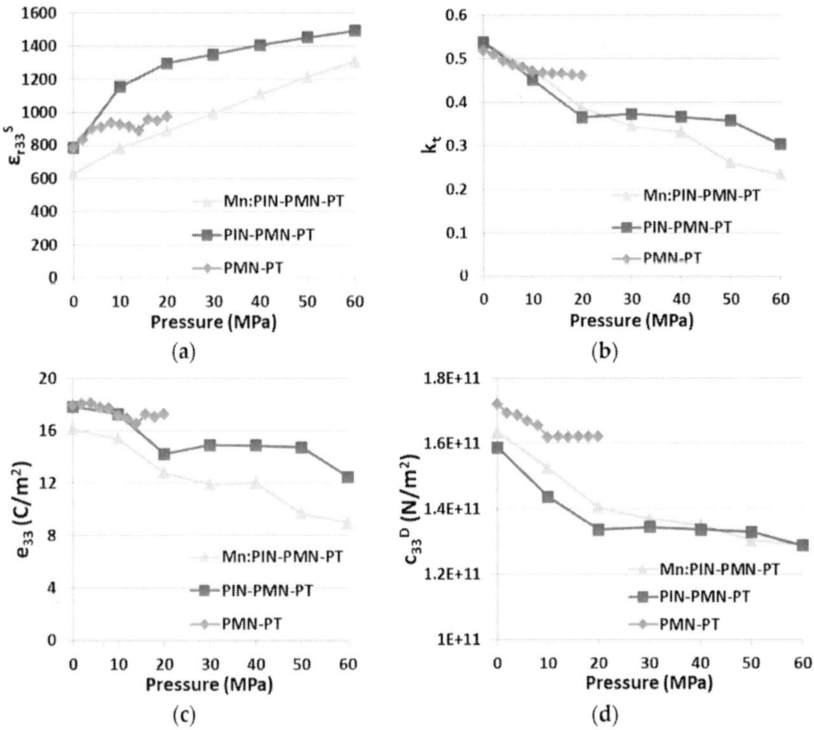

Figure 5. Effect of *P* on three generations of piezocrystals (**a**) ε_{r33}^{S}; (**b**) k_t; (**c**)e_{33}; and (**d**) c_{33}^{D}. (PMN-PT results adapted from [15]).

Despite the differences in the magnitudes of the variations for the three generations, the trends are in reasonable agreement. For PMN-PT, perturbations are evident for 10 MPa ≤ *P* ≤ 14 MPa, which may be termed the phase transition zone induced by uniaxial compressive pressure. For PIN-PMN-PT, this pressure range is extended to around 20 MPa. For the Mn:PIN-PMN-PT sample, however, there is no obvious evidence of a phase transition zone. This suggests that the

Mn acceptor dopant has significantly hardened the material to allow it to work under more demanding conditions than can be sustained with Generations I (PMN-PT) and II (PIN-PMN-PT) piezocrystals.

Details of the variations between PIN-PMN-PT and Mn:PIN-PMN-PT are listed in Table 3, with pressure ranging from 0 to 60 MPa. The variations of PMN-PT are not included because of the different, more limited pressure test range, $P < 20$ MPa, to which it was subjected.

Table 3. Variation with P for two generations of piezocrystals, 0 MPa $\leq P \leq$ 60 MPa (Reference P = 0 MPa).

(0 MPa < P < 60 MPa)	ε_{r33}^{S} (%)	k_t (%)	e_{33} (%)	c_{33}^{D} (%)
PIN-PMN-PT	+90.8	−43.7	−29.9	−18.9
Mn:PIN-PMN-PT	+107.8	−56.4	−44.2	−21.0

+: maximum variation above the reference; −: minimum variation below the reference.

4.1.4. Property Variation with Environmental Temperature and Uniaxial Pressure

To look at the dual influences from environmental temperature and uniaxial pressure simultaneously, Figure 6and Figure 7 present the variations for the four key piezoelectric properties in PIN-PMN-PT and Mn:PIN-PMN-PT, respectively. Variations in the characteristics are shown with environmental conditions of T and P, for 20 °C $\leq T \leq$ 100 °C with a step of 20 °C and for $0 \leq P \leq 60$ MPa with a step of 10 MPa.

As illustrated in Figure 6, for 20 °C $\leq T \leq$ 80 °C, both k_t and c^{D}_{33} of PIN-PMN-PT decrease with P, while ε_{r33}^{S}, on the contrary, increases with increasing T and P, and e_{33} decreases with P and increases with T in general. At T= 100 °C, the opposite trend was observed, with k_t, c^{D}_{33} and e_{33} rising, and ε_{r33}^{S} falling in value following increased P. It is believed that the combined effect of applied cycles of P and increases in T has induced a phase transition at lower temperature, with the piezocrystal transferring from the rhombohedral to orthorhombic phase, as suggested by McLaughlin [41].

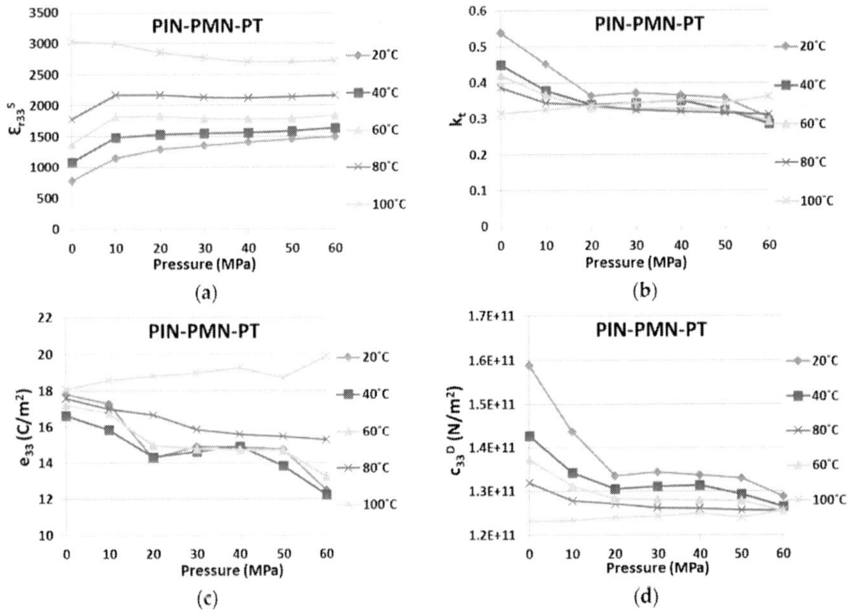

Figure 6. Property variation caused by environmental temperature and uniaxial pressure for PIN-PMN-PT on (**a**) ε_{r33}^{S}; (**b**) k_t; (**c**) e_{33}; and (**d**) c_{33}^{D}.

In comparison, Mn:PIN-PMN-PT in Figure 7 behaves similarly to PIN-PMN-PT, except that the same general trends are maintained for all four properties, even at $T = 100°C$. Again, this provides evidences that Generation III piezocrystals are more resilient to T and P, and thus likely to be more suitable for high power applications. However, as indicated in Table 3 and Table 6, the absolute variations in Mn:PIN-PMN-PT are bigger than for PIN-PMN-PT. This is because the deteriorating trends shown for Mn:PIN-PMN-PT with T and P, especially with$P$, did not recover after the phase transition.

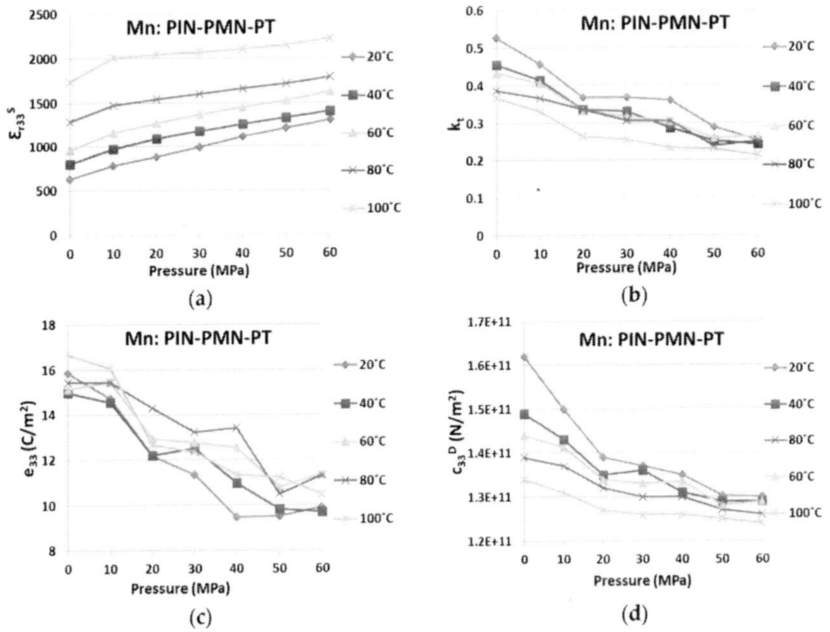

Figure 7. Property variation caused by environmental temperature and uniaxial pressure for Mn:PIN-PMN-PT on (**a**) ε_{r33}^{S}; (**b**) k_t; (**c**) e_{33}; and (**d**)c_{33}^{D}.

4.2. Electric Field Induced Temperature Rise

In high power applications such as active SONAR, conventional piezoelectric materials are subject to high cyclic electric fields to generate large displacement outputs. In turn, the intrinsic mechanical, piezoelectric and dielectric losses generate considerable heat, leading to an increase in *T* and consequently affecting the material's phase structure, piezoelectric properties and performance stability. This performance deterioration is exacerbated when the cyclic electric field is applied continuously for long periods of time, especially in industrial applications, such as ultrasonic cleaning, cutting and welding. In the context of integration of piezocrystals in such applications, it is necessary to carefully investigate these materials at different operating temperatures, and hence drive voltages.

Both temporal and spectral responses were examined for performance assessment. First, characterisation of the temporal response was conducted on a 10 mm × 10 mm × 1 mm PMN-PT TE plate, with AC driving voltages of 80, 90, 100 and 110 V_{pp}, with an RF power amplifier (31002A, Electronics & Innovation, Rochester, NY, USA). The driving voltage was set at the plate's resonant frequency of 2.19 MHz, and was applied continuously for 5 min. The thermal response was recorded with a thermal imaging camera (TIM-160, Micro-Epsilon Ltd., Ortenburg, Germany), as shown in Figure 8.

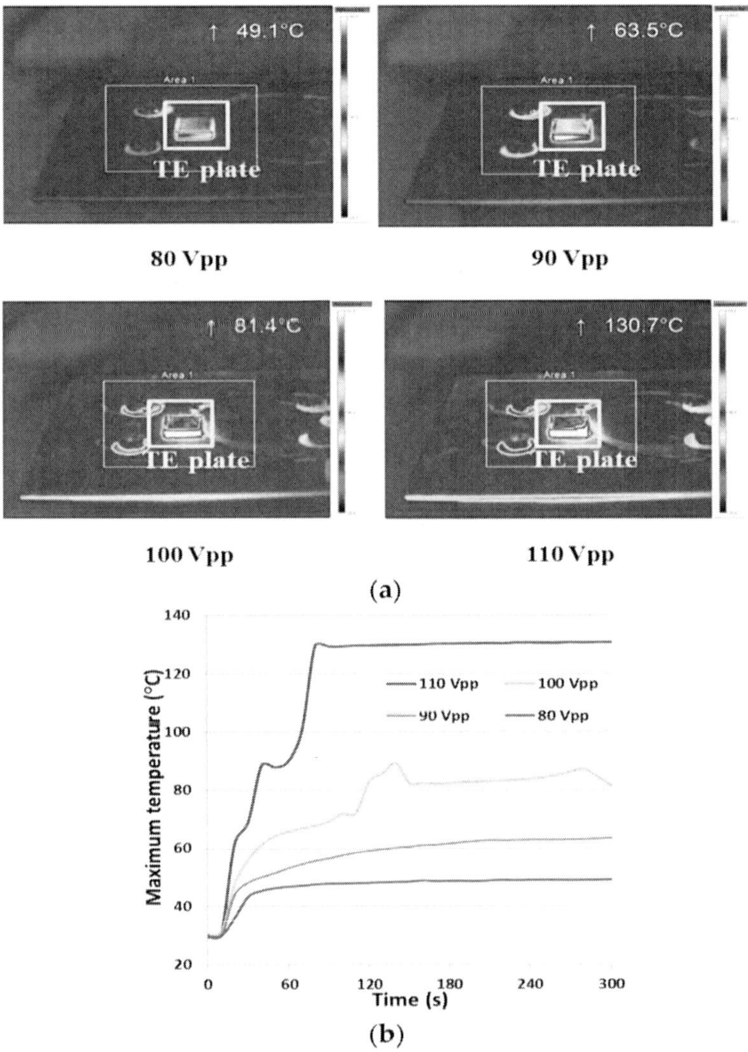

Figure 8. (**a**) Thermal mapping and (**b**) Maximum spot temperature of a PMN-PT TE plate driven with different voltages for 5 min.

From the thermal maps of the TE plate sample in Figure 8a, it can be seen that the maximum electric field-induced temperatures were 49.1, 63.5, 81.4 and 130.7 °C respectively for 80 V_{pp}, 90 V_{pp}, 100 V_{pp}, and 110 V_{pp}sinusoidal excitation for 5 min. The temporal response curves in Figure 8b show that at 80 V_{pp} and 90 V_{pp}, T rose monotonically and then stabilised; however, at 100 and 110 V_{pp}, T experienced particularly sharp increases at 110 s and at 60 s, corresponding to temperatures of 68.9 and 96.9 °C respectively. These increases may indicate a phase transition, as reported recently as T_{RT} = 97 °C for PMN-PT [42].

Subsequently, characterisation of electric field-induced temperature increases was performed with respect to the spectral response, measuring the electrical impedance of the samples at induced temperatures maintained at stable levels. This spectral characterisation was performed on PIN-PMN-PT and Mn:PIN-PMN-PT TE plates with dimensions 5 mm × 5 mm × 0.5 mm. The electric field-induced temperature was maintained at successive temperatures of 40, 60, 80, 100 and 120 °C with the plate driven at its TE resonance with the AC driving voltage controlled adaptively. Impedance spectra were measured in the range 10 Hz–10 MHz with 800 sampling points, using a 10:1 passive voltage probe (N2862B, Agilent Technologies, Santa Clara, CA, USA) and a current probe (P6021A, Tektronix, Plano, TX, USA). As with the use of commercial impedance analyser systems, fixture compensation was performed to ensure accurate measurement calibration.

In Figure 9b, Mn:PIN-PMN-PT shows significant changes in the resonant impedance peak at 120 °C with an average driving voltage of 26 V_{pp}. For PIN-PMN-PT in Figure 9a, the corresponding peak changes significantly at the lower temperature of 100 °C, with an average driving voltage of 28 V_{pp}. These changes in impedance peaks may indicate the critical phase transition point for piezocrystals. For PIN-PMN-PT, the TE plate sample was depolarised at 120 °C with an average driving voltage of 84 V_{pp}. The Mn:PIN-PMN-PT sample exhibited better thermal stability at high electric field than PIN-PMN-PT sample. It is also notable that T = 120 °C is far below T_C(=191 °C [1]) for PIN-PMN-PT; the effects we observed may be a result of the combination of high E and T.

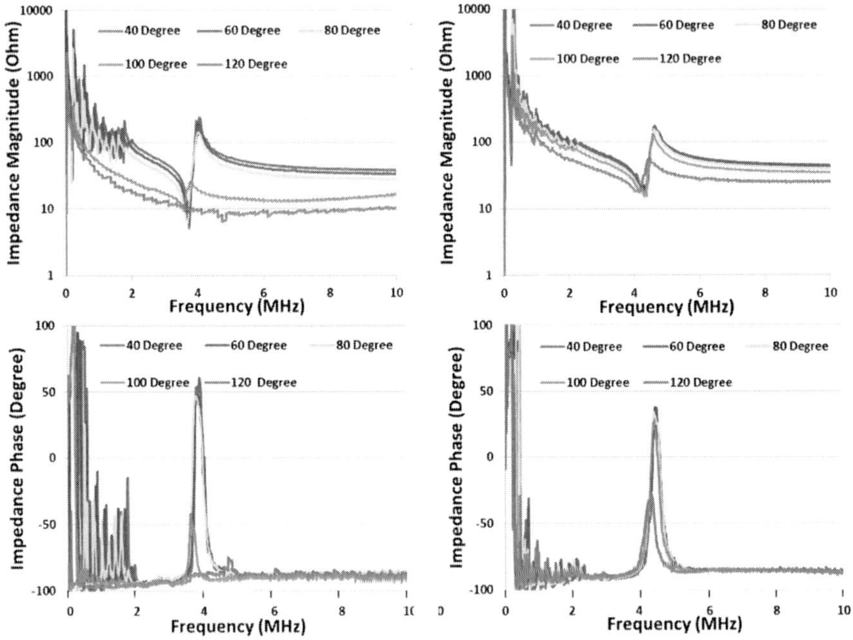

(a) (b)

Figure 9. Impedance spectra of (**a**) PIN-PMN-PT and (**b**) Mn:PIN-PMN-PT when the electric field induced temperature was maintained as 40, 60, 80, 100 and 120 °C.

4.3. Laser Doppler Vibrometry under Ambient Conditions

LDV is based on the use of the inverse piezoelectric effect to measure sample surface vibration amplitude with applied electrical field. Not needing physical contact with the sample, LDV avoids the interference and energy loss from ultrasonic coupling at the interface that is unavoidable in most other measurements. It can be performed at a single measurement spot or by scanning a surface to obtain a mode shape.

In the work reported here, LDV was first used to measure the mode shape of a d_{36} FS plate by scanning the four side surfaces around the plate, with both vibration amplitude and phase data acquired. Based on the measured data, the FS mode shape can be obtained by sweeping through the phases. In Figure 10a, the red dashed lines represent the original shape of the plate; the blue asterisks represent the LDV-measured vibration at 10° phase; the blue solid lines represent the curves fitted to the measured data points, illustrating the plate shape profile for FS vibration. The measured FS mode shape had a good agreement with the FS mode shape simulated with the PZFlex FEA software package, with displacements mapped along *x* and *y* directions, as shown in Figure 10b.

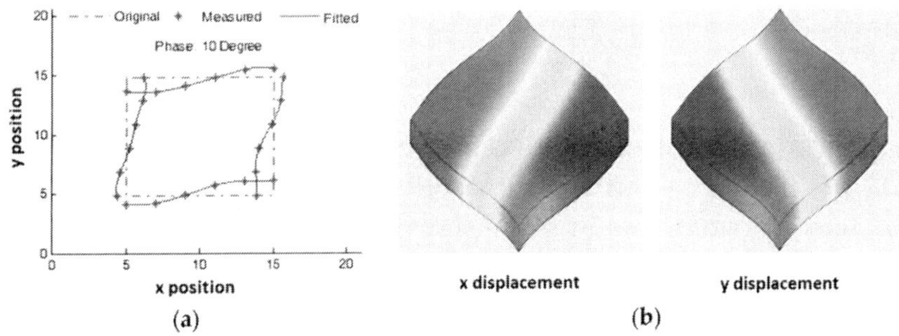

Figure 10. (**a**) Experimentally measured mode shape of a d_{36} FS plate; and (**b**) corresponding simulated mode shapes with displacements along *x* and *y*directions mapped in colours.

LDV was also used to measure piezoelectric constants, d_{ij}. For a 10 mm × 3 mm × 1 mm Mn:PIN-PMN-PT LTE plate with 45° rotation about the *z* axis, the piezoelectric constant can be determined with:

$$d_{31} = \frac{S_1}{E_3} = \frac{A_1/l}{V_3/t} \tag{5}$$

where S_1 represents the strain along the 1st direction, measured as the ratio between the vibration amplitude, $A_1$and the plate's length, *l*, and E_3 represents the electric field along the 3rd direction, calculated as the ratio of applied voltage, V_3, to the thickness, *t*. The LDV was set up as shown in Figure 11a, with a Mn:PIN-PMN-PT LTE plate mounted on a spring-loaded electrode fixture and the beam focused on the side of the LTE plate.

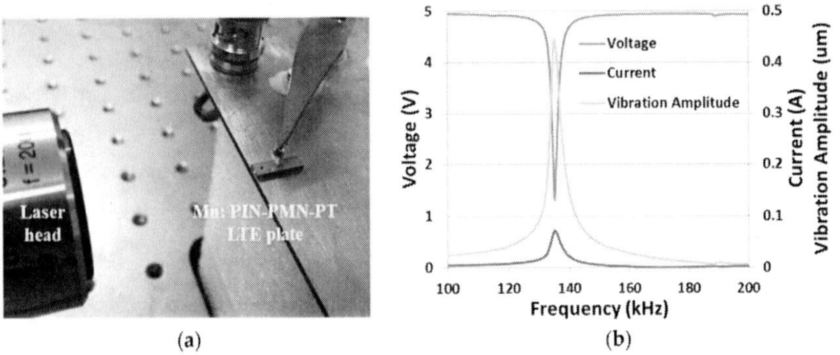

Figure 11. (**a**) Laser vibrometry characterisation setup with a Mn:PIN-PMN-PT LTE plate with (**b**) responses of voltage, current, and vibration amplitude when a constant AC 5 V driving voltage was swept from 100 to 200 kHz.

An AC signal of constant 5 V amplitude was swept spectrally across the LTE resonance from 100 to 200 kHz. The vibration amplitude, V_3 and current are shown in Figure 11b. V_3 dropped and the current increased at resonance because of the low electric impedance; the vibration amplitude increase approximately matched the increase in current. Applying Equation (5) off resonance, d_{31} = 457 pC/N was measured at 100 kHz for this sample, a value similar to the figure of 474 pC/N provided by the supplier (TRS Technologies, State College, PA, USA).

To compare the LDV measurement with different methodologies, a study was conducted with Mn:PIN-PMN-PT samples with different vibration modes, as shown in Table 4. All the measurements were performed locally under ambient conditions, except for the material supplier's data. There is reasonable agreement between the different techniques, with no single technique consistently appearing to over or underestimate the coefficient values. Equally, however, this data set offers no gold standard measurement techniques in close agreement with each other.

Table 4. Characterisation of piezoelectric constant d_{ij} on Mn:PIN-PMN-PT under ambient conditions with different vibration modes measured with different techniques.

(pC/N)	Mode	Berlincourt Method	IEEE Method	LDV	Manufacturer Data
d_{33}	LE	1053	1492	1160	-
	TE	1062	-	-	-
d_{31}	LTE	478	458	402	467
$d_{31}^{45z\Delta}$	LTE	518	493	457	474
d_{15}	TS	156	120	156	121
d_{36}	FS	-	-	1109	1128

$d_{31}^{45z\Delta}$: d_{31} value for LTE sample rotated 45° about z axis [12].

5. DISCUSSION

Table 5 summarises the basic temperature and pressure characteristics of the three generations of piezocrystal as considered in this paper. Where comparison is possible with the literature, it is generally in good agreement, except for a much higher figure for T_{RT} for Mn:PIN-PMN-PT, as measured approximately in the present work. In consideration of high power transducer operation, this suggests that Generation II and III piezocrystals (PIN-PMN-PT and Mn:PIN-PMN-PT respectively) have usefully raised temperature limits. However, material suppliers may care to look into the issue of phase transitions under the influence of pressure for widest acceptance of piezocrystals.

Table 5. Basic comparison between three generations of piezocrystals.

Crystal	T_C ◇ (°C)	T_C * (°C)	T_{RT} ◇ (°C)	T_{RT} * (°C)
PMN-PT	135	≈130	96	≈100
PIN-PMN-PT	191	>180	125	≈130
Mn:PIN-PMN-PT	193	>180	119	≈150
Crystal	**σ_{RO} ◇ (MPa)**	**σ_{RO} * (MPa)**	**E_c ◇ (kV/cm)**	**E_{int} ◇ (kV/cm)**
PMN-PT	≈15 MPa	≈10 MPa	2.3	0.0
PIN-PMN-PT	≈20 MPa	≈20 MPa	5.0	0.0
Mn:PIN-PMN-PT	>21 MPa	>20 MPa	6.0	1.0

◇ Literature values [1,28,34,40,41]; * Measured values in this study.

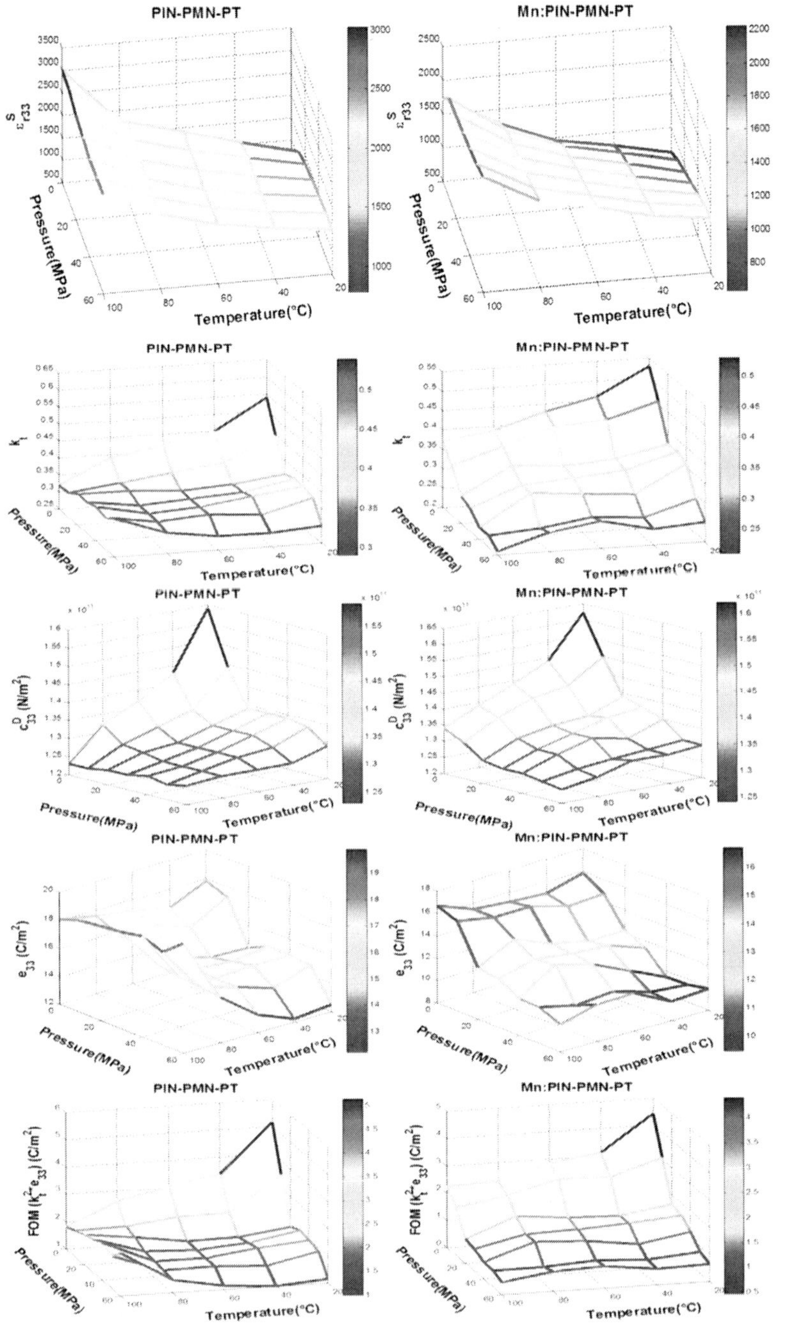

Figure 12. A summary of the variation of key parameters of PIN-PMN-PT and Mn:PIN-PMN-PT with varying temperature and pressure on ε^S_{r33}, k_t,c^D_{33}, e_{33} and one figure of merit ($k_t^2e_{33}$).

Figure 12 summarises the variation in the key material parameters with temperature and pressure, presented in a format to allow easier interpretation and comparison. As expected from many previous reports in the literature, ε^S_{r33} increases with *T*; however, changes in *P* have less effect. Electromechanical coupling, k_t, and stiffness, c^D_{33}, are significantly higher at ambient *T* and *P*, and they generally fall as either and both of these parameters increases. In both cases, the reduction is likely to be sufficient to require a device designer to take it into account if the operating envelope of the device approaches the limits of *T* and *P* considered here. The most complicated characteristics are exhibited by e_{33}. Mn:PIN-PMN-PT shows relatively little change with *T* but a generally falling trend with *P*. PIN-PMN-PT shows similar characteristics up to 60 °C then a recovery to the ambient temperature values for all values of *P*. This may be caused by a phase transition mediated by *T* and *P*. However, further work is required to confirm this.

Table 6. (**a**) Detail of FOM $k_t^2 e_{33}$ (C/m^2) for PIN-PMN-PT with variation in *T* and *P*; (**b**) Detail of FOM $k_t^2 e_{33}$ (C/m^2) for Mn:PIN-PMN-PT with variation in *T* and *P*.

(a)

T/P	0 MPa	10 MPa	20 MPa	30 MPa	40 MPa	50 MPa	60 MPa	Variation
20 °C	5.16	3.52	1.89	2.07	1.99	1.88	1.15	−77.8%
40 °C	3.36	2.26	1.65	1.74	1.82	1.46	1.00	−70.3%
60 °C	3.04	2.23	1.62	1.59	1.58	1.56	1.14	−62.4%
80 °C	2.62	2.01	1.89	1.67	1.60	1.55	1.48	−43.5%
100 °C	1.80	1.96	2.12	2.24	2.39	2.20	2.60	44.7%
Variation	−65.1%	−44.4%	12.2%	8.3%	20.6%	17.1%	127.2%	**−49.5%**

(b)

T/P	0 MPa	10 MPa	20 MPa	30 MPa	40 MPa	50 MPa	60 MPa	Variation
20 °C	4.42	3.09	1.67	1.56	1.24	0.80	0.64	−85.6%
40 °C	3.12	2.50	1.38	1.39	0.91	0.62	0.57	−81.6%
60 °C	2.84	2.56	1.44	1.29	1.16	0.73	0.80	−71.9%
80 °C	2.32	2.08	1.62	1.25	1.26	0.60	0.72	−68.9%
100 °C	2.25	1.77	0.89	0.82	0.62	0.60	0.47	−79.1%
Variation	−49.2%	−42.7%	−46.6%	−47.5%	−50.0%	−25.0%	−26.3%	**−89.4%**

The FOM, $k_t^2 e_{33}$, is finally shown in Figure 12 for PIN-PMN-PT and Mn:PIN-PMN-PT and the values are detailed in Table 6a,b respectively, with the addition of variances calculated from ambient to peak *T* and *P*. Somewhat contrary to expectation, this suggests that

PIN-PMN-PT generally exhibits less variance in FOM than Mn:PIN-PMN-PT, crucially at a combination of the highest values of *T* and *P*. We believe this tends to support our proposal that a phase transition is responsible for the variation under the extreme of the conditions considered here. It is also notable that the values for FOM reduce very significantly as *T* and *P* increase.

As summarised in Table 5 and Table 6, deterioration in piezocrystal functional performance under a combination of temperature, stress and electric field has been noted and has remained of research interest since investigations started in PMN-PT. Early characterisation was done on LE and LTE modes with a series of *T* and *P* values applied and phase transitions were examined to explain the performance variations. A satisfactory dimensional height:width:depth ratio was suggested as 3:1:1 to define a LE mode bar for the application of pressure [40,41]. Following that work, with the development of PIN-PMN-PT, characterisation was done under conditions of elevated *T*, *P* and *E* with special attention focused on the stress-induced ferroelectric rhombohedral (F_R) to ferroelectric orthorhombic (F_O) phase transition [28]. The relevant value, σ = 20 MPa, is consistent with the measurements in our study. Another study was reported more recently, including the newly developed Mn:PIN-PMN-PT, with uniaxial pressure and different electric fields, and reductions in the piezoelectric constant, d_{32}, were demonstrated in both PIN-PMN-PT and Mn:PIN-PMN-PT [34]. The results reported here add to the evidence of such performance reductions and reinforce the need for functional characterisation of new piezocrystal materials to allow device designers to estimate the effects of their behaviour on practical performance.

6. CONCLUSIONS AND FURTHER WORK

The use of piezocrystals to exploit their potential high performance brings with it the need to consider performance variations contingent on both their growth processes and their intrinsic behavior. This paper has reported application-oriented research to determine trends in the latter when the three generations of these materials are subject to ranges of temperature, pressure and drive voltages (electric fields) representative of practical use. This research has been based on a measurement system including an oven to apply elevated temperatures and a material testing system for the application of uniaxial pressure, as shown in Figure 2, for measuring the passive response of the piezocrystals to their environment. Through high drive components, the system also allows the response to active transducer driving to be determined, e.g., in terms of increases in temperature because of drive conditions. Furthermore, it allows comparison of different measurement methods, such as electrical impedance spectroscopy and laser Doppler vibrometry.

Using this system, it has been shown that Generation I, PMN-PT, is relatively sensitive to temperature, though with marginally the highest peak performance under ambient conditions. Given the low coercive field values reported in the literature, this suggests it is most suitable for applications demanding low average power levels and with low excitation fields to avoid the need for DC bias. This matches practical experience, with PMN-PT now as the material of choice for biomedical imaging. Generation II, PIN-PMN-PT, and Generation III, Mn:PIN-PMN-PT, materials have been developed specifically for applications with greater demands in terms of temperature and pressure, such as are typically found in power ultrasound, including SONAR and the oil and gas industry. Here, we have confirmed that samples of both generations have higher phase transition and Curie temperatures than Generation I PMN-PT. However, pressure in the range to 60 MPa appears to have a highly deleterious effect on parameters such as TE mode coupling coefficient, k_t, and Figure of Merit, FOM = $k_t^2 e_{33}$. In addition, our specific measurements suggest relatively little or no advantage from Mn-doping in terms of immunity to environmental temperature and pressure. However, we have not considered coercive field, E_c, or mechanical quality factor, Q_m, in detail, and both of these parameters have been reported to be significantly higher and useful for performance assessment in active transduction.

In considering work to further elucidate the behaviour of piezocrystals for practical configurations, we noted that the temperature investigation presented in Section 4.1 ranged upwards from 20 °C. However, it may be useful to determine the behaviour at lower temperatures for applications such as underwater SONAR. Although commercial temperature-controlled cryostats can provide very wide temperature ranges and environmental chambers are available, their configurations are generally unsuitable for simultaneous application of pressure in the range up to 60 MPa, representing full ocean depth. Initial trials of low temperature characterisation were carried out in our lab based on the system described in this paper, cooling the load applicator bars in a −80 °C freezer and then assembling them with further cooling packs to allow characterisation to start at 0 °C. Further investigation of the use of LN_2 is under consideration to make this arrangement less cumbrous.

Another issue that has been considered in a preliminary way is the effect of pressure cycling and fatigue. Measurements were performed on PIN-PMN-PT and Mn:PIN-PMN-PT; comparison with a hard piezoceramic is also of interest. For both piezocrystals, property variations are particularly evident for $0 \leq P \leq 25$ MPa; after only three pressure cycles, k_t was found to have suffered a permanent reduction in both materials while the properties of the hard ceramic were much more stable under loading and there was no long term change upon completion of a small number of pressure cycles. This clearly merits further consideration.

Benefitting from the increase in computational power owing to Moore's law, FEA has largely superseded analytical methods for commercial transducer design. This technique helps to verify design validity but generally requires the full elastoelectric matrix as input data. Measurement of this matrix under varying conditions of temperature, pressure and electric field allows the transducer to be simulated under a range of operating conditions, in principle. However, whilst measurement is feasible over a range of temperatures, it is difficult at elevated pressure because of the likelihood of mechanical damage, especially for the LE mode bar. In addition, the generation of a high electric field may be difficult because the thickness of this sample demands a very high applied voltage. In this context, techniques similar to those reported by Cao *et al.* may be helpful, using the elastoelectric matrix of a known piezoelectric material as an initial condition and applying optimisation techniques to reduce the number of samples required to obtain the matrix for the new material. Additional techniques combining simulation and modelling, such as those reported by Gallagher and Lynch to characterise piezocrystal properties across the phase transition, can reduce the number of measurements and combination of external conditions needed to fully characterise the material [44]. Such techniques are also beginning to find favour in commercial FEA packages, and it is expected that these will be combined with the external application of temperature and pressure in due course.

ACKNOWLEDGMENTS

Aspects of this work have been supported by UK EPSRC projects EP/K020250 and EP/K020013. Zhen Qiu and Xiaochun Liao were supported by the Scottish Universities Physics Alliance. Tingyi Jiang was supported by the China Scholarship Council. Christine E.M. Démoré is supported by a Royal Society of Edinburgh Caledonian Research Fund Biomedical Personal Research Fellowship. The authors are also grateful for the assistance of Mr. Aleksandrs Bolhovitins and Mr. Romans Poltarjonoks in experimental equipment development and validation, and for continuing financial support from Thales Underwater Systems (UK).

AUTHOR CONTRIBUTIONS

Xiaochun Liao, Zhen Qiu and Tingyi Jiang carried out experimental work and contributed to discussion, writing and reviewing of the text. Muhammad R. Sadiq planned and supervised development of the experimental equipment, carried out experimental work and contributed to reviewing of the text. Zhihong Huang provided resources and contributed to discussion and reviewing of the text. Christine E.M. Demore contributed to experimental protocol development, discussion and reviewing of the text. Sandy Cochran provided resources and contributed to discussion, writing and reviewing of the text. The contributions of Zhen Qiu and Sandy Cochran were made whilst they were employed by the Institute for Medical Science and Technology, University of Dundee.

REFERENCES

1. Zhang, S.; Li, F.; Jiang, X.; Kim, J.; Luo, J.; Geng, X. Advantages and challenges of relaxo-$PbTiO_3$ferroelectric crystals for electroacoustic transducers—A review. *Prog. Mater. Sci.* **2015**, *68*, 1–66.
2. Zhou, Q.; Lam, K.H.; Zheng, H.; Qiu, W.; Shung, K.K. Piezoelectric single crystal ultrasonic transducers for biomedical applications. *Prog. Mater. Sci.* **2014**, *66*, 87–111.

3. Qiu, Y.; Wang, H.; Demore, C.; Hughes, D.; Glynne-Jones, P.; Gebhardt, S.; Bolhovitins, A.; Poltarjonoks, R.; Weijer, K.; Schönecker, A.; *et al.* Acoustic devices for particle and cell manipulation and sensing. *Sensors* **2014**, *14*, 14806–14838.

4. Jiang, X.; Kim, J.; Kim, K. Relaxor-PT single crystal piezoelectric sensors. *Crystals* **2014**, *4*, 351.

5. Qiu, Z.; Habeshaw, R.; Fortinec, J.; Huang, Z.; Demore, C.; Cochran, S. New piezocrystal material in the development of a 96-element array transducer for Mr-guided focused ultrasound surgery. In Proceedings of the 2012 International Society for Therapeutic Ultrasound Symposium, Heidelberg Germany, 10–13 June 2012.

6. Gallego-Juárez, J.A.; Graff, K.F. *Power Ultrasonics: Applications of High Intensity Ultrasound*; Woodhead Publishing: Cambridge, UK, 2015.

7. Stansfield, D. *Underwater Electroacoustic Transducers: A Handbook for Users and Designers*; Bath University Press: Bath, UK, 1991.

8. Zhang, S.; Li, F. High performance ferroelectric relaxor-$PbTiO_3$ single crystals: Status and perspective. *J. Appl. Phys.* **2012**, *111*.

9. Sun, E.; Cao, W. Relaxor-based ferroelectric single crystals: Growth, domain engineering, characterization and applications. *Prog. Mater. Sci.* **2014**, *65*, 124–210.

10. Mitsuyoshi, M.; Yoshihito, T.; Yosuke, I. Development of large-diameter single-crystal PMN-PT with high energy conversion efficiency. *JFE Tech. Rep.* **2005**, *6*, 46–53.

11. Berlincourt, D.A.; Curran, D.R.; Jaffe, H. 3-piezoelectric and piezomagnetic materials and their function in transducers. In *Physical Acoustics*; Mason, W.P., Ed.; Academic Press: Waltham, MA, USA, 1964; pp. 169–270.

12. 176-1987-IEEE Standard on Piezoelectricity. Available online: http://ieeexplore.ieee.org/stamp/stamp.jsp?arnumber=26560 (accessed on 30 November 2015).

13. Li, S.; Zheng, L.; Jiang, W.; Sahul, R.; Gopalan, V.; Cao, W. Characterization of full set material constants of piezoelectric

materials based on ultrasonic method and inverse impedance spectroscopy using only one sample. *J. Appl. Phys.* **2013**, *114*.

14. Tang, L.; Cao, W. Temperature dependence of self-consistent full matrix material constants of lead zirconate titanate ceramics. *Appl. Phys. Lett.* **2015**, *106*.

15. Qiu, Z.; Sadiq, M.R.; Demore, C.; Parker, M.F.; Marin, P.; Mayne, K.; Cochran, S. Characterization of piezocrystals for practical configurations with temperature- and pressure-dependent electrical impedance spectroscopy. *IEEE Trans. Ultrason. Ferroelectr. Freq. Control* **2011**, *58*, 1793–1803.

16. Lee, H.J.; Ural, S.O.; Chen, L.; Uchino, K.; Zhang, S. High power characteristics of lead-free piezoelectric ceramics. *J. Am. Ceram. Soc.* **2012**, *95*, 3383–3386.

17. Cochran, S.; Démoré, C.E.M.; Courtney, C.R.P. 6-modelling ultrasonic-transducer performance: One-dimensional models. In *Ultrasonic Transducers*; Nakamura, K., Ed.; Woodhead Publishing: Sawston, UK, 2012; pp. 187–219.

18. Zhang, S.; Li, F.; Luo, J.; Xia, R.; Hackenberger, W.; Shrout, T.R. Field stability of piezoelectric shear properties in PIN-PMN-PT crystals under large drive field. *IEEE Trans. Ultrason. Ferroelectr. Freq. Control***2011**, *58*, 274–280.

19. Chen, R.; Wu, J.; Lam, K.H.; Yao, L.; Zhou, Q.; Tian, J.; Han, P.; Shung, K.K. Thermal-independent properties of PIN-PMN-PT single-crystal linear-array ultrasonic transducers. *IEEE Trans. Ultrason. Ferroelectr. Freq. Control* **2012**, *59*, 2777–2784.

20. Zhang, S.; Li, F.; Luo, J.; Sahul, R.; Shrout, T.R. Relaxor-$PbTiO_3$ single crystals for various applications.*IEEE Trans. Ultrason. Ferroelectr. Freq. Control* **2013**, *60*, 1572–1580.

21. Standard Draft for Relaxor-Based Single Crystals for Transducer and Actuator Applications. Available online: http://www.ieee-uffc.org/main/standards/CrystalStandard_V8_for_distribution.pdf (accessed on 30 November 2015).

22. Liu, X.; Zhang, S.; Luo, J.; Shrout, T.R.; Cao, W. A complete set of material properties of single domain

$0.26Pb(In_{(1/2)}Nb_{(1/2)})O_3$–$0.46Pb(Mg_{(1/3)}Nb_{(2/3)})O_3$–$0.28PbTiO_3$ single crystals. *Appl. Phys. Lett.* **2010**, *96*.

23. Luo, J.; Hackenberger, W.; Zhang, S.; Shrout, T.R. Elastic, piezoelectric and dielectric properties of PIN-PMN-PT crystals grown by bridgman method. In Proceedings of the 2008 IEEE International Ultrasonics Symposium (IUS), Beijing, China, 2–5 November 2008; pp. 261–264.

24. Sun, E.; Zhang, S.; Luo, J.; Shrout, T.R.; Cao, W. Elastic, dielectric, and piezoelectric constants of $Pb(In_{1/2}Nb_{1/2})O_3$–$Pb(Mg_{1/3}Nb_{2/3})O_3$–$PbTiO_3$ single crystal poled along [011]c. *Appl. Phys. Lett.* **2010**, *97*.

25. Zheng, L.; Sahul, R.; Zhang, S.; Jiang, W.; Li, S.; Cao, W. Orientation dependence of piezoelectric properties and mechanical quality factors of $0.27Pb(In_{1/2}Nb_{1/2})O_3$–$0.46Pb(Mg_{1/3}Nb_{2/3})O_3$–$0.27PbTiO_3$:Mn single crystals. *J. Appl. Phys.* **2013**, *114*.

26. Zhang, S.; Li, F.; Jiang, W.; Luo, J.; Meyer, R.J.; Cao, W.; Shrout, T.R. Face shear piezoelectric properties of relaxor-$PbTiO_{(3)}$ single crystals. *Appl. Phys. Lett.* **2011**, *98*.

27. Huo, X.; Zhang, S.; Liu, G.; Zhang, R.; Luo, J.; Sahul, R.; Cao, W.; Shrout, T.R. Elastic, dielectric and piezoelectric characterization of single domain PIN-PMN-PT: Mn crystals. *J. Appl. Phys.* **2012**, *112*.

28. Finkel, P.; Robinson, H.; Stace, J.; Amin, A. Study of phase transitions in ternary lead indium niobate-lead magnesium niobate-lead titanate relaxor ferroelectric morphotropic single crystals. *Appl. Phys. Lett.* **2010**,*97*.

29. Wooldridge, J.; Stewart, M.; Vecchini, C.; Cain, M.G.; Gutmann, M.; Reece, M. Temperature and frequency dependence of electric field-induced phase transitions in PMN–0.32PT. *J. Am. Ceram. Soc.* **2014**, *97*, 2111–2115.

30. Yuan, Z.; Seyit, O.U.; Kenji, U. Loss factor characterization methodology for piezoelectric ceramics. *IOP Conf. Ser. Mater. Sci. Eng.* **2011**, *18*.

31. Uchino, K.; Hirose, S. Loss mechanisms in piezoelectrics: How to measure different losses separately. *IEEE Trans. Ultrason. Ferroelectr. Freq. Control* **2001**, *48*, 307–321.

32. Snook, K.A.; Rehrig, P.W.; Hackenberger, W.S.; Jiang, X.; Meyer, J.R.J.; Markley, D. Advanced piezoelectric single crystal based transducers for naval sonar applications. In Proceedings of the 2005 IEEE Ultrasonics Symposium, Rotterdam, The Netherlands, 18–21 September 2005; pp. 263–271.

33. Meyer, R.J.; Montgomery, T.C.; Hughes, W.J. Tonpilz transducers designed using single crystal piezoelectrics. In Proceedings of the OCEANS '02 MTS/IEEE, Biloxi, MS, USA, 29–31 October 2002; Volume 4, pp. 2328–2333.

34. Zhang, S.; Taylor, S.; Li, F.; Luo, J.; Meyer, R.J. Piezoelectric property of relaxor-$PbTiO_3$ crystals under uniaxial transverse stress. *Appl. Phys. Lett.* **2013**, *102*.

35. Stewart, M.; Cain, M. Direct piezoelectric measurement: The berlincourt method. In *Characterisation of Ferroelectric Bulk Materials and Thin Films*; Cain, M.G., Ed.; Springer: Dordrecht, The Netherlands, 2014; Volume 2, pp. 37–64.

36. Jian, T.; Pengdi, H.; Payne, D.A. Measurements along the growth direction of PMN-PT crystals: Dielectric, piezoelectric, and elastic properties. *IEEE Trans. Ultrason. Ferroelectr. Freq. Control* **2007**, *54*, 1895–1902.

37. Sherrit, S.; Mukherjee, B.K. Characterization of piezoelectric materials for transducers. Available online: http://arxiv.org/ftp/arxiv/papers/0711/0711.2657.pdf (accessed on 30 November 2015).

38. McSkimin, H.J. Notes and references for the measurement of elastic moduli by means of ultrasonic waves.*J. Acoust. Soc. Am.* **1961**, *33*, 606–615.

39. Zhu, S.; Jiang, B.; Cao, W. Characterization of piezoelectric materials using ultrasonic and resonant techniques. *Med. Imaging Ultrason. Transducer Eng.* **1998**, *3341*, 154–162.

40. McLaughlin, E.A.; Liu, T.; Lynch, C.S. Relaxor ferroelectric PMN-32%PT crystals under stress and electric field loading: I-32 mode measurements. *Acta Mater.* **2004**, *52*, 3849–3857.

41. McLaughlin, E.A.; Liu, T.; Lynch, C.S. Relaxor ferroelectric PMN-32%PT crystals under stress, electric field and temperature loading: II-33-mode measurements. *Acta Mater.* **2005**, *53*, 4001–4008.

42. Tian, J.; Han, P. Growth and characterization on PMN-PT-based single crystals. *Crystals* **2014**, *4*, 331–341.

43. Zhang, S.; Liu, G.; Jiang, W.; Luo, J.; Cao, W.; Shrout, T.R. Characterization of single domain $Pb(In_{0.5}Nb_{0.5})O_3$-$Pb(Mg_{1/3}Nb_{2/3})O_3$-$PbTiO_3$ crystals with monoclinic phase. *J. Appl. Phys.* **2011**, *110*.

44. Gallagher, J.A.; Lynch, C.S. Combining experiments and modeling to characterize field driven phase transformations in relaxor ferroelectric single crystals. *Acta Mater.* **2015**, *89*, 41–49.

Chapter 5

Postbuckling Investigations of Piezoelectric Microdevices Considering Damage Effects

Zhigang Sun [1,*] **and Xianqiao Wang** [2,*]

[1]*College of Energy and Power Engineering and State Key Laboratory of Mechanics and Control of Mechanical Structures, Nanjing University of Aeronautics and Astronautics, Nanjing 210016, China*

[2]*College of Engineering, University of Georgia, Athens, GA 30602, USA*

ABSTRACT

Piezoelectric material has been emerging as a popular building block in MEMS devices owing to its unique mechanical and electrical material properties. However, the reliability of MEMS devices under buckling deformation environments remains elusive and needs to be further explored. Based on the Talreja's tensor valued internal state damage variables as well as the Helmhotlz free energy of piezoelectric material, a constitutive model of piezoelectric materials with damage is presented. The Kachanvo damage evolution law under in-plane compressive loads is employed. The model is applied to the specific case of the postbuckling analysis of the piezoelectric plate with damage. Then, adopting von Karman's plate theory, the nonlinear governing equations of the piezoelectric plates with initial geometric deflection including damage effects under in-plane compressive loads are established. By using the finite difference method and the Newmark scheme, the damage evolution for damage accumulation is developed and the finite difference procedure for postbuckling equilibrium path is simultaneously employed. Numerical results show the postbuckling behaviors of initial flat and deflected piezoelectric plates with damage or no damage under different sets of electrical loading conditions. The effects of applied voltage, aspect ratio of plate, thick-span ratio of plate, damage as well as initial geometric deflections on the postbuckling behaviors of the piezoelectric plate are discussed.

Keywords: piezoelectric plates; tensor valued internal state variables; postbuckling behavior; damage evolution; initial deflections

1. INTRODUCTION

The use of piezoelectric materials in intelligent structures has received considerable attention in recent years due to the intrinsic direct and converse piezoelectric effects. Piezoelectric materials have been used as sensors or actuators for the control of the active shape or vibration of structures. Defects such as microcracks, voids, dislocations and delamination are introduced in piezoelectric materials during the manufacturing and poling process. The existence of these defects greatly affects the electric, dielectric, elastic, mechanical and piezoelectric properties of the piezoelectric materials, especially the service life of piezoelectric structures. When subjected to mechanical and electrical loads, these defects may grow in size and cracks may propagate leading to premature mechanical or electrical fatigue failure. Therefore, it is important to understand the growth of these defects, the damage accumulation and the overall effect of these defects on the average mechanical and electrical properties of piezoelectric structures.

Damage in fiber-reinforced composite materials has been extensively investigated, and many theories have been established and used to predict the life of composite structures. Based on the framework of irreversible thermodynamics with internal state variables, Talreja [1] developed a phenomenological theory for composite laminated plates. In his study, the Helmholtz free energy was expanded into a polynomial in terms of elastic strains and damage variables to obtain the stiffness-damage relations. Utilizing a continuum mechanics approach, Allen *et al.* [2,3] developed a model for predicting the thermomechanical constitution of initially elastic composites subjected to both monotonic and cyclic fatigue loading. Valliappan *et al.* [4] established the elastic constitutive equations for anisotropic damage mechanics, and the implementation of these constitutive equations in the finite element analysis was explained. By defining damage variables as the material stiffness reduction, Ladeveze and Dantec [5] formulated the constitutive equations and the corresponding damage evolution laws of the elementary ply for laminated composites that can be used to describe the matrix micro-cracking and fiber/matrix debonding. Schapery and Sicking [6] discussed the homogenized constitutive equations for the mechanical behavior of unidirectional fiber composites with growing damage, and the emphasis was on resin matrices reinforced with high modulus elastic fiber. Zhang *et al.* [7] investigated a computational model for the damage evolution of engineering materials under dynamic loading, and two models for dynamic damage evolution of materials in general anisotropic damage state were presented. Moore and Dillard [8] have observed time dependent growth of transverse cracks in graphite/epoxy and Kevlar/epoxy cross-ply laminated at room temperature. Luo and Daniel [9] have shown that the macroscopic mechanical behavior of unidirectional fiber-reinforced brittle matrix composites can be correlated explicitly with the microscopic deformation and damage.

Modeling and analysis of multilayer piezoelectric beams and plates have reached a relative maturity as attested by the numerous papers. Mindlin [10] presented the theory of piezoelectric crystals plate considering shear and bending. Chandrashekhara and Tiersten [11] developed general piezoelectric nonlinear theory and detailed the vibration equations of different piezoelectric crystals. Chandrashkhara, Tenneti [12] and Zhou *et al.* [13] investigated the dynamic control of laminated piezoelectric plates by the finite element (FE) method. Wang and Rogers [14] presented a model for laminated plates with spatially distributed piezoelectric patches. Tzou and Gadre [15] analyzed thin laminates coupled with shell actuators for distributed vibration control. Xu *et al.* [16] analyzed the free vibration of laminated piezothermoelectric plate based on the 3D theory. Mitchell and Reddy [17] proposed the theory of the laminated piezoelectric plates by using classical plate theory and simple third-order theory, respectively. Noor and Peters [18] presented postbuckling analysis of multilayered composite plates subjected to combined axial and thermal loads and investigated the effects of mechanical and thermal loading on postbuckling behaviors of composite plates. Based on transverse shear-deformable theory, Librescu and Souza [19] gave the postbuckling analysis of geometrically imperfect flat panels under combined thermal and compressive edge loadings. Shen [20] investigated the postbuckling behaviors of laminated plates with piezoelectric actuators under complex loading conditions based on Reddy's higher order shear deformation plate theory. Oh *et al.* [21] studied thermal postbuckling behavior of laminated plates with top and/or bottom actuators subjected to thermal and electrical loads. In their analysis all the static/dynamic behaviors of laminated plates were investigated without considering the damage effects which would greatly influence the mechanical behaviors of smart structures. To the best of the author's knowledge, up to now postbuckling analysis of piezoelectric structures considering the damage effects has rarely been investigated and reported. For example, Aydin [22] studied the dynamic characteristics of functionally graded beams with open edge cracks, in which an analytical method was proposed to determine the free vibration of beams with any number of cracks. Mao *et al.*[23] studied the creep buckling and post-buckling of laminated piezoelectric viscoelastic functionally graded material (FGM) plates by adopting the Boltzmann superposition principle, and the nonlinear creep buckling governing equations of the laminated piezoelectric viscoelastic FGM plates with initial deflection were derived on the basis of the Reddy's higher-order shear deformation plate theory. Hamed [24] presented a nonlinear theoretical model for their bending and creep buckling analysis. The model accounted for the viscoelasticity of the materials using differential-type constitutive relations that were based on the linear Boltzmann's principle of superposition. Xu *et al.* [25] presented a high performance and simple structure bi-stable piezoelectric energy harvester based on simply supported piezoelectric buckled beam. Cottone *et al.* [26] investigated an approach for piezoelectric beams by exerting an increasing axial compression and demonstrated that the numerical model and experimental results were in good qualitative agreement. A constitutive model of fully coupled electro-magneto-thermo-elastic multiphase composites has been proposed by Aboudi [27]. In his works, the linear displacement, electric potential and magnetic potential are adopted, which can't predict the micro fields precisely. Bansal and Pindera [28] proposed a unified macro-and micromechanics failure model with method of cells as finite-volume direct averaging micromechanics (FVDAM). Based on the FVDAM theory, Sun *et al.* [29] built a unified macro- and micro-mechanics constitutive model of fully coupled fields in composite materials by high-order displacement, electric potential and magnetic potential.

In the present study, a new constitutive model for piezoelectric materials using the Talreja's tensor valued internal state damage variables and the Helmhotlz free energy of piezoelectric material is presented. This model is then applied to a specific case of postbuckling analysis of piezoelectric plates under in-plane compressive loads. By adopting von Karman's plate theory and using the finite difference and the Newmark scheme, the damage evolution for damage accumulation is developed and the finite difference procedure for postbuckling equilibrium path is simultaneously employed. In the numerical examples, the effects of variation in the load parameters, damage influences and geometric parameters of the plate on postbuckling equilibrium paths are discussed.

2. BASIC EQUATIONS

2.1. Constitutive Equations for Damaged Piezoelectric Materials

Consider a representative volume element of a piezoelectric solid with a multitude of damage entities in the form of microcracks, as shown in Figure 1. As discussed in Talreja, two vectors are needed to define each damage entity. These are the damage influence vector $\vec{a}_i$ and the $\vec{n}_i$ unit normal to the damage entity surface. The damage influence vector represents an appropriately chosen effect of the damage entity on the surrounding medium. With these two vectors, a damage entity tensor d_{ij} is formed by taking an integral of the diad $\vec{a}_i\ \vec{n}_i$ over the surface of the damaged entity:

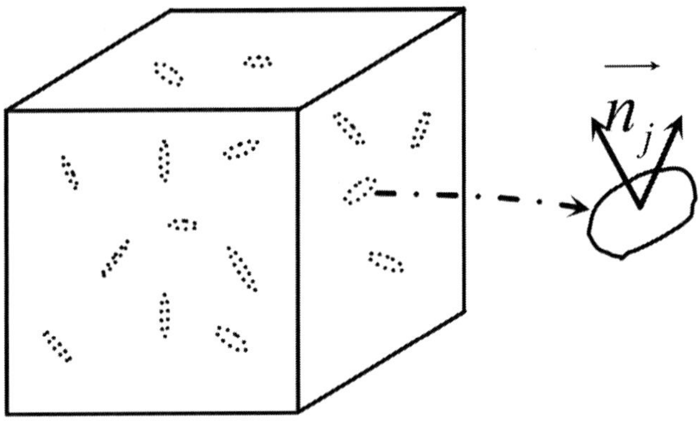

Figure 1. Representative volume element with internal damage variables for piezoelectric materials.

$$d_{ij} = \int_S \vec{a_i}\, \vec{n_j} dS \tag{1}$$

where *S* is the damage entity surface.

Now if there are *n* distinct damage modes in the representative volume element (e.g., intralaminar cracks in different orientations, *etc.*) denoted by *k* = 1,2,...*n*, a damage tensor can be defined for each mode as:

$$\omega_{ij}^k = \frac{1}{V_r} \sum_{\vartheta_k} (d_{ij})_{\vartheta_k} \tag{2}$$

where V_r is the volume of the representative volume element and ϑ_k represents the number of damage entities in the *k*th damage mode. The tensor ωkij is an unsymmetrical tensor in general. However, we can represent the vector ai→ along the normal and tangential directions at any point on the surface of the damage entity and write:

$$d_{ij} = d_{ij}^1 + d_{ij}^2 \tag{3}$$

where $d_{ij}^1 = \int_S a\vec{n_i}\, \vec{n_j} dS$ and $d_{ij}^2 = \int_S b\vec{m_i}\, \vec{n_j} dS$. in which, *a* and *b* are the magnitudes of the normal and tangential projections of vector $\vec{a_i}$ respectively, and vectors $\vec{n_i}$ and $\vec{m_i}$ are unit normal and tangential vectors, respectively. Thus the damage tensor ω_{ij} can be written as:

$$\omega_{ij}^{k} = \omega_{ij}^{1k} + \omega_{ij}^{2k} \tag{4}$$

where $\omega_{ij}^{1k} = \sum_{\vartheta_k} (d_{ij}^{1})_{\vartheta_k} / V_r$ and $\omega_{ij}^{2k} = \sum_{\vartheta_k} (d_{ij}^{2})_{\vartheta_k} / V_r$.

Physically, the damage tensor ω_{ij}^{1k} represents the effects of crack opening on the surrounding medium whereas the damage tensor ω_{ij}^{2k} represents the effects of sliding between the two crack faces. In many situations, the sliding between the crack faces can be negligible, e.g., for intralaminar cracks constrained by stiff plies, and hence we assume $\omega_{ij}^{2k} \equiv 0$. This implies $\omega_{ij}^{k} \equiv \omega_{ij}^{1k}$ which is a symmetric tensor.

For the case of damaged piezoelectric material without temperature effect where the damage is represented by internal state variables, the Helmholtz free energy of piezoelectric material can be written as a function of the transformed elastic strains, the electric field vector and damage internal variables, that is:

$$H = H(\varepsilon_{ij}, E_i, \omega_{ij}^{k}) \tag{5}$$

The transformed stress components σ_{ij} and the electric displacement components G_i at any fixed damage state are now given by:

$$\sigma_{ij} = \frac{\partial H(\varepsilon_{ij}, E_i, \omega_{ij}^{k})}{\partial \varepsilon_{ij}}$$

$$G_i = -\frac{\partial H(\varepsilon_{ij}, E_i, \omega_{ij}^{k})}{\partial E_i} \tag{6}$$

When the damage induced by the cracks in the piezoelectric material has the orthotropic property, the irreducible integrity bases for a scalar polynomial function of two symmetric second rank tensors can be expressed as [30]:

ε11,ε22,ε33,ε232,ε312,ε122,ε12ε23ε31,ωk11,ωk22,ωk33,(ωk23)2,(ωk31)2,(ωk12)2,ωk12ωk23ωk31ε23ωk23,ε31ωk31,ε12ωk12,ωk23ε12ε13,ωk31ε32ε12,ωk12ε13ε23,(k=1,2,·······n)ε23ωk12ωk13,ε31ωk32ωk12,ε12ωk13ωk23E1,E2,E3

$$\begin{aligned}
&\varepsilon_{11},\ \varepsilon_{22},\ \varepsilon_{33},\ \varepsilon_{23}^{\ 2},\ \varepsilon_{31}^{\ 2},\ \varepsilon_{12}^{\ 2},\ \varepsilon_{12}\varepsilon_{23}\varepsilon_{31}, \\
&\omega_{11}^{k},\ \omega_{22}^{k},\ \omega_{33}^{k},\ (\omega_{23}^{k})^2,\ (\omega_{31}^{k})^2,\ (\omega_{12}^{k})^2,\ \omega_{12}^{k}\omega_{23}^{k}\omega_{31}^{k} \\
&\varepsilon_{23}\omega_{23}^{k},\ \varepsilon_{31}\omega_{31}^{k},\ \varepsilon_{12}\omega_{12}^{k},\ \omega_{23}^{k}\varepsilon_{12}\varepsilon_{13},\ \omega_{31}^{k}\varepsilon_{32}\varepsilon_{12},\ \omega_{12}^{k}\varepsilon_{13}\varepsilon_{23},\quad (k=1,2,\cdots\cdots n) \\
&\varepsilon_{23}\omega_{12}^{k}\omega_{13}^{k},\ \varepsilon_{31}\omega_{32}^{k}\omega_{12}^{k},\ \varepsilon_{12}\omega_{13}^{k}\omega_{23}^{k} \\
&E_1,\ E_2,\ E_3
\end{aligned} \tag{7}$$

where n is the number of the cracks' direction in the material. For a piezoelectric single-layer plate, the local coordinate system $o-1-2-3$ is selected, in which 1,2 denote the two principal direction of the piezoelectric plate, 3 is vertical to the midsurface. According to the Kirchhoff hypothesis for plate $\varepsilon_{13} = \varepsilon_{23} = 0$ and applying Voigt notation to describe strains and damage variables, the bases of invariants can be further written as:

$$\begin{aligned}
&\varepsilon_1,\ \varepsilon_2,\ \varepsilon_3,\ \varepsilon_6^{\ 2},\ \omega_1^{\ k},\ \omega_2^{\ k},\ \omega_3^{\ k},\ (\omega_4^{\ k})^2,\ (\omega_5^{\ k})^2,\ (\omega_6^{\ k})^2, \\
&\omega_4^{\ k}\omega_5^{\ k}\omega_6^{\ k},\ \varepsilon_6\omega_6^{\ k},\ \varepsilon_6\omega_4^{\ k}\omega_5^{\ k} \qquad ,\ (k=1,2,\cdots\cdots n) \\
&E_1,\ E_2,\ E_3
\end{aligned} \tag{8}$$

Using the above stated irreducible integrity bases, the Helmholtz free energy of piezoelectric materials can be expressed as a quadratic expression of the strains or the electric field intensity, a mixture quadratic expression of strains and electric field intensity and a linear expression of damage variables [31] as follows:

$$
\begin{aligned}
H = {} & C_1^0\varepsilon_1^2 + C_2^0\varepsilon_1\varepsilon_2 + C_3^0\varepsilon_2^2 + C_4^0\varepsilon_6^2 + C_5^0\varepsilon_1^2 + C_6^0\varepsilon_1\varepsilon_3 + C_7^0\varepsilon_2\varepsilon_3 - (\kappa_1^0E_1^2 + \kappa_2^0E_2^2 + \\
& \kappa_3^0E_3^2 + \kappa_4^0E_1E_2 + \kappa_5^0E_2E_3 + \kappa_6^0E_3E_1) - (e_1^0E_1\varepsilon_1 + e_2^0E_1\varepsilon_2 + e_3^0E_1\varepsilon_3 + e_4^0E_2\varepsilon_1 + \\
& e_5^0E_2\varepsilon_2 + e_6^0E_2\varepsilon_3 + e_7^0E_3\varepsilon_1 + e_8^0E_3\varepsilon_2 + e_9^0E_3\varepsilon_3) + \sum_{k=1}^{n}(C_1^k\varepsilon_1^2\omega_1^k + C_2^k\varepsilon_1^2\omega_2^k + C_3^k\varepsilon_1^2\omega_3^k \\
& + C_4^k\varepsilon_2^2\omega_1^k + C_5^k\varepsilon_2^2\omega_2^k + C_6^k\varepsilon_2^2\omega_3^k + C_7^k\varepsilon_3^2\omega_1^k + C_8^k\varepsilon_3^2\omega_2^k + C_9^k\varepsilon_3^2\omega_3^k + C_{10}^k\varepsilon_6^2\omega_1^k + \\
& C_{11}^k\varepsilon_6^2\omega_2^k + C_{12}^k\varepsilon_6^2\omega_3^k + C_{13}^k\varepsilon_1\varepsilon_2\omega_1^k + C_{14}^k\varepsilon_1\varepsilon_2\omega_2^k + C_{15}^k\varepsilon_1\varepsilon_2\omega_3^k + C_{16}^k\varepsilon_2\varepsilon_3\omega_1^k + C_{17}^k\varepsilon_2\varepsilon_3\omega_2^k \\
& + C_{18}^k\varepsilon_2\varepsilon_3\omega_3^k + C_{19}^k\varepsilon_3\varepsilon_1\omega_1^k + C_{20}^k\varepsilon_3\varepsilon_1\omega_2^k + C_{21}^k\varepsilon_3\varepsilon_1\omega_3^k + C_{22}^k\varepsilon_6\varepsilon_1\omega_6^k + C_{23}^k\varepsilon_6\varepsilon_2\omega_6^k + \\
& C_{24}^k\varepsilon_6\varepsilon_3\omega_6^k) - \sum_{k=1}^{n}(\kappa_1^kE_1^2\omega_1^k + \kappa_2^kE_1^2\omega_2^k + \kappa_3^kE_1^2\omega_3^k + \kappa_4^kE_2^2\omega_1^k + \kappa_5^kE_2^2\omega_2^k + \kappa_6^kE_3^2\omega_3^k + \\
& \kappa_7^kE_3^2\omega_1^k + \kappa_8^kE_3^2\omega_2^k + \kappa_9^kE_3^2\omega_3^k + \kappa_{10}^kE_1E_2\omega_1^k + \kappa_{11}^kE_1E_2\omega_2^k + \kappa_{12}^kE_1E_2\omega_3^k + \kappa_{13}^kE_2E_3\omega_1^k \\
& + \kappa_{14}^kE_2E_3\omega_2^k + \kappa_{15}^kE_2E_3\omega_3^k + \kappa_{16}^kE_3E_1\omega_1^k + \kappa_{17}^kE_3E_1\omega_2^k + \kappa_{18}^kE_3E_1\omega_3^k) - \sum_{k=1}^{n}(e_1^kE_1\varepsilon_1\omega_1^k \\
& + e_2^kE_1\varepsilon_1\omega_2^k + e_3^kE_1\varepsilon_1\omega_3^k + e_4^kE_1\varepsilon_2\omega_1^k + e_5^kE_1\varepsilon_2\omega_2^k + e_6^kE_1\varepsilon_2\omega_3^k + e_7^kE_1\varepsilon_3\omega_1^k + e_8^kE_1\varepsilon_3\omega_2^k \\
& + e_9^kE_1\varepsilon_3\omega_3^k + e_{10}^kE_2\varepsilon_1\omega_1^k + e_{11}^kE_2\varepsilon_1\omega_2^k + e_{12}^kE_2\varepsilon_1\omega_3^k + e_{13}^kE_2\varepsilon_2\omega_1^k + e_{14}^kE_2\varepsilon_2\omega_2^k + e_{15}^kE_2\varepsilon_2\omega_3^k \\
& + e_{16}^kE_2\varepsilon_3\omega_1^k + e_{17}^kE_2\varepsilon_3\omega_2^k + e_{18}^kE_2\varepsilon_3\omega_3^k + e_{19}^kE_3\varepsilon_1\omega_1^k + e_{20}^kE_3\varepsilon_1\omega_2^k + e_{21}^kE_3\varepsilon_1\omega_3^k + e_{22}^kE_3\varepsilon_2\omega_1^k \\
& + e_{23}^kE_3\varepsilon_2\omega_2^k + e_{24}^kE_3\varepsilon_2\omega_3^k + e_{25}^kE_3\varepsilon_3\omega_1^k + e_{26}^kE_3\varepsilon_3\omega_2^k + e_{27}^kE_3\varepsilon_3\omega_3^k + e_{28}^kE_1\varepsilon_6\omega_6^k + e_{29}^kE_2\varepsilon_6\omega_6^k \\
& + e_{30}^kE_3\varepsilon_6\omega_6^k) + P_0 + P_1\left(\varepsilon_p, \omega_q^k\right) + P_2\left(\omega_q^k\right) + P_3\left(E_i, \omega_q^k\right)
\end{aligned}
\tag{9}
$$

where $C_i^0(i=1,2,\cdots 7)$ are the elastic material constants without damage, $e_i^0(i=1,2,\cdots 9)$ are the piezoelectric constants without damage, $\kappa_i^0(i=1,2,\cdots 6)$ are the permittivity matrix constants without damage, $C_i^k(i=1,2,\cdots 24)$ are the material constants with damage; $e_i^k(i=1,2,\cdots 30)$ are the piezoelectric constants with damage, $\kappa_i^k(i=1,2,\cdots 18)$ are the permittivity matrix constants with damage, ρ is the density of piezoelectric material, P_0 is a constant, P_1 is a linear function of strains, P_2 is a linear function of damage variables and P_3 is a linear function of the electric field intensity. Then the stresses and the electric displacements can be expressed as:

$$
\begin{aligned}
\sigma_p &= \frac{\partial H}{\partial \varepsilon_p} = [C_{pq}^0 + \sum_{k=1}^{n} C_{pq}^k]\varepsilon_q - [e_{pm}^0 + \sum_{k=1}^{n} e_{pm}^k]E_m \\
G_m &= -\frac{\partial H}{\partial E_m} = [e_{qm}^0 + \sum_{k=1}^{n} e_{qm}^k]^T \varepsilon_q + [\kappa_{mn}^0 + \sum_{k=1}^{n} \kappa_{mn}^k]E_n
\end{aligned}
\tag{10}
$$

where C^0_{pq}, C^k_{pq}, κ^0_{mn} and κ^k_{mn} are all symmetric matrixes having the forms as follows:

$$[C^0_{pq}] = \begin{bmatrix} 2C^0_1 & C^0_2 & C^0_6 & 0 \\ & 2C^0_3 & C^0_7 & 0 \\ & & 2C^0_5 & 0 \\ & & & 2C^0_4 \end{bmatrix} \quad (11)$$

$$[C^k_{pq}] = \begin{bmatrix} 2C^k_1\omega^k_1 + 2C^k_2\omega^k_2 + 2C^k_3\omega^k_3 & C^k_{13}\omega^k_1 + C^k_{14}\omega^k_2 + C^k_{15}\omega^k_3 & C^k_{19}\omega^k_1 + C^k_{20}\omega^k_2 + C^k_{21}\omega^k_3 & C^k_{22}\omega^k_6 \\ & 2C^k_4\omega^k_1 + 2C^k_5\omega^k_2 + 2C^k_6\omega^k_3 & C^k_{16}\omega^k_1 + C^k_{17}\omega^k_2 + C^k_{18}\omega^k_3 & C^k_{23}\omega^k_6 \\ & & 2C^k_7\omega^k_1 + 2C^k_8\omega^k_2 + 2C^k_9\omega^k_3 & C^k_{24}\omega^k_6 \\ & & & 2C^k_{10}\omega^k_1 + 2C^k_{11}\omega^k_2 + 2C^k_{12}\omega^k_3 \end{bmatrix} \quad (12)$$

$$[e^0_{pm}] = \begin{bmatrix} e^0_1 & e^0_4 & e^0_7 \\ e^0_2 & e^0_5 & e^0_8 \\ e^0_3 & e^0_6 & e^0_9 \\ 0 & 0 & 0 \end{bmatrix} \quad (13)$$

$$[\kappa^0_{mn}] = \begin{bmatrix} 2\kappa^0_1 & \kappa^0_4 & \kappa^0_6 \\ & 2\kappa^0_2 & \kappa^0_5 \\ & & 2\kappa^0_3 \end{bmatrix} \quad (14)$$

$$[e^k_{pm}] = \begin{bmatrix} e^k_1\omega^k_1 + e^k_2\omega^k_2 + e^k_3\omega^k_3 & e^k_{10}\omega^k_1 + e^k_{11}\omega^k_2 + e^k_{12}\omega^k_3 & e^k_{19}\omega^k_1 + e^k_{20}\omega^k_2 + e^k_{21}\omega^k_3 \\ e^k_4\omega^k_1 + e^k_5\omega^k_2 + e^k_6\omega^k_3 & e^k_{13}\omega^k_1 + e^k_{14}\omega^k_2 + e^k_{15}\omega^k_3 & e^k_{22}\omega^k_1 + e^k_{23}\omega^k_2 + e^k_{24}\omega^k_3 \\ e^k_7\omega^k_1 + e^k_8\omega^k_2 + e^k_9\omega^k_3 & e^k_{16}\omega^k_1 + e^k_{17}\omega^k_2 + e^k_{18}\omega^k_3 & e^k_{25}\omega^k_1 + e^k_{26}\omega^k_2 + e^k_{27}\omega^k_3 \\ e^k_{28}\omega^k_6 & e^k_{29}\omega^k_6 & e^k_{30}\omega^k_6 \end{bmatrix} \quad (15)$$

$$[\kappa^k_{mn}] = \begin{bmatrix} 2\kappa^k_1\omega^k_1 + 2\kappa^k_2\omega^k_2 + 2\kappa^k_3\omega^k_3 & \kappa^k_{10}\omega^k_1 + \kappa^k_{11}\omega^k_2 + \kappa^k_{12}\omega^k_3 & \kappa^k_{16}\omega^k_1 + \kappa^k_{17}\omega^k_2 + \kappa^k_{18}\omega^k_3 \\ & 2\kappa^k_4\omega^k_1 + 2\kappa^k_5\omega^k_2 + 2\kappa^k_6\omega^k_3 & \kappa^k_{13}\omega^k_1 + \kappa^k_{14}\omega^k_2 + \kappa^k_{15}\omega^k_3 \\ & & 2\kappa^k_7\omega^k_1 + 2\kappa^k_8\omega^k_2 + 2\kappa^k_9\omega^k_3 \end{bmatrix} \quad (16)$$

Assuming that there is only one damage mode in the representative volume element, the relations of the strains, the stresses, the electric field intensity and the electric displacements in Equation (10) can be simplified as:

$$\begin{aligned} \sigma_p &= [C^0_{pq} + C^1_{pq}]\varepsilon_q - [e^0_{pm} + e^1_{pm}]E_m \\ G_m &= [e^0_{qm} + e^1_{qm}]^T \varepsilon_q + [\kappa^0_{mn} + \kappa^1_{mn}]E_n \end{aligned} \tag{17}$$

where C^0_{pq}, e^0_{pm} and κ^0_{mn} are the same as before. C^k_{pq}, e^k_{pm}, κ^k_{mn} are replaced by C^1_{pq}, e^1_{pm}, κ^1_{mn}.

In present study, consider that the matrix cracks in the piezoelectric plate are parallel to the coordinate plane 2–3, all damage variables except ω_1 are zero, then the coefficient matrixes in Equations (12), (15) and (16) can be simplified as:

$$[C^1_{pq}] = \begin{bmatrix} 2C_1\omega_1 & C_{13}\omega_1 & C_{19}\omega_1 & 0 \\ & 2C_4\omega_1 & C_{16}\omega_1 & 0 \\ & & 2C_7\omega_1 & 0 \\ & & & 2C_{10}\omega_1 \end{bmatrix} \tag{18}$$

$$[e^1_{pm}] = \begin{bmatrix} e_1\omega_1 & e_{10}\omega_1 & e_{19}\omega_1 \\ e_4\omega_1 & e_{13}\omega_1 & e_{22}\omega_1 \\ e_7\omega_1 & e_{16}\omega_1 & e_{25}\omega_1 \\ 0 & 0 & 0 \end{bmatrix} \tag{19}$$

$$[\kappa^1_{mn}] = \begin{bmatrix} 2\kappa_1\omega_1 & \kappa_{10}\omega_1 & \kappa_{16}\omega_1 \\ & 2\kappa_4\omega_1 & \kappa_{13}\omega_1 \\ & & 2\kappa_7\omega_1 \end{bmatrix} \tag{20}$$

Due to the fact the cracks are parallel to the coordinate plane 2 – 3, the effect of the damage on the stiffness of the plate in this coordinate plane 2 – 3 can be neglected, which means the component C_{11}^{1} and C_{11}^{3} of stiffness matrix due to damage effect are negligible. Then matrix (18) can be further simplified as:

$$[C_{pq}^{1}]=\begin{bmatrix} 2C_1\omega_1 & C_{13}\omega_1 & C_{19}\omega_1 & 0 \\ & 0 & C_{16}\omega_1 & 0 \\ & & 0 & 0 \\ & & & 2C_{10}\omega_1 \end{bmatrix} \tag{21}$$

Letting σ_3=0 based on plane-stress assumption and using Equation (17), the constitutive relations with damage of the piezoelectric plate for the plane stress problems are obtained as follows:

$$\begin{aligned} \sigma_p &= [C_{pq}]\varepsilon_q - [e_{pm}]E_m = [C_{pq}^{0} + C_{pq}^{1}]\varepsilon_q - [e_{pm}^{0} + e_{pm}^{1}]E_m \\ D_m &= [e_{qm}]^T \varepsilon_q + [\kappa_{mn}]E_n = [e_{qm}^{0} + e_{qm}^{1}]^T \varepsilon_q + [\kappa_{mn}^{0} + \kappa_{mn}^{1}]E_n \quad (p,q=1,2,6 \quad m,n=1,2,3) \end{aligned} \tag{22}$$

where:

$$[C_{pq}^{0}]=\begin{bmatrix} 2C_1^0 - \dfrac{\left(C_6^0\right)^2}{2C_5^0} & C_2^0 - \dfrac{C_6^0 C_7^0}{2C_5^0} & 0 \\ & 2C_3^0 - \dfrac{\left(C_7^0\right)^2}{2C_5^0} & 0 \\ & & 2C_4^0 \end{bmatrix} \underline{\underline{def}} \begin{bmatrix} C_{11}^{\bullet} & C_{12}^{\bullet} & 0 \\ & C_{22}^{\bullet} & 0 \\ & & C_{66}^{\bullet} \end{bmatrix} \tag{23}$$

$$[C_{pq}^{1}]=\begin{bmatrix} \left(2C_1 - \dfrac{C_6^0 C_{19}}{C_5^0}\right)\omega_1 & \left(C_{13} - \dfrac{C_7^0 C_{19} + C_6^0 C_{16}}{2C_5^0}\right)\omega_1 & 0 \\ & -\dfrac{C_7^0 C_{16}}{C_5^0}\omega_1 & 0 \\ & & 2C_{10}\omega_1 \end{bmatrix} \underline{\underline{def}} \begin{bmatrix} \alpha_{11} & \alpha_{12} & 0 \\ & \alpha_{22} & 0 \\ & & \alpha_{66} \end{bmatrix}\omega_1 \tag{24}$$

$$[e_{pm}^{0}]=\begin{bmatrix} e_1^0-\dfrac{C_6^0e_3^0}{2C_5^0} & e_4^0-\dfrac{C_6^0e_6^0}{2C_5^0} & e_7^0-\dfrac{C_6^0e_9^0}{2C_5^0} \\ e_2^0-\dfrac{C_7^0e_3^0}{2C_5^0} & e_5^0-\dfrac{C_7^0e_6^0}{2C_5^0} & e_8^0-\dfrac{C_7^0e_9^0}{2C_5^0} \\ 0 & 0 & 0 \end{bmatrix} \underline{\underline{def}} \begin{bmatrix} e_{11}^{*} & e_{12}^{*} & e_{13}^{*} \\ e_{21}^{*} & e_{22}^{*} & e_{23}^{*} \\ 0 & 0 & 0 \end{bmatrix} \tag{25}$$

$$[e_{pm}^{1}]=\begin{bmatrix} \left(e_1-\dfrac{C_6^0e_7+C_{19}e_3^0}{2C_5^0}\right)\omega_1 & \left(e_{10}-\dfrac{C_6^0e_{16}+C_{19}e_6^0}{2C_5^0}\right)\omega_1 & \left(e_{19}-\dfrac{C_6^0e_{25}+C_{19}e_9^0}{2C_5^0}\right)\omega_1 \\ \left(e_4-\dfrac{C_7^0e_7+C_{16}e_3^0}{2C_5^0}\right)\omega_1 & \left(e_{13}-\dfrac{C_6^0e_{16}+C_{16}e_6^0}{2C_5^0}\right)\omega_1 & \left(e_{22}-\dfrac{C_7^0e_{25}+C_{16}e_9^0}{2C_5^0}\right)\omega_1 \\ 0 & 0 & 0 \end{bmatrix} \underline{\underline{def}} \begin{bmatrix} \beta_{11} & \beta_{12} & \beta_{13} \\ \beta_{21} & \beta_{22} & \beta_{23} \\ 0 & 0 & 0 \end{bmatrix}\omega_1 \tag{26}$$

$$[\kappa_{mn}^{0}]=\begin{bmatrix} 2\kappa_1^0+\dfrac{\left(e_3^0\right)^2}{2C_5^0} & \kappa_4^0+\dfrac{e_3^0e_6^0}{2C_5^0} & \kappa_6^0+\dfrac{e_3^0e_9^0}{2C_5^0} \\ & 2\kappa_2^0+\dfrac{\left(e_6^0\right)^2}{2C_5^0} & \kappa_5^0+\dfrac{e_6^0e_9^0}{2C_5^0} \\ & & 2\kappa_3^0+\dfrac{\left(e_9^0\right)^2}{2C_5^0} \end{bmatrix} \underline{\underline{def}} \begin{bmatrix} \kappa_{11}^{*} & \kappa_{12}^{*} & \kappa_{13}^{*} \\ & \kappa_{22}^{*} & \kappa_{23}^{*} \\ & & \kappa_{33}^{*} \end{bmatrix} \tag{27}$$

$$[\kappa_{mn}^{1}]=\begin{bmatrix} \left(2\kappa_1+\dfrac{e_3^0e_7}{C_5^0}\right)\omega_1 & \left(\kappa_{10}+\dfrac{e_3^0e_{16}+e_6^0e_7}{2C_5^0}\right)\omega_1 & \left(\kappa_{16}+\dfrac{e_3^0e_{25}+e_9^0e_7}{2C_5^0}\right)\omega_1 \\ & \left(2\kappa_4+\dfrac{e_6^0e_{17}}{C_5^0}\right)\omega_1 & \left(\kappa_{13}+\dfrac{e_6^0e_{25}+e_9^0e_{16}}{2C_5^0}\right)\omega_1 \\ & & \left(2\kappa_7+\dfrac{e_9^0e_{25}}{C_5^0}\right)\omega_1 \end{bmatrix} \underline{\underline{def}} \begin{bmatrix} \gamma_{11} & \gamma_{12} & \gamma_{13} \\ & \gamma_{22} & \gamma_{23} \\ & & \gamma_{33} \end{bmatrix}\omega_1 \tag{28}$$

In the present research, the Kachanvo damage evolution law [32] is adopted for an arbitrary point *i* of the piezoelectric plate with damage:

$$\frac{\partial\omega_1^i}{\partial t}=\begin{cases} B\left(\dfrac{\left|\sigma_{eq}^i\right|}{1-\omega_1^i}-\mu\sigma_f\right)^m & \left|\sigma_{eq}^i\right|\geq\mu(1-\omega_1^i)\sigma_f \\ 0 & \left|\sigma_{eq}^i\right|<\mu(1-\omega_1^i)\sigma_f \end{cases} \tag{29}$$

where B, m and μ are the material constants, σ_{eq} is an equivalent stress which is based on certain failure criterion, σ_f is the limit stress.

The relations between the electric fields E_x, E_y, E_z and the electric potential ϕ in the Cartesian coordinate system are defined by:

$$E_x = -\phi_{,x}, \quad E_y = -\phi_{,y}, \quad E_z = -\phi_{,z} \tag{30}$$

For the piezoelectric plate, only thickness direction electric field E_z is dominant. If the voltage applied to the piezoelectric plate with piezoelectric effect in the thickness only, then:

$$E_z = V_e / h \tag{31}$$

where V_e is the applied voltage across the thickness of piezoelectric plates.

2.2. Basic Equations of Piezoelectric Plates

Now, consider a thin piezoelectric plate with transverse cracks subjected to uniformly distributed in-plane compressive load P in the x-direction, having thickness h, length a in the x-direction, width b in the y-direction shown in Figure 2. The reference surface defined by z = 0 is set on the middle surface of the undeformed plate.

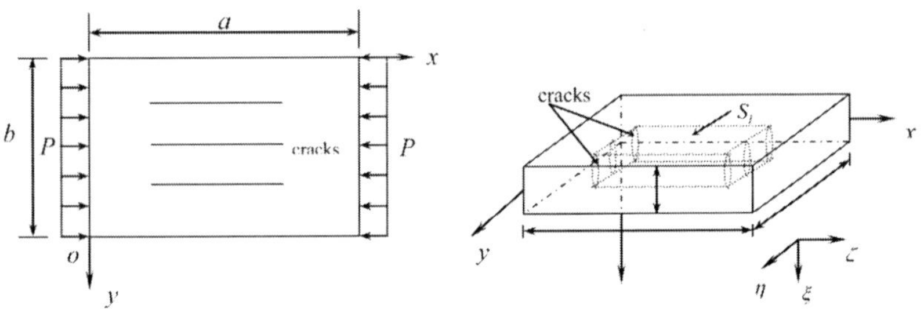

Figure 2. Geometric configuration of a piezoelectric plate with transverse cracks under the uniform compressive in-plane loads.

Setting u, v and w as the displacement components of an arbitrary point on the mid-surface along the direction of x, y and z, respectively, and denote $\bar{w}$ as the initial geometric deflection. According to classical nonlinear theory, the strain components ε_x^0, ε_y^0 and γ_{xy}^0 of the mid-surface [33] can be written as:

$$\varepsilon_x^0 = u_{,x} + \frac{1}{2} w_{,x}^{\ 2} + \bar{w}_{,x} w_{,x}, \quad \varepsilon_y^0 = v_{,y} + \frac{1}{2} w_{,y}^{\ 2} + \bar{w}_{,y} w_{,y}, \quad \gamma_{xy}^0 = u_{,y} + v_{,x} + w_{,x} w_{,y} + \bar{w}_{,x} w_{,y} + \bar{w}_{,y} w_{,x} \tag{32}$$

and the curvatures κ_x, κ_y and κ_{xy} of the mid-surface as:

$$\kappa_x = -w_{,xx}, \quad \kappa_y = -w_{,yy}, \quad \kappa_{xy} = -2w_{,xy} \tag{33}$$

then the nonlinear strain-displacement relations are expressed as follows:

$$\varepsilon_x = \varepsilon_x^0 + z\kappa_x, \quad \varepsilon_y = \varepsilon_y^0 + z\kappa_y, \quad \gamma_{xy} = \gamma_{xy}^0 + z\kappa_{xy} \tag{34}$$

Suppose the damage variable remains constant through the thickness of plate. Denoting N_x, N_y, N_{xy} as the membrane stress resultants and M_x, M_y, M_{xy} as the stress couples of the plate, according to the classical nonlinear plate theory, the nonlinear governing equations of the piezoelectric plate with initial geometric deflection [33] can be written as:

$$\begin{aligned}
&N_{x,x} + N_{xy,y} = 0 \\
&N_{xy,x} + N_{y,y} = 0 \\
&M_{x,xx} + 2M_{xy,xy} + M_{y,yy} + N_x(w_{,xx} + \bar{w}_{,xx}) + 2N_{xy}(w_{,xy} + \bar{w}_{,xy}) + N_y(w_{,yy} + \bar{w}_{,yy}) = 0
\end{aligned} \tag{35}$$

Using Equations (22) and (34), the following constitutive equations can be obtained:

$$\begin{Bmatrix} N_x \\ N_y \\ N_{xy} \end{Bmatrix} = \int_{-h/2}^{h/2} \begin{Bmatrix} \sigma_x \\ \sigma_y \\ \sigma_{xy} \end{Bmatrix} dz = \begin{bmatrix} A_{11} & A_{12} & 0 \\ & A_{22} & 0 \\ & & A_{66} \end{bmatrix} \begin{Bmatrix} \varepsilon_x^0 \\ \varepsilon_y^0 \\ \gamma_{xy}^0 \end{Bmatrix} - \begin{Bmatrix} N_x^p \\ N_y^p \\ N_{xy}^p \end{Bmatrix} \tag{36}$$

$$\begin{Bmatrix} M_x \\ M_y \\ M_{xy} \end{Bmatrix} = \int_{-h/2}^{h/2} z \begin{Bmatrix} \sigma_x \\ \sigma_y \\ \sigma_{xy} \end{Bmatrix} dz = \begin{bmatrix} D_{11} & D_{12} & 0 \\ & D_{22} & 0 \\ & & D_{66} \end{bmatrix} \begin{Bmatrix} \kappa_x \\ \kappa_y \\ \kappa_{xy} \end{Bmatrix} - \begin{Bmatrix} M_x^p \\ M_y^p \\ M_{xy}^p \end{Bmatrix} \tag{37}$$

where N_x^p, N_x^p, N_{xy}^p and M_x^p, M_y^p, M_{xy}^p represent the component resultants and couples due to the piezoelectric effect, respectively. The stiffness coefficients A_{ij} and D_{ij} of the piezoelectric plate are defined as follows:

$$A_{ij} = \int_{-h/2}^{h/2} C_{ij} dz, \quad D_{ij} = \int_{-h/2}^{h/2} z^2 C_{ij} dz \quad (i, j = 1, 2, 6) \tag{38}$$

The resultants and couples due to the piezoelectric effect can be written as:

$$\begin{aligned} & N_x^p = \int_{-h/2}^{h/2} e_{31} \frac{V_e}{h} dz, \quad N_y^p = \int_{-h/2}^{h/2} e_{32} \frac{V_e}{h} dz, \quad N_{xy}^p = 0 \\ & M_x^p = 0, \qquad M_y^p = 0, \qquad M_{xy}^p = 0 \end{aligned} \tag{39}$$

Introducing the following dimensionless parameters:

$$\begin{gathered} \xi = \frac{x}{a}, \ \eta = \frac{y}{b}, \ U = \frac{u}{a}, \ V = \frac{v}{b}, \ W = \frac{w}{h}, \ \overline{W} = \frac{w}{h}, \\ \bar{A}_{11} = \frac{A_{11}}{C_{11}^{*} h}, \ \bar{A}_{12} = \frac{A_{12}}{C_{11}^{*} h}, \ \bar{A}_{22} = \frac{A_{22}}{C_{11}^{*} h}, \ \bar{A}_{66} = \frac{A_{66}}{C_{11}^{*} h}, \ \lambda_1 = \frac{h}{a}, \ \lambda_2 = \frac{h}{b}, \ \bar{D}_{11} = \frac{D_{11}}{C_{11}^{*} h^3}, \\ \bar{D}_{12} = \frac{D_{12}}{C_{11}^{*} h^3}, \ \bar{D}_{22} = \frac{D_{22}}{C_{11}^{*} h^3}, \ \bar{D}_{66} = \frac{D_{66}}{C_{11}^{*} h^3}, \ \overline{C}_{12} = \frac{C_{12}}{C_{11}^{*}}, \ \overline{C}_{11} = \frac{C_{11}}{C_{11}^{*}}, \ \bar{\sigma}_{eq} = \frac{\sigma_{eq}}{C_{11}^{*}}, \ \bar{\sigma}_f = \frac{\sigma_f}{C_{11}^{*}}, \\ \overline{\alpha}_{11} = \frac{\alpha_{11}}{C_{11}^{*}}, \ \overline{\alpha}_{12} = \frac{\alpha_{12}}{C_{11}^{*}}, \ \overline{\alpha}_{22} = \frac{\alpha_{22}}{C_{11}^{*}}, \ \overline{\alpha}_{66} = \frac{\alpha_{66}}{C_{11}^{*}}, \ \overline{e_{31}} = \frac{e_{31}}{e_{31}^{*}}, \ \overline{e_{32}} = \frac{e_{32}}{e_{31}^{*}}, \ \overline{\beta}_{31} = \frac{\beta_{31}}{e_{31}^{*}}, \ \overline{\beta}_{32} = \frac{\beta_{32}}{e_{31}^{*}}, \\ \bar{z} = \frac{z}{h}, \ \tau = \frac{t}{h}\sqrt{\frac{C_{11}^{*}}{\rho}}, \ \overline{P} = \frac{P}{C_{11}^{*}}, \ \overline{V}_e = \frac{e_{31}^{*} V_e}{C_{11}^{*} h} \end{gathered} \tag{40}$$

By using Equations (22), (32–39) and (40), the dimensionless nonlinear governing equations of piezoelectric plate with initial geometric deflection under compressive loads in-plane including damage effects are obtained and expressed in terms of U, V and W as follow:

$$
\begin{aligned}
&\bar{A}_{11,\xi}(U_{,\xi}+\tfrac{1}{2}\lambda_1^2W_{,\xi}^2+\lambda_1^2\bar{W}_{,\xi}W_{,\xi})+\bar{A}_{11}(U_{,\xi\xi}+\lambda_1^2W_{,\xi}W_{,\xi\xi}+\lambda_1^2\bar{W}_{,\xi\xi}W_{,\xi}+\lambda_1^2\bar{W}_{,\xi}W_{,\xi\xi})+\bar{A}_{12,\xi}(V_{,\eta}\\
&+\tfrac{1}{2}\lambda_2^2W_{,\eta}^2+\lambda_2^2\bar{W}_{,\eta}W_{,\eta})+\bar{A}_{12}(V_{,\xi\eta}+\lambda_2^2W_{,\eta}W_{,\xi\eta}+\lambda_2^2\bar{W}_{,\xi\eta}W_{,\eta}+\lambda_2^2\bar{W}_{,\eta}W_{,\xi\eta})+\tfrac{\lambda_2}{\lambda_1}\bar{A}_{66,\eta}(\tfrac{\lambda_2}{\lambda_1}U_{,\eta}+\\
&\tfrac{\lambda_1}{\lambda_2}V_{,\xi}+\lambda_1\lambda_2W_{,\xi}W_{,\eta}+\lambda_1\lambda_2\bar{W}_{,\xi}W_{,\eta}+\lambda_1\lambda_2W_{,\xi}\bar{W}_{,\eta})+\bar{A}_{66}(\tfrac{\lambda_2^2}{\lambda_1^2}U_{,\eta\eta}+V_{,\xi\eta}+\lambda_2^2W_{,\eta}W_{,\xi\eta}+\lambda_2^2W_{,\xi}W_{,\eta\eta}+\\
&\lambda_2^2\bar{W}_{,\xi\eta}W_{,\eta}+\lambda_2^2\bar{W}_{,\eta\eta}W_{,\xi}+\lambda_2^2\bar{W}_{,\xi}W_{,\eta\eta}+\lambda_2^2\bar{W}_{,\eta}W_{,\xi\eta})-\overline{e_{31}}_{,\xi}\bar{V}_e=0\\
&\bar{A}_{12,\eta}(U_{,\xi}+\tfrac{1}{2}\lambda_1^2W_{,\xi}^2+\lambda_1^2\bar{W}_{,\xi}W_{,\xi})+\bar{A}_{12}(U_{,\xi\eta}+\lambda_1^2W_{,\xi}W_{,\xi\eta}+\lambda_1^2\bar{W}_{,\xi\eta}W_{,\xi}+\lambda_1^2\bar{W}_{,\xi}W_{,\xi\eta})+\bar{A}_{22,\eta}(V_{,\eta}\\
&+\tfrac{1}{2}\lambda_2^2W_{,\eta}^2+\lambda_2^2\bar{W}_{,\eta}W_{,\eta})+\bar{A}_{22}(V_{,\eta\eta}+\lambda_2^2W_{,\eta}W_{,\eta\eta}+\lambda_2^2W_{,\eta}\bar{W}_{,\eta\eta}+\lambda_2^2\bar{W}_{,\eta}W_{,\eta\eta})+\tfrac{\lambda_1}{\lambda_2}\bar{A}_{66,\xi}(\tfrac{\lambda_2}{\lambda_1}U_{,\eta}+\\
&\tfrac{\lambda_1}{\lambda_2}V_{,\xi}+\lambda_1\lambda_2W_{,\xi}W_{,\eta}+\lambda_1\lambda_2\bar{W}_{,\xi}W_{,\eta}+\lambda_1\lambda_2W_{,\xi}\bar{W}_{,\eta})+\bar{A}_{66}(U_{,\xi\eta}+\tfrac{\lambda_1^2}{\lambda_2^2}V_{,\xi\xi}+\lambda_1^2W_{,\eta}W_{,\xi\xi}+\lambda_1^2W_{,\xi}W_{,\xi\eta}+\\
&\lambda_1^2\bar{W}_{,\xi\xi}W_{,\eta}+\lambda_1^2\bar{W}_{,\xi}W_{,\xi\eta}+\lambda_1^2\bar{W}_{,\xi\eta}W_{,\xi}+\lambda_1^2\bar{W}_{,\eta}W_{,\xi\xi})-\overline{e_{32}}_{,\eta}\bar{V}_e=0\\
&-\lambda_1^4(\bar{D}_{11,\xi\xi}W_{,\xi\xi}+2\bar{D}_{11,\xi}W_{,\xi\xi\xi}+\bar{D}_{11}W_{,\xi\xi\xi\xi})-\lambda_1^2\lambda_2^2(\bar{D}_{12,\xi\xi}W_{,\eta\eta}+2\bar{D}_{12,\xi}W_{,\xi\eta\eta}+\bar{D}_{12}W_{,\xi\xi\eta\eta})-4\lambda_1^2\lambda_2^2(\\
&\bar{D}_{66,\xi\eta}W_{,\xi\eta}+\bar{D}_{66,\xi}W_{,\xi\eta\eta}+\bar{D}_{66,\eta}W_{,\xi\xi\eta}+\bar{D}_{66}W_{,\xi\xi\eta\eta})-\lambda_1^2\lambda_2^2(\bar{D}_{12,\eta\eta}W_{,\xi\xi}+2\bar{D}_{12,\eta}W_{,\xi\xi\eta}+\bar{D}_{12}W_{,\xi\xi\eta\eta})-\\
&\lambda_2^4(\bar{D}_{22,\eta\eta}W_{,\eta\eta}+2\bar{D}_{22,\eta}W_{,\eta\eta\eta}+\bar{D}_{22}W_{,\eta\eta\eta\eta})+\lambda_1^2[\bar{A}_{11}(U_{,\xi}+\tfrac{1}{2}\lambda_1^2W_{,\xi}^2+\lambda_1^2\bar{W}_{,\xi}W_{,\xi})+\bar{A}_{12}(V_{,\eta}+\tfrac{1}{2}\lambda_2^2W_{,\eta}^2\\
&+\lambda_2^2\bar{W}_{,\eta}W_{,\eta})-\overline{e_{31}}\bar{V}_e](W_{,\xi\xi}+\bar{W}_{,\xi\xi})+2\lambda_1\lambda_2\bar{A}_{66}(\tfrac{\lambda_2}{\lambda_1}U_{,\eta}+\tfrac{\lambda_1}{\lambda_2}V_{,\xi}+\lambda_1\lambda_2W_{,\xi}W_{,\eta}+\lambda_1\lambda_2\bar{W}_{,\xi}W_{,\eta}+\lambda_1\lambda_2W_{,\xi}\bar{W}_{,\eta})(W_{,\xi\eta}\\
&+\bar{W}_{,\xi\eta})+\lambda_2^2[\bar{A}_{12}(U_{,\xi}+\tfrac{1}{2}\lambda_1^2W_{,\xi}^2+\lambda_1^2\bar{W}_{,\xi}W_{,\xi})+\bar{A}_{22}(V_{,\eta}+\tfrac{1}{2}\lambda_2^2W_{,\eta}^2+\lambda_2^2\bar{W}_{,\eta}W_{,\eta})-\overline{e_{32}}\bar{V}_e](W_{,\eta\eta}+\bar{W}_{,\eta\eta})=0
\end{aligned}
\tag{41}
$$

Suppose the boundary of the piezoelectric plate is simply movably supported, the dimensionless boundary conditions can be expressed as:

$$\xi=0,1:\; V=0,\; \bar{A}_{11}(U_{,\xi}+\tfrac{1}{2}\lambda_1^2W_{,\xi}^2+\lambda_1^2\bar{W}_{,\xi}W_{,\xi})-\overline{e_{31}^1}\bar{V}_e=-\bar{P} \tag{42}$$

$$W=0,\; W_{,\xi\xi}=0$$

$$\eta=0,1:\; U=0,\; \bar{A}_{22}(V_{,\eta}+\tfrac{1}{2}\lambda_2^2W_{,\eta}^2+\lambda_2^2\bar{W}_{,\eta}W_{,\eta})-\overline{e_{32}^1}\bar{V}_e=0$$

$$W=0,\; W_{,\eta\eta}=0$$

The dimensionless damage evolution equation of the piezoelectric plate subjected to the uniformly in-plane compressive load can be written respectively as follows:

$$\frac{\partial\omega_1^i}{\partial\tau}=\begin{cases}\bar{B}\left(\dfrac{\left|\bar{\sigma}_{eq}^i\right|}{1-\omega_1^i}-\mu\bar{\sigma}_f\right)^m & \left|\bar{\sigma}_{eq}^i\right|\geq\mu(1-\omega_1^i)\bar{\sigma}_f\\ 0 & \left|\bar{\sigma}_{eq}^i\right|<\mu(1-\omega_1^i)\bar{\sigma}_f\end{cases} \tag{43}$$

Taking the mid-surface normal stress of the piezoelectric plate as the equivalent stress $\bar{\sigma}_{eq}$ that is parallel to the fibrous direction, it can be presented as:

$$\bar{\sigma}_{eq} = \overline{C}_{11}(U_{,\xi} + \tfrac{1}{2}\lambda_1^2 W_{,\xi}^{\ 2} + \lambda_1^2 \overline{W_{,\xi}} W_{,\xi}) + \overline{C}_{12}(V_{,\eta} + \tfrac{1}{2}\lambda_2^2 W_{,\eta}^{\ 2} + \lambda_2^2 \overline{W_{,\eta}} W_{,\eta}) - \overline{e_{31}^1}\,\overline{V_e} \tag{44}$$

3. SOLUTION METHODOLOGY

Suppose the dimensionless initial geometric deflection is taken as:

$$\overline{W} = \overline{W_0} \sin p\pi\xi \sin q\pi\eta \tag{45}$$

where p and q are the mode number in the ξ-direction and η-direction of the piezoelectric plate, respectively. Since the load and the structure are symmetric, only one quarter of the plate needs be considered. So the domain of the problem is selected as $0 \le \xi \le 1/2$, $0 \le \eta \le 1/2$.

To seek the approximate solutions of the governing Equations (41) which satisfied the boundary conditions(42), the unknown functions U,V and W are separated both for space and for time. The finite difference method is used for space, and the partial derivatives with respect to the space coordinate variables are replaced by difference form. The time τ is equally divided into small time segments $\Delta\tau$, and the whole equations are iterated to seek solutions. At each step of the iteration, the nonlinear items in the equations and the boundary conditions are linearized. For example, at the step J, the nonlinear items may be transformed to:

$$(x \cdot y)_J = (x)_J \cdot (y)_{J_p} \tag{46}$$

where $(y)_{Jp}$ is the average value of those obtained in the preceding two iterations. For the initial step of the iteration, it can be determined by using the quadratic extrapolation, *i.e.*:

$$(y)_{J_p} = A(y)_{J-1} + B(y)_{J-2} + C(y)_{J-3} \tag{47}$$

and for the different step of the iteration, the coefficients A, B and C can be expressed as follows:

$$\begin{aligned} J=1: &\quad A=1, B=0, C=0 \\ J=2: &\quad A=2, B=-1, C=0 \\ J\geq 3: &\quad A=3, B=-3, C=1 \end{aligned} \tag{48}$$

Moreover, using the Newmark scheme, the inertia in Equation (41) can be expressed as follows:

$$\begin{aligned} (W_{,\tau\tau})_J &= \frac{4(W_J - W_{J-1})}{(\Delta\tau)^2} - \frac{4(W_{,\tau})_{J-1}}{\Delta\tau} - (W_{,\tau\tau})_{J-1} \\ (W_{,\tau})_J &= (W_{,\tau})_{J-1} + \tfrac{1}{2}[(W_{,\tau\tau})_{J-1} + (W_{,\tau\tau})_J]\Delta\tau \\ W_J &= W_{J-1} + (W_{,\tau})_{J-1}\Delta\tau + \tfrac{1}{4}[(W_{,\tau\tau})_{J-1} + (W_{,\tau\tau})_J](\Delta\tau)^2 \end{aligned} \tag{49}$$

For every time step, the iteration lasts until the difference of the present value and the former is smaller than 0.1%, then continue the calculation of the next step.

4. NUMERICAL RESULTS

4.1. Comparison Study

To ensure the accuracy and effectiveness of the present method, a test example was calculated for postbuckling analysis of isotropic rectangular plate with initial geometric deflection. Comparison of postbuckling response curves for isotropic rectangular plate with initial geometric deflection is shown in Figure 3. The boundaries of the plate are clamped movable edges. W_0 denotes the center deflection of the plate. The close agreements between the present results and those of reference [33] demonstrate the present method is accurate and effective.

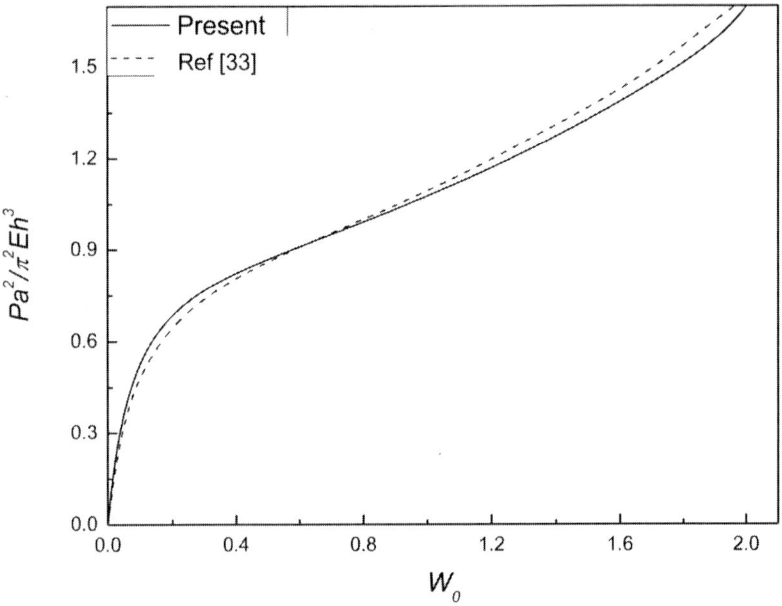

Figure 3. Comparison of postbuckling response curves for isotropic rectangular plate with initial deflection ($\bar{W}_0$ =0.1, v = 1/3).

4.2. Parametric Study

To study the piezo-effects and damage effects on the postbuckling behavior of the plates, several numerical examples were solved for initial flat and deflected plate. A piezoelectric plate consisting of the PZT-5A including initial damage is considered for postbuckling analysis. The material properties of PZT-5A are given as follows:

$$\begin{aligned}
&E_{11} = E_{22} = 61.0\text{GPa},\ E_{33} = 53.2\text{GPa},\quad \upsilon_{12} = 0.35,\quad \upsilon_{13} = \upsilon_{23} = 0.38 \\
&G_{12} = 22.6\text{GPa},\ G_{13} = G_{23} = 21.1\text{GPa},\quad \rho = 7.75\times 10^3 (\text{kg/m}^3) \\
&e_{31} = e_{32} = 7.209\text{C/m}^2,\ e_{33} = 15.118\text{C/m}^2,\ e_{15} = e_{24} = 12.322\text{C/m}^2 \\
&\kappa_{11} = \kappa_{22} = 1.53\times 10^{-8}\,\text{F/m},\ \kappa_{33} = 1.5\times 10^{-8}\,\text{F/m}
\end{aligned} \tag{50}$$

When the effect of damage is omitted and the linear strain-displacement relations are adopted, the dimensionless governing equation corresponding to Equation (41) is presented as:

$$-\lambda_1^4 \overline{D}_{11} W_{,\xi\xi\xi\xi} - 2\lambda_1^2 \lambda_2^2 \overline{D}_{12} W_{,\xi\xi\eta\eta} - 4\lambda_1^2 \lambda_2^2 \overline{D}_{66} W_{,\xi\xi\eta\eta} - \lambda_2^4 \overline{D}_{22} W_{,\eta\eta\eta\eta}$$
$$-\lambda_1^2 (\overline{P} + \overline{e_{31}} \overline{V}_e)(W_{,\xi\xi} + \overline{W}_{,\xi\xi}) = 0$$

(51)

The corresponding dimensionless boundary conditions of the simply movable supported plate can be written as:

$$\begin{aligned} \xi = 0,\ 1: &\qquad W = 0,\quad W_{,\xi\xi} = 0 \\ \eta = 0,\ 1: &\qquad W = 0,\quad W_{,\eta\eta} = 0 \end{aligned} \qquad (52)$$

Considering a harmonic displacement solution for this buckling problem (51), the displacement that satisfies with the boundary conditions (52) can be expressed as:

$$W(\xi, \eta) = W_0 \sin m\pi\xi \sin n\pi\eta \qquad (53)$$

Substituting Equation (53) into the Equation (51), the buckling load $\overline{P}_{mn}$ of a perfect plate and the relation of the center deflection and the compressive load in-plane of the plate with initial geometric imperfection can be obtained, respectively, as:

$$\overline{P}_{mn} = \pi^2 (\lambda_1^2 m^2 \overline{D}_{11} + 2\lambda_2^2 n^2 \overline{D}_{12} + \frac{\lambda_2^4 n^4}{\lambda_1^2 m^2} \overline{D}_{22} + 4\lambda_2^2 n^2 \overline{D}_{66}) - \overline{e_{31}} \overline{V}_e$$

(54)

$$W_0(\overline{P}) = \frac{(P_{mn} + \overline{e_{31}} \overline{V}_e) \overline{W}_0}{P_{mn} - \overline{P}} \qquad (55)$$

The least critical buckling load $\overline{P}_{cr}$ of the piezoelectric plate is determined by applying Equation (54) for the buckling mode $(m, n) = (1, 1)$ $(m, n) = (1, 1)$, as:

$$\overline{P}_{cr} = \pi^2(\lambda_1^2\overline{D}_{11} + 2\lambda_2^2\overline{D}_{12} + \frac{\lambda_2^4}{\lambda_1^2}\overline{D}_{22} + 4\lambda_2^2\overline{D}_{66}) - \overline{e}_{31}\overline{V}_e \quad (56)$$

When the geometric parameters are given as $\lambda_1 = \lambda_2 = 0.1$, $\overline{W}_0 = 0.1$, the critical buckling load $\overline{P}_{cr}$ is obtained as 0.03290 and 0.03272 by using the Equation (56) and the algorithm in the present paper, respectively. Figure 4shows the relations of the center deflection W_0 of the piezoelectric plate and the in-plane compressive load $\overline{P}$ without considering the damage effect. It can be seen that the nonlinearity of the plate has great influence on the postbuckling paths of the piezoelectric plate with initial deflection.

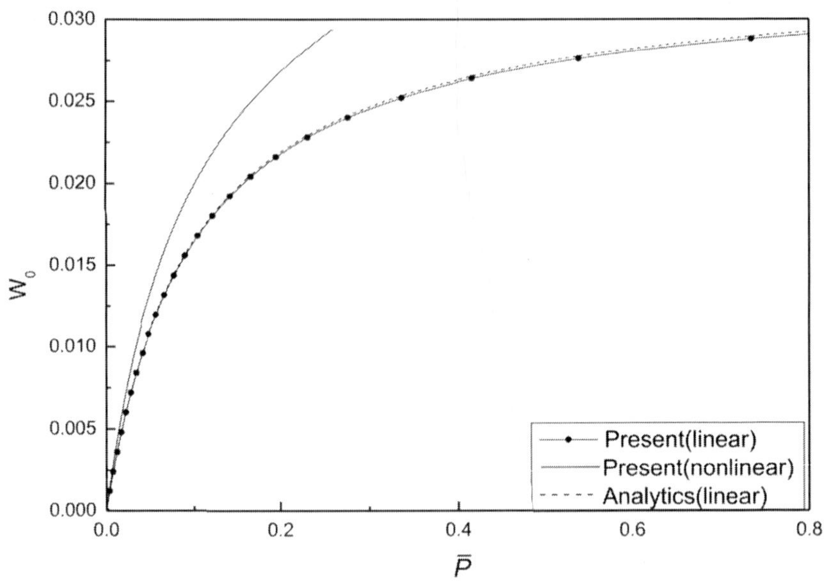

Figure 4. Response curves of the centre deflection *vs.* compressive loads in plane without damage.

A parametric study has been carried out and typical results are shown in Figures 5, 6, 7, 8, 9, 10, 11, 12, 13 and14. It should be appreciated that in all figures W_0, $\overline{W}$, $\overline{V}_e$ and $\overline{P}_{cr}$ denote the dimensionless maximum deflection, the dimensionless maximum initial deflection of the plate, the dimensionless applied voltage acted upon the plate and

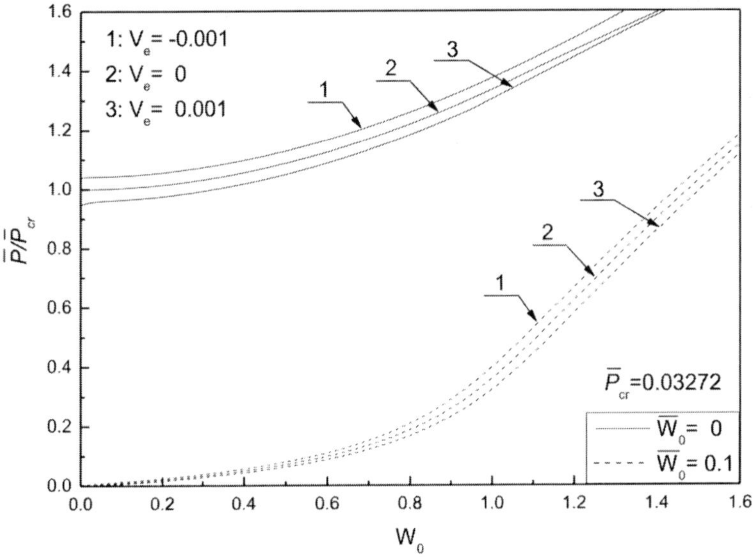

Figure 5. Comparisons of postbuckling response curves for initially flat and deflected piezoelectric plate without damage under different electrical loads.

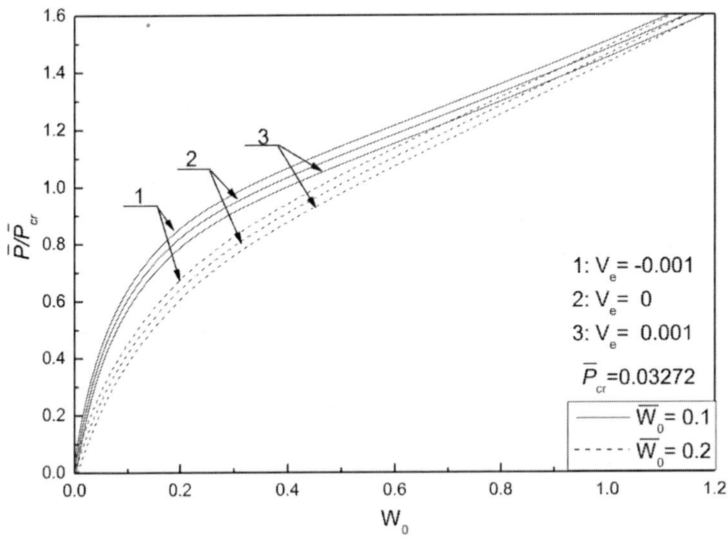

Figure 6. Effect of electrical loads on postbuckling response curves of piezoelectric plate without damage under two different initial deflections.

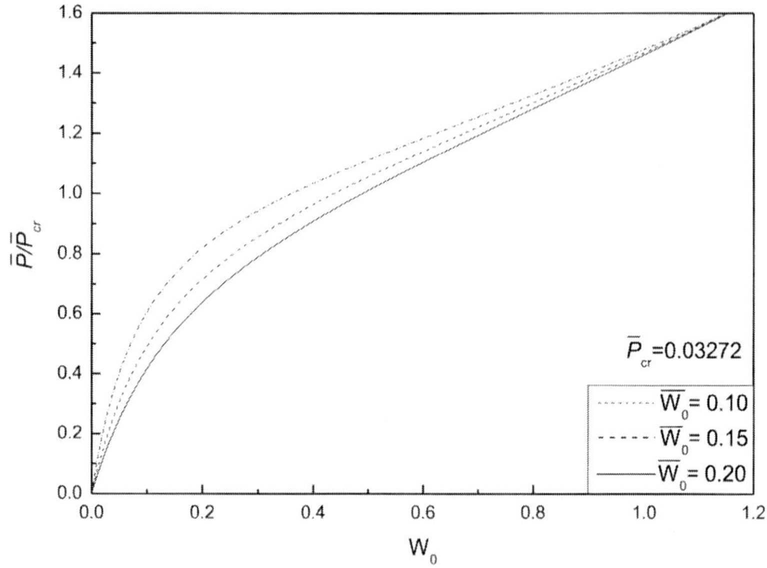

Figure 7. Effect of initial geometric deflections on the postbuckling response curves of piezoelectric plate without damage.

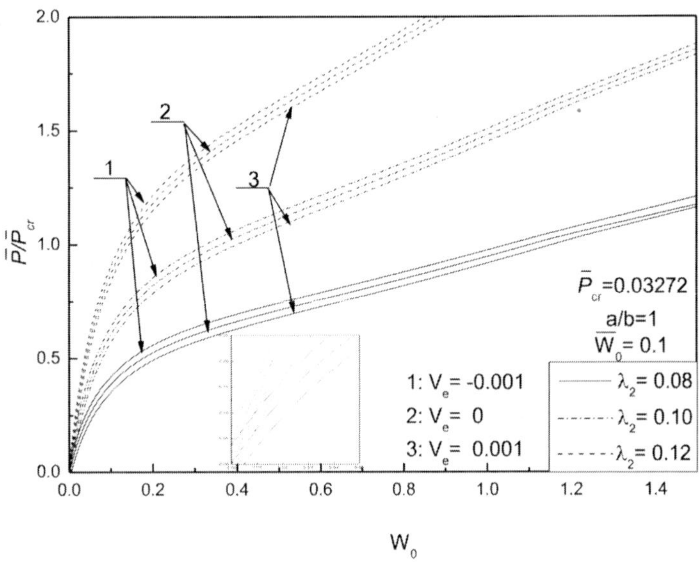

Figure 8. Effect of thick-span ratio on postbuckling response curves of piezoelectric plate without damage under different electrical loads (the inset figure is a zoom-in snapshot of the region around orgin point to depict the difference of three cases).

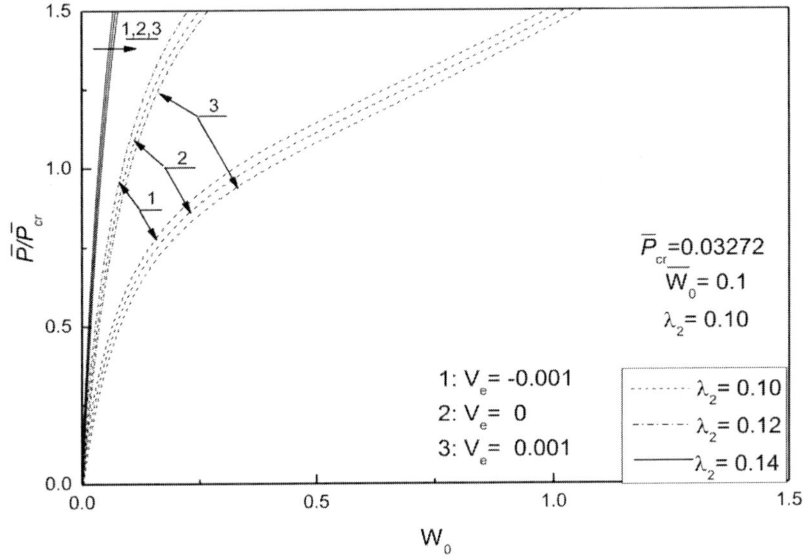

Figure 9. Effect of aspect ratio on postbuckling response curves of piezoelectric plate without damage under different electrical loads.

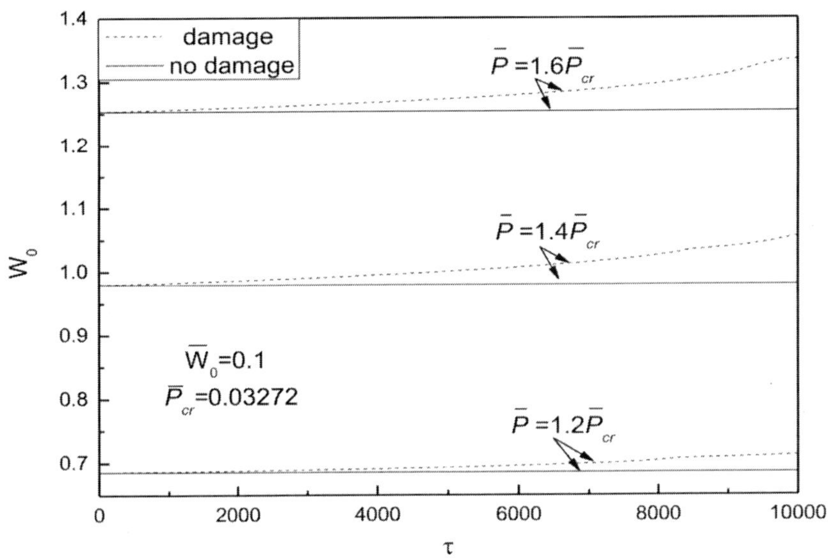

Figure 10. Effect of external loads on postbuckling response curves of piezoelectric plate with damage and initial deflection.

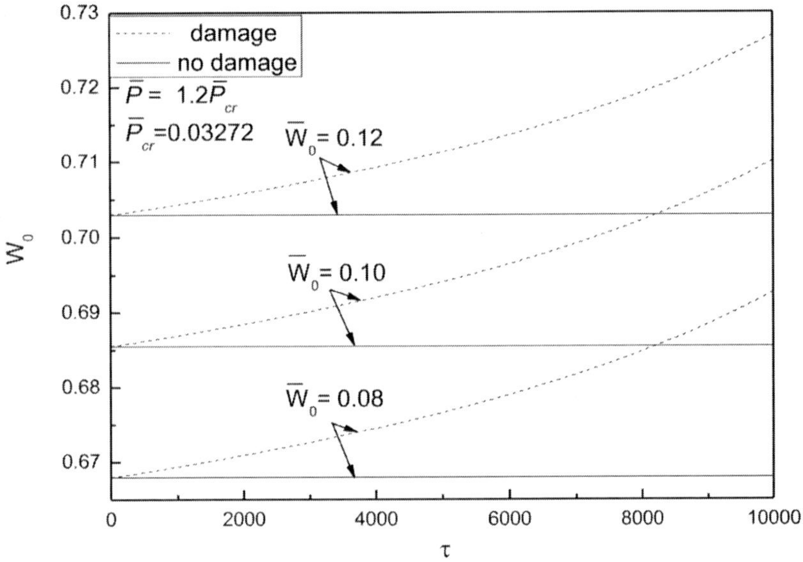

Figure 11. Effect of initial deflections on the postbuckling response curves of piezoelectric plate with damage.

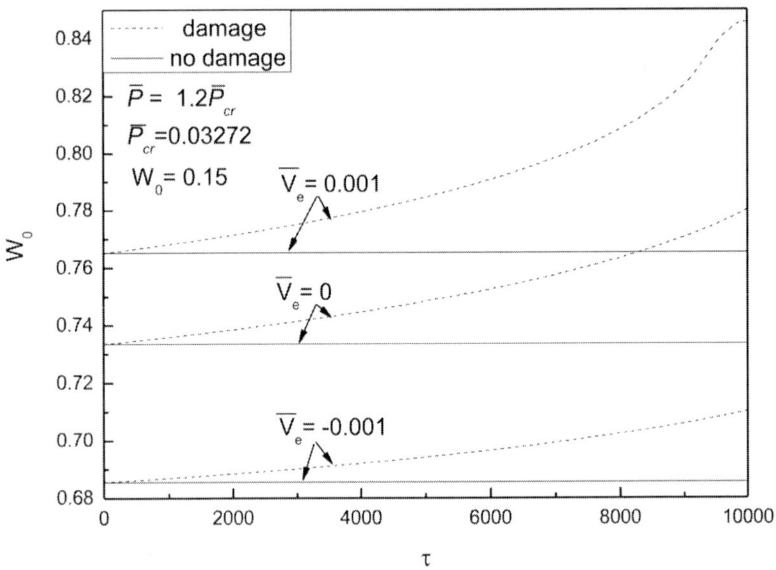

Figure 12. Effect of electrical loads on postbuckling response curves of piezoelectric plate with damage and initial deflection.

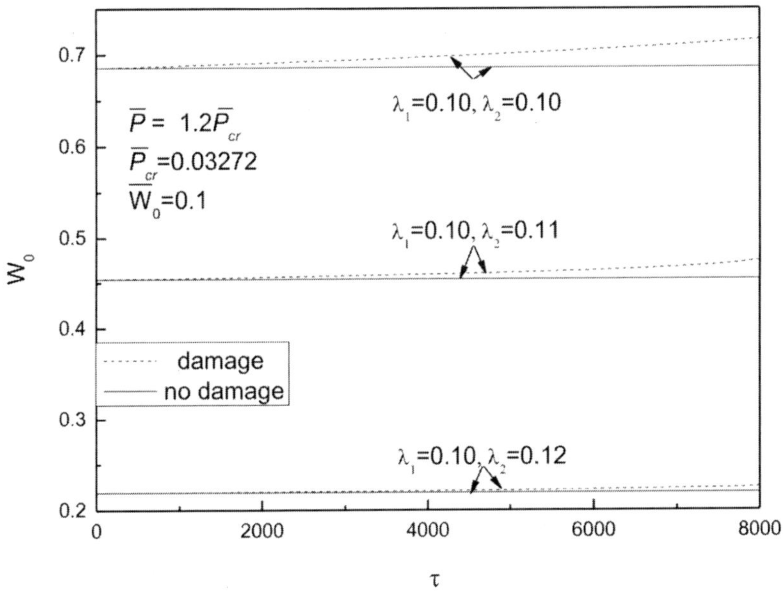

Figure 13. Effect of aspect ratio on postbuckling response curves of piezoelectric plate with damage and initial deflection.

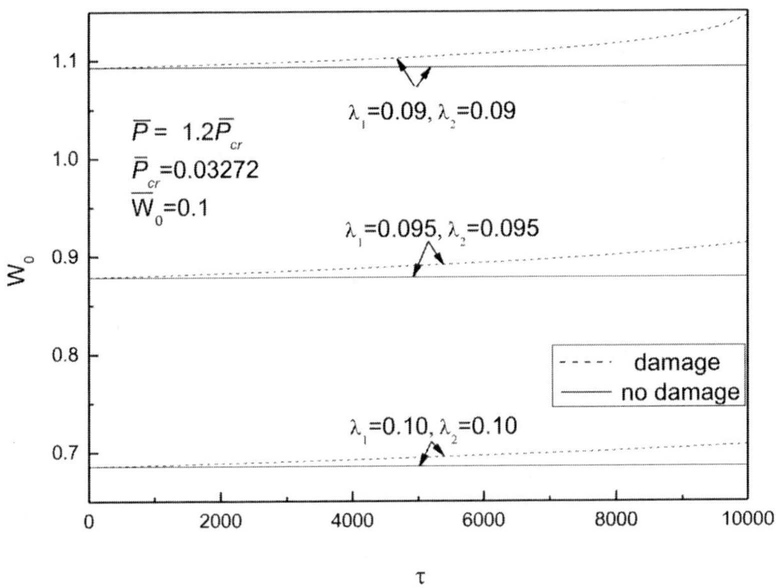

Figure 14. Effect of thick-span ratio on postbuckling response curves of piezoelectric plate with damage and initial deflection.

the dimensionless least critical buckling load. In all examples the least critical buckling load is taken as $\overline{P_{cr}}$=0.03272, which was calculated without considering piezo-effects and with the geometric parameters $\lambda_1 = \lambda_2 = 0.01$.

When the damage effect is in consideration, the material parameters related to damage in all examples are taken as:

$\bar{B}=0.5$, $n=1.1$, $\bar{\sigma}_f=4\times10^{-2}$, $\overline{\alpha_{11}}=-0.1$, $\overline{\alpha_{12}}=-0.04$, $\overline{\alpha_{22}}=-0.1$, $\overline{\alpha_{66}}=-0.04$, $\overline{\beta_{31}}=-0.1$, $\overline{\beta_{32}}=-0.1$, $\mu=0.6$.

Figure 5 shows the postbuckling response curves for an initially flat and deflected piezoelectric plate without damage under different electrical loads and Figure 6 shows the effects of electrical loads on the postbuckling response curve of a piezoelectric plate without damage under two initial deflection conditions, respectively. The geometric parameters are given as $\lambda_1 = \lambda_2 = 0.1$. Three electrical load conditions, referred as 1, 2 and 3, are considered. It can be seen that the negative control voltage results in the increase of the buckling load and the decrease of postbuckled deflection at the same compressive loads. In contrast, the positive control voltage decreases the buckling load and induces larger postbuckled deflections. It can be concluded that the positive control voltage acting upon the piezoelectric plate is equivalent to a compressive piezoelectric force acting in the in-plane direction of the plate to some certain extent, which leads to the smaller buckling loads.

Figure 7 shows the effect of initial deflections on the postbuckling response curve of piezoelectric plate without damage. The geometric parameters are given as $\lambda_1 = \lambda_2 = 0.1$. It can be seen that the larger the initial deflections of the plate, the larger the postbuckled deflection of the plate under the same compressive load, and that the postbuckled deflections of the plate under different initial deflections will reach the same value with the increase of postbuckling loads.

Figure 8 shows the effect of thick-span ratio of the plate on postbuckling response curves of piezoelectric plate without damage under different electrical loads. It can be found that the buckling loads increase with the increase of the thick-span ratio of plate, and that the control voltage has a small effect on the postbuckling behaviors of the plate with lower thick-span ratio.

Figure 9 shows the effect of aspect ratio of the plate on postbuckling response curves of piezoelectric plate without damage under different electrical loads. As expected, these results show that the buckling loads are increased by increasing the aspect ratio of the plate. It can also be found that the effect of control voltage is more pronounced for the square plate than for the rectangular plate.

Figure 10 shows the effect of external loads on postbuckling response curves of piezoelectric plate with damage and initial deflection. The electrical load is taken as V_e =0. It can be concluded that the larger the external compressive loads, the quicker the development of damage and the more obvious the effect of damage on the postbuckling deflection. From Figure 11, the conclusion that the change of initial deflection has a small effect on the postbuckled deflection and the damage effect varies slightly can be drawn.

Figure 12 shows the effect of electrical loads on the postbuckling response curves of piezoelectric plate with damage and initial deflection. It can be seen that the control voltage has a notable effect on the postbuckled deflection and the effect of damage varies greatly. The negative control voltage results in a smaller rate of damage development of the plate than that of the same plate without electrical loads, so it can be concluded that the positive control voltage can increase the rate of degradation of the stiffness of the piezoelectric plate. This can be explained as follows: the effect of positive control voltage on the damaged piezoelectric plate is equivalent to a tensile force acting in the in-plane direction of the plate, resulting in the acceleration of the piezoelectric plate degradation process. Therefore, it demonstrates a prominent effect on the deformation of piezoelectric plate with damage under the same loading capacity.

Figure 13 shows the effect of aspect ratio on postbuckling response curves of the piezoelectric plate with damage and initial deflections and Figure 14 shows the effect of thick-span ratio on postbuckling response curves of the piezoelectric plate with damage and initial deflections. The electrical load is taken as V_e =0. From Figure 13, it can be seen that the bigger the aspect ratio of the plate, the smaller the postbuckled deflection of the plate and the smaller the effect of damage. Figure 14 indicates that the larger the thick-span ratio, the smaller the postbuckled deflection of the plate and the smaller the effect of damage.

5. CONCLUSIONS

This paper presents an approach to investigate the postbuckling analysis of piezoelectric plates including damage effects using Talreja's tensor valued internal state damage variables and the Kachanvo damage evolution equation. The effects of applied voltage, plate aspect ratio, thick-span ratio, damage as well as initial geometric deflections on the postbuckling behaviors of the piezoelectric plate are investigated. Numerical results show that the nonlinearity of structure has a great influence on the postbuckling paths of the piezoelectric plate. The negative control voltage results in the increase of the buckling loads and the decrease of postbuckled deflections under the same in-plane compressive loads, whereas the positive control voltage decreases the buckling loads and induces larger postbuckled deflections. The buckling loads increase with the increase of the thick-span ratio of the plate, and the control voltage has a small effect on the postbuckling behaviors of the plate with lower thick-span ratio. When the damage and damage evolution are considered, the postbuckled deflection of the plate will gradually grow with the increase of the time until the damage reaches a characteristic damage state. The external in-plane compressive loads and the applied control voltage have great effects on the postbuckled deflections of the plate and the damage development. The negative control voltage can decrease the degradation rate of the stiffness of the piezoelectric structures and will provide a control mean for the damaged smart structures.

ACKNOWLEDGMENTS

Partial support for this project provided by National Natural Science Foundation of China (51075204); Aeronautical Science Foundation of China (2012ZB52026); the Fundamental Research Funds for the Central Universities (NS2014024) is gratefully acknowledged.

AUTHOR CONTRIBUTIONS

Zhigang Sun carried out the theory and calculation. Xianqiao Wang provided the thought. Zhigang Sun wrote the paper. Xianqiao Wang reviewed and edited the manuscript. All authors read and approved the manuscript

REFERENCES

1. Talreja, R. A continuum mechanics characterization of damage in composite materials. *Proc. R. Soc. Lond. A***1985**, *399*, 195–216.
2. Allen, D.H.; Harris, C.E.; Groves, S.E. A thermomechanical constitutive theory for elastic composites with distributed damage-I. Theoretical development. *Int. J. Solids Struct.* **1987**, *23*, 1301–1318.
3. Allen, D.H.; Harris, C.E.; Groves, S.E. A thermomechanical constitutive theory for elastic composites with distributed damage-II. Application to matrix cracking in laminated composites. *Int. J. Solids Struct.* **1987**,*23*, 1319–1338.
4. Valliappan, S.; Murti, V.; Zhang, W.H. Finite element analysis of anisotropic damage mechanics problems.*Eng. Fract. Mech.* **1990**, *35*, 1061–1071.
5. Ladeveze, P.; Dantec, E.L. Damage modelling of the elementary ply for laminated composites. *Compos. Sci. Technol.* **1992**, *43*, 257–267.
6. Schapery, R.A.; Sicking, D.L. On nonlinear constitutive equations for elastic and viscoelastic composites with growing damage. Proceedings of the 7th International Conference on Mechanical Behavior of Materials-ICM7, The Hague, The Netherlands, 28 May–2 June 1995; pp. 45–76.
7. Zhang, W.H.; Jin, T.; Chen, Y.M. Behavior of dynamic response for damage materials. *J. Vib. Eng.* **2000**, *13*, 415–425. (In Chinese).
8. Moore, R.H.; Dillard, D.A. Time-dependent matrix cracking in cross-ply laminates. *Compos. Sci. Technol.***1990**, *39*, 1–12.
9. Luo, J.J.; Daniel, I.M. A cylinder model for characterization of deformation and damage development in a unidirectional composite. *Compos. Sci. Technol.* **2000**, *60*, 2791–2802.

10. Mindlin, R.D. Forced thickness-shear and flexural vibration of piezoelectric crystal plates. *J. Appl. Phys.* **1952**, *23*, 83–88.

11. Tiersten, H.F. *Linear Piezoelectric Plate Vibrations*; Plenum Press: New York, NY, USA, 1978.

12. Chandrashekhara, K.; Thenneti, R. Termally induced vibration suppression of laminated plates with piezoelectric sensors and actuators. *Smart Mater. Struct.* **1955**, *4*, 281–290.

13. Zhou, X.; Chattopadhyay, A.; Gu, H.Z. Dynamic response of smart composites using a coupled thermo-piezoelectric mechanical mode. *AIAA J.* **2000**, *38*, 1939–1948.

14. Wang, B.; Rogers, A. Laminate plate theory for spatially distributed induced strain actuators. *J. Compos. Mater.* **1991**, *25*, 433–452.

15. Tzou, H.S.; Gadre, M. Theoretical analysis of a multi-layered thin shell coupled with piezoelectric shell actuators for distributed vibration controls. *J. Sound Vib.* **1989**, *132*, 433–450.

16. Xu, K.M.; Noor, A.K.; Tang, Y.Y. Three-dimensional solutions for free vibration of initially stressed thermoelectroelastic multilayered plates. *Comput. Methods Appl. Mech. Eng.* **1997**, *141*, 125–139.

17. Mitchell, J.A.; Reddy, J.N. A refined hybrid plate theory for composite laminates with piezoelectric laminate. *Int. J. Solid Struct.* **1996**, *32*, 2345–2367.

18. Noor, A.K.; Peters, J.M. Postbuckling of multilayered composite plates subjected to combined axial and thermal loads. *Finite Elem. Anal. Des.* **1992**, *11*, 91–104.

19. Librescu, L.; Souza, M.A. Post-buckling of geometrically imperfect shear-deformable flat panels under combined thermal and compressive edge loadings. *ASME J. Appl. Mech.* **1993**, *60*, 526–533.

20. Shen, H.S. Postbuckling of shear deformable laminated plates with piezoelectric actuators under complex loading conditions. *Int. J. Solids Struct.* **2001**, *38*, 7703–7721.

21. Oh, I.K.; Han, J.H.; Lee, I. Postbuckling and vibration characteristics of piezolaminated composite plate subjected to thermo-piezoelectric loads. *J. Sound Vib.* **2000**, *233*, 19–40.

22. Aydin, K. Free vibration of functionally graded beams with arbitrary number of surface cracks. *Eur. J. Mech. A Solids* **2013**, *42*, 112–124.

23. Mao, Y.Q.; Fu, Y.M.; Dai, H.L. Creep buckling and post-buckling analysis of the laminated piezoelectric viscoelastic functionally graded plates. *Eur. J. Mech. A Solids* **2011**, *30*, 547–558.

24. Hamed, E. Bending and creep buckling response of viscoelastic functionally graded beam-columns.*Compos. Struct.* **2012**, *94*, 3043–3051.

25. Xu, C.; Liang, Z.; Ren, B.; Di, W.; Luo, H.; Wang, D.; Wang, K.; Chen, Z. Bi-stable energy harvesting based on a simply supported piezoelectric buckled beam. *J. Appl. Phys.* **2013**, *114*, 114507.

26. Cottone, F.; Gammaitoni, L.; Vocca, H.; Ferrari, M.; Ferrari, V. Piezoelectric buckled beams for random vibration energy harvesting. *Smart Mater. Struct.* **2012**, *21*, 035021:1–035021:11.

27. Aboudi, J. Micromechanical Prediction of the Effective Behavior of Fully Coupled Electro-Magneto-Thermo-Elastic Multiphase Composites. *Smart Mater. Struct.* **2001**, *10*, 867–877.

28. Bansal, Y.; Pindera, M. Finite-volume direct averaging micromechanics of heterogeneous materials with elastic-plastic phases. *Int. J. Plast.* **2006**, *22*, 775–825.

29. Sun, Z.G.; Niu, X.N.; Huang, S.W.; Song, Y.D. A Unified Macro- and Micro-mechanics Constitutive Model of Fully Coupled Fields. *Mech. Compos. Mater.* **2014**. in press.

30. Wang, C.C.; Truesdell, C. Introduction to Rational Elasticity. In *Mechanics of Continua*; Noordhoff International Pub. Co.: Leyden, The Netherlands, 1973; p. p. 556.

31. Robert, E. *Mechanics of Continua*, 2nd ed.; Robert, E. Krieger Pub Co. Huntington, NY, USA, 1980; p. p. 592.

32. Kachanov, L.M. *Introduction to Continuum Damage Mechanics*; Martinus Nijhoff Publishers: Dordrecht, The Netherlands, 1986.

33. Chia, C.Y. *Nonlinear Analysis of Plates*; McGraw-Hill International Book Company: New York, NY, USA, 1980.

Chapter 6

Direct Ink Writing of Three-Dimensional (K, Na)NbO_3-Based Piezoelectric Ceramics

Yayun Li[1,*], Longtu Li[1] and Bo Li[2]

[1]State Key Laboratory of New Ceramics and Fine Processing, School of Materials Science and Engineering, Tsinghua University, Beijing 100084, China

[2]Research Institute for Advanced Materials, Graduate School at Shenzhen, Tsinghua University, Shenzhen 518055, China

ABSTRACT

A kind of piezoelectric ink was prepared with Li, Ta, Sb co-doped (K, Na)NbO_3(KNN) powders. Piezoelectric scaffolds with diameters at micrometer scale were constructed from this ink by using direct ink writing method. According to the micro-morphology and density test, the samples sintered at 1100 °C for 2 h have formed ceramics completely with a high relative density of 98%. X-ray diffraction (XRD) test shows that the main phase of sintered samples is orthogonal $(Na_{0.52}K_{0.4425}Li_{0.0375})(Nb_{0.87}Sb_{0.07}Ta_{0.06})O_3$. The piezoelectric constant d_{33} of 280 pC/N, dielectric constant ε of 1775, remanent polarization P_r of 18.8 $\mu C/cm^2$ and coercive field E_c of 8.5 kV/cm prove that the sintered samples exhibit good electrical properties. The direct ink writing method allows one to design and rapidly fabricate piezoelectric structures in complex three-dimensional (3D) shapes without the need for any dies or lithographic masks, which will simplify the process of material preparation and offer new ideas for the design and application of piezoelectric devices.

Keywords: (K, Na)NbO_3 (KNN); direct ink writing; 3D structures

1. INTRODUCTION

Environmentally-friendly lead-free piezoelectric ceramics have attracted much attention in recent years. In 2004, Saito *et al.* [1] reported a big breakthrough of a high d_{33} (416 pC/N) in the Li^+-, Ta^{5+}-, and Sb^{5+}-modified (K, Na)NbO_3 (KNN) textured ceramics as a promising candidate for lead-free piezoelectric ceramics. Up to now, studies about KNN are focusing on the improvement of the materials and the structure designs. Material improvement mainly concerns the substitution behavior [2,3], phase transitional behavior [4,5,6,7] and adulteration of the materials [8,9,10,11,12]. For the structure design refers to the KNN-based piezoelectric devices such as the sensors, transducers and actuators [13,14,15,16,17], most of which are fabricated by the 1–3 KNN composite [18,19,20] or the single crystalline KNN nanostructures [21,22,23,24]. Few works about the three-dimensional (3D) KNN structures, however, have been reported in the fabrication of piezoelectric devices. Direct ink writing (DIW) technique is a powerful fabrication method in building 3D KNN piezoelectric scaffolds due to its favorable performance. In this method, subtle 3D KNN architectures can be constructed without the need for any dies or lithographic masks [25], which is more convenient than the dice and fill method to fabricate 1–3 type KNN products [26]. DIW shows a potential for application in catalytic materials [27], tissue engineering [28] and especially the composites [29].

In this paper, we prepared methyl methacrylate based suspensions of KNN inks and fabricated 3D piezoelectric woodpile structures with various structural parameters by using DIW technique. The phase composition, the microstructure and the piezoelectric parameters of the sintered samples were observed.

2. EXPERIMENTAL SECTION

The KNN-based powders with enhanced piezoelectricity by co-additive of Li, Sb, and Ta were prepared by a two-step calcining and ball milling route. The average particle size of the powder was about 0.5 μm with the specific surface area of 10.1 m^2/g. Methyl methacrylate (AR, Beijing Yili Fine Chemicals, Beijing, China) and pentaerythritol three acrylate (Alfa Aesar, Beijing, China) were mixed as a solvent. A certain amount of KNN powders were added into the mixture with ultrasound treatment and the final suspension had a solid content fraction of 56 wt%. The plot of viscosity as a function of the shear stress and the plot of shear stress on shear rate were measured by a controlled stress rheometer (Physica MRC 300 Modular Compact Rheometer, Physica Paar, Berlin, Germany) under the pressure from 1 Pa to 1000 Pa at room temperature.

The direct ink writing equipment (A3200, Aerotech Inc., Pittshurgh, PA, USA) consists of three parts (Figure 1). The first one is the computer-controlled system which designs the structure and translates the mobile paths into code languages. The second one is the 3D motion platform which can move under the instruction of the computer. The last one is the extrusion system which can eject the inks under a proper pressure. In this experiment, we designed three woodpile structures with different sizes and changed those into code languages. The 3D platform constructed all the KNN woodpile structures after receiving the motion instructions.

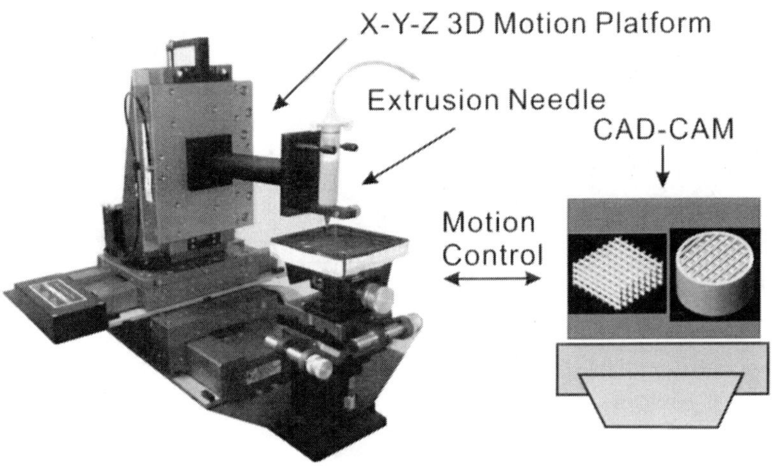

Figure 1. Experimental set-up for the direct write printing process.

Those green samples were sintered at varying temperatures for 2 h with a heating rate of 3 °C/min. The macrostructures of the samples were characterized by optical microscope (AXio Imager.Z1m, Carl Zeiss Shanghai Co. Ltd., Shanghai, China). The microstructures of the sintered samples were analyzed by using scanning electron microscopy (Leo-1530, Oberkochen, Germany). The density of the sintered sample was measured by the Archimedes technique with deionized water as the suspension fluid. The dielectric properties were measured by an impedance analyzer (HP4278A, Santa Clara, CA, USA) and the d_{33}measurements were performed by using a ZJ-3A quasi-static d_{33} tester (Institute of Acoustics, Chinese Academy of Sciences, Beijing, China). The hysteresis loop was tested by a ferroelectric tester (RT 6000HVA, Radiant Technologies Inc., Albuquerque, NM, USA) under a maximum electric field of 2.5 kV/mm.

3. RESULTS AND DISCUSSION

Nanosize KNN powders with little soft-agglomerates is shown in Figure 2a. The X-ray diffraction (XRD) pattern of the KNN powders (Figure 2b) shows the main phase is $(Na_{0.52}K_{0.4425}Li_{0.0375})(Nb_{0.87}Sb_{0.07}Ta_{0.06})O_3$ which is prepared by the co-additive of Li, Sb, and Ta in the $(K, Na)NbO_3$ powders.

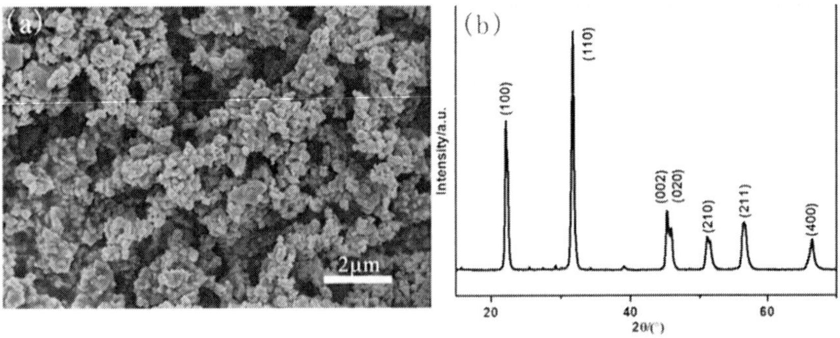

Figure 2. (**a**)The SEM image of the KNN powders; (**b**) the XRD pattern of the KNN powders.

Figure 3a shows the plot of shear stress as a function on shear rate, which is close to that of the classic shear thinning fluid. When the shear stress less than 15 Pa, the ink shows an elastic deformation behavior, and the deformation can be restored after unloading (Figure 3b). Then the viscosity decreases rapidly and experiences a plastic deformation to reach a steady state at 50 Pa. This deformation type cannot be restored after unloading. The changing trend of the plot indicates that the ink has a shear-thinning behavior.

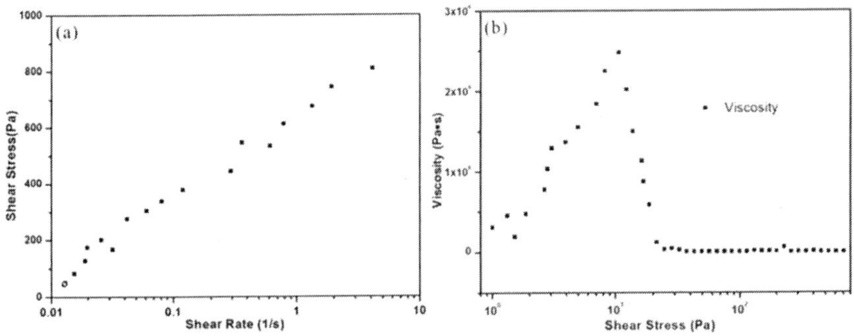

Figure 3. (**a**) Plot of the shear stress as a function of shear rate; (**b**) plot of the viscosity as a function of shear stress.

The optimum solid content of the ink is related to the specific surface area of the original powders, and the KNN ink with a solid content fraction of 56 wt% is beneficial for the fabrication process and can be sintered with a high relative density. For example, the ink formed by TiO_2 nanoparticle powders (specific surface area of 50 m^2/g) has a proper solid content of 42.5 wt% [30]. The ink formed by PLZT powders (specific surface area of 4.735 m^2/g) has an appropriate solid content about 70 wt% [31]. Both of the above mentioned two inks can be sintered with a high relative density. In this paper, we use the KNN powders with a specific surface area of 10.1 m^2/g to form the ink with a final solid content of 56 wt%. We can conclude that the powders with a larger specific surface area can be developed to an ink with a smaller solid content fraction.

Here, we design 3D KNN woodpile structures with various shape sizes, filament diameters and rod-to-rod distances. Figure 4a shows the top view of the three structures with shape sizes of 10 mm × 10 mm or 12 mm × 12 mm. Figure 4b shows the sample with the filament diameter of 400 μm and rod-to-rod distance of 1 mm. Figure 4c shows the sample with the filament diameter of 250 μm and rod-to-rod distance of 600 μm. Figure 4d shows the sample with the filament diameter of 300 μm and rod-to-rod distance of 500 μm. It is obvious that the direct ink writing (DIW) method has an advantage in the structure design and size precision control.

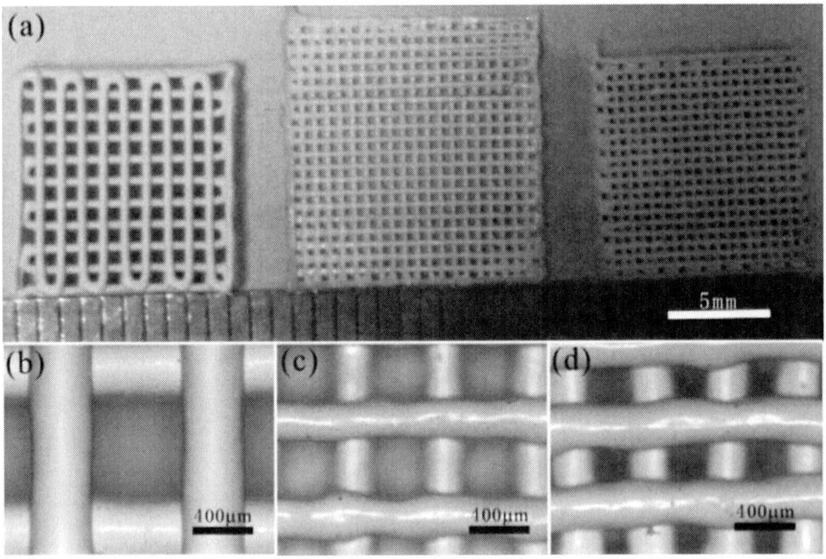

Figure 4. Optical images of the woodpile structures: (**a**) top view of three samples with different sizes; (**b**) 10 mm × 10 mm sample with a filament diameter of 400 μm; (**c**) 12 mm × 12 mm sample with a filament diameter of 250 μm; (**d**) 10 mm × 10 mm sample with filament a diameter of 300 μm.

A series of sintered experiments have been performed to find out the optimum sintering process. We designed three sintering temperatures, namely: 1050 °C, 1100 °C and 1130 °C.Figure 5a shows the sample sintered in 1050 °C for 2 h, we can see that this sample has not been sintered completely with the non-uniform grain size and lots of pores. Figure 5c shows the SEM image of the sample sintered under 1130 °C for 2 h, which is over sintered with an abnormal grain growth. From Figure 5b we can conclude that the KNN sample sintered at 1100 °C for 2 h has a better performance with the uniform grain size and less pores. The density of the sintered sample was measured by the Archimedes method with deionized water as the suspension fluid. Moreover, the testing density is 4.52 g/cm^3, while the theoretical density of KNN is about 4.60 g/cm^3. The relative density (98%) is the ratio of the testing density and the theoretical density. Figure 5d shows the top view of the sample (on the right-hand side ofFigure 4a) sintered at 1100 °C for 2 h, which shows that the deformation did not occur during the sintering process.

Figure 5. SEM photos of KNN samples sintered to varying maximum temperatures: (**a**) 1050 °C; (**b**) 1100 °C; (**c**) 1130 °C; (**d**) the top view of the sintered 10 mm × 10 mm sample.

Figure 6 shows the XRD pattern of the KNN sample sintered at 1100 °C. The sintered sample agrees well with the origin powders, indicating that no significant phase transition occurred during the sintering process.

The sintered sample with the original size of 10 mm×10 mm and filament diameter of 400 μm was printed with the low temperature silver paste on the top and bottom surface and covered with aluminum foil. Then the sample was poled at 2.5 kV/mm for 20 min in

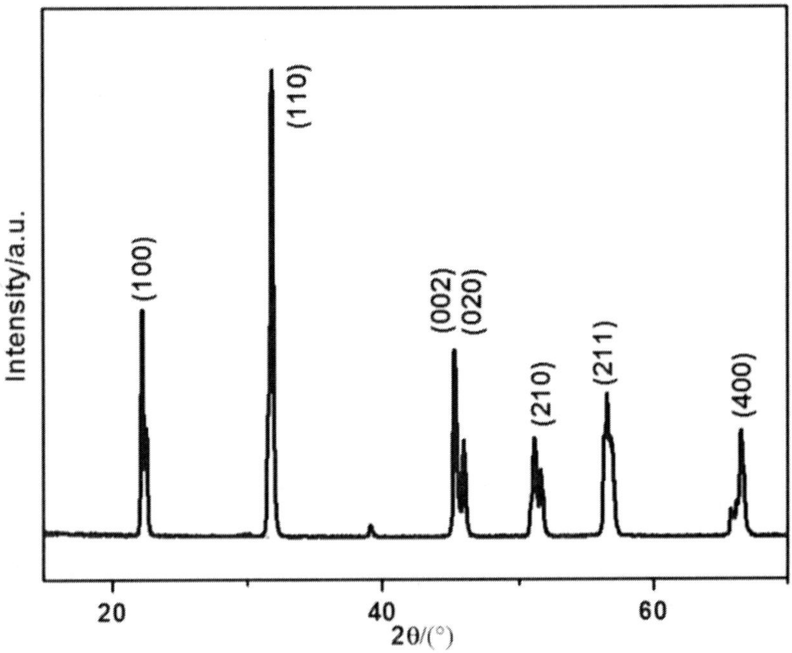

Figure 6. XRD pattern of the sample sintered at 1100 °C.

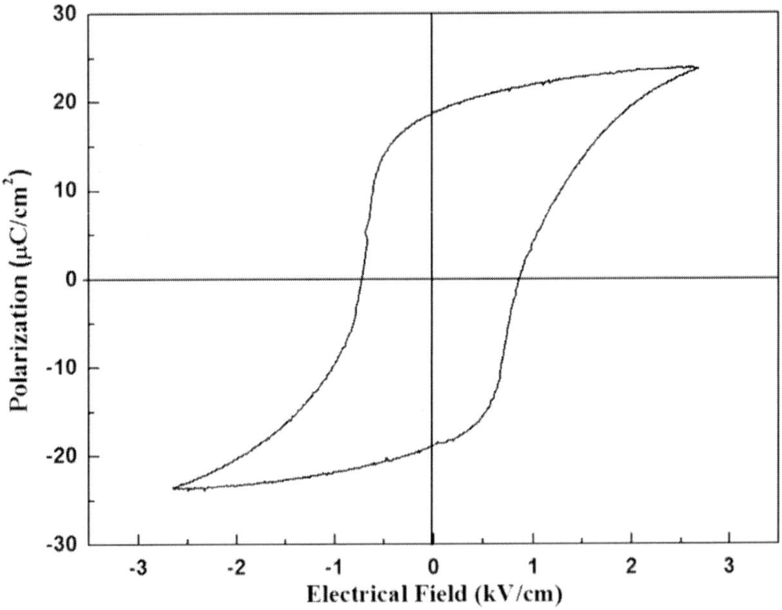

Figure 7. The hysteresis loop of the sintered KNN sample.

Table 1. Dielectric and piezoelectric properties of the sintered KNN sample.

Electrical properties	$\varepsilon_{33}/\varepsilon_0$	d_{33} (pC/N)	P_r (μC/cm^2)	E_c (kV/cm)
KNN sample	1775	280	18.8	8.5

a silicone oil bath at 100 °C. The hysteresis loop was tested under a maximum electric field of 2.5 kV/mm (Figure 7), and the value of remanent polarization P_r and coercive field E_c were gained. Table 1 shows the dielectric and piezoelectric properties of the sintered 3D KNN sample. We can see that sintered sample has good electrical properties with piezoelectric constant d_{33} of 280 pC/N, dielectric constant of 1775, remanent polarization P_r of 18.8 μC/cm^2 and coercive field E_c of 8.5 kV/cm.

4. CONCLUSIONS

We prepared KNN woodpile structures with a relatively high piezoelectric constant by using direct ink writing method along with the ink methyl methacrylate based suspensions of KNN. Three woodpile structures with various shape sizes, filament diameters and rod-to-rod distances have been constructed. Samples sintered in 1100 °C for 2 h have a relative density of 98%. The piezoelectric constant d_{33} of 280 pC/N, dielectric constant ε of 1775, remanent polarization P_r of 18.8 μC/cm^2 and coercive field E_c of 8.5 kV/cm show that the sintered samples exhibit good electrical properties. Compared with the other methods such as dice and fill method and injection molding, the direct ink writing method allows one to design and rapidly fabricate piezoelectric structures in complex 3D shapes without the need for any dies or lithographic masks, which will simplify the processing of the material preparation and offer ideas for the design and application of new piezoelectric devices.

ACKNOWLEDGMENTS

This work was financially supported by National Science Foundation of China (Grant Nos. 50902081, 61077029 and 51032003), National 973 project (Grant No. 2009CB623304), Shenzhen Science and technology plan under JSE201007200050A, and by the Project of Cooperation from Industry, Education and Academy supported by Guangdong Province (Grant No. 2009A090100021).

AUTHOR CONTRIBUTIONS

Longtu Li and Bo Li provide some theoretical guidances about KNN. Renlong Gao offers some experimental guidances on this paper. Yayun Li wrote the manuscript.

REFERENCES

1. Saito, Y.; Takao, H.; Tani, T.; Nonoyama, T.; Takatori, K.; Homma, T.; Nagaya, T.; Nakamura, M. Lead-free piezoceramics. *Nature* **2004**, *432*, 84–87.
2. Ha, J.; Ryu, J.; Lee, H. Substitution behavior of $x(Na_{0.5}K_{0.5})NbO_3$-$(1-x)BaTiO_3$ ceramics for multilayer ceramic capacitors by a near edge X-ray absorption fine structure analysis.*Appl. Phys. Lett.* **2014**, *104*, 262904:1–262904:4.
3. Lu, L.; Gong, Y.Q.; Gong, L.J.; Dong, H.D.; Dong, H.; Yi, X.F.; Zheng, X.J. Low-temperature hydro/solvothermal synthesis of Ta-modified $K_{0.5}Na_{0.5}NbO_3$ powders and piezoelectric properties of corresponding ceramics. *Mater. Des.* **2012**, *33*, 362–366.
4. Toshio, K.; Yuan, Y.; Fumito, S. Mechanisms of texture development in lead-free piezoelectric ceramics with perovskite structure made by the templated grain growth process. *Materials* **2010**, *3*, 4965–4978.
5. Dai, Y.J.; Zhang, X.W.; Zhou, G.Y. Phase transitional behavior in $K_{0.5}Na_{0.5}NbO_3$-$LiTaO_3$ceramics. *Appl. Phys. Lett.* **2007**, *90*, 262903:1–262903:3.
6. Zuo, R.Z.; Fu, J.; Lv, D.Y.; Liu, Y. Antimony tuned rhombohedral-orthorhombic phase transition and enhanced piezoelectric properties in sodium potassium niobate. *J. Am. Ceram. Soc.* **2010**, *93*, 2783–2787.

7. Fisher, J.G.; Rout, D.; Moon, S.K.; Kang, S.J.L. Structural changes in potassium sodium niobate ceramics sintered in different atmospheres. *J. Alloys Compd.* **2009**, *479*, 467–472.

8. Fu, J.; Zuo, R.Z.; Wang, X.H.; Li, L.T. Phase transition characteristics and piezoelectric properties of compositionally optimized alkaline niobate based ceramics. *J. Alloys Compd.* **2009**, *486*, 790–794.

9. Wang, X.P.; Wu, J.G.; Xiao, D.Q.; Zhu, J.G.; Cheng, X.J.; Zheng, T.; Zhang, B.Y.; Lou, X.J.; Wang, X.J. Giant piezoelectricity in potassium-sodium niobate lead-free ceramics. *J. Am. Ceram. Soc.* **2014**, *136*, 2905–2910.

10. Lee, Y.; Yoo, J.; Lee, K.; Kim, I.; Song, J.; Park, Y.W. Dielectric and piezoelectric characteristics of the non-stoichiometric (Na, K)NbO_3 ceramics doped with CuO. *J. Alloys Compd.* **2010**, *506*, 872–876.

11. Wang, Y.L.; Damjanovic, D.; Klein, N.; Setter, N. High-temperature instability of Li- and Ta- modified (K, Na)NbO_3 piezoceramics. *J. Am. Ceram. Soc.* **2008**, *91*, 1962–1970.

12. Ming, B.Q.; Wang, J.F.; Qi, P.; Zang, G.Z. Piezoelectric properties of (Li, Sb, Ta) modified (Na, K)NbO_3 lead-free ceramics. *J. Appl. Phys.* **2007**, *101*, 054103:1–054103:4.

13. Kwok, K.W.; Hon, S.F.; Lin, D. Lead-free self-focused piezoelectric transducers for viscous liquid ejection. *Sens. Actuators A Phys.* **2011**, *168*, 168–171.

14. Wu, D.W.; Chen, R.M.; Zhou, Q.F.; Shung, K.K.; Lin, D.M.; Chan, H.L.W. Lead-free KNLNT piezoelectric ceramics for high-frequency ultrasonic transducer application.*Ultrasonic* **2009**, *49*, 395–398.

15. Hong, J.; Yoo, J.; Lee, K.; Lee, S.; Song, H. Characteristics of acoustic emission sensor using lead-free (LiNaK)(NaTaSb)O_3 ceramics for fluid leak detection at power plant valves. *Jpn. J. Appl. Phys.* **2008**, *47*, 2192–2194.

16. Oh, Y.; Noh, J.; Yoo, J.; Kang, J.; Hwang, L.; Hong, J. Dielectric and piezoelectric properties of CeO_2-added Nonstoichiometric $(Na_{0.5}K_{0.5})_{0.97}(Nb_{0.96}Sb_{0.04})O_3$ ceramics for piezoelectric energy

harvesting device applications. *IEEE Trans. Ultrason. Ferroelectr. Freq. Control* **2011**, *58*, 1860–1866.

17. Lee, T.; Kwok, K.W.; Li, H.L.; Chan, H.L.W. Lead-free alkaline niobate-based transducer for ultrasonic wirebonding applications. *Sens. Actuators A Phys.* **2009**, *150*, 267–271.

18. Shen, Z.Y.; Li, J.F.; Chen, R.M.; Zhou, Q.F.; Shung, K.K. Microscale 1–3 type (Na, K)NbO_3-based Pb-free piezocomposites for high-frequency ultrasonic transducer applications. *J. Am. Ceram. Soc.* **2011**, *94*, 1346–1349.

19. Zhou, D.; Lam, K.H.; Chen, Y.; Zhang, Q.H.; Chiu, Y.C.; Luo, H.; Dai, J.Y.; Chan, H.L.W. Lead-free piezoelectric single crystal based 1–3 composites for ultrasonic transducer applications. *Sens. Actuators A Phys.* **2012**, *182*, 95–100.

20. Xu, Y.; Li, J.F.; Ma, J.; Nan, C.W. Magnetoelectricity of lateral 1–3 type composites with $CoFe_2O_4$ ferromagnetic microstrips embedded in (K, Na)NbO_3-based piezoelectric substrate. *J. Appl. Phys.* **2011**, *110*, 044104:1–044104:4.

21. Li, L.H.; Deng, J.X.; Chen, J.; Yu, R.B.; Qiao, L.J.; Xing, X.R. Piezoelectric and ferroelectric properties of $0.96(Na,K)(Nb_{0.9}Ta_{0.1})O_{3-0.04}LiSbO_3$ ceramics synthesized by molten salt method. *J. Alloys Compd.* **2009**, *471*, 428–431.

22. Urban, J.J.; Spanier, J.E.; Lian, O.Y.; Yun, W.S.; Park, H. Single-crystalline barium titanate nanowires. *Adv. Mater.* **2003**, *15*, 423–426.

23. Farooq, M.U.; Fisher, J.G. Growth of $(K_{0.5}Na_{0.5})NbO_3SrTiO_3$ lead-free piezoelectric single crystals by the solid state crystal growth method and their characterization.*Ceram. Int.* **2014**, *40*, 3199–3207.

24. Hou, Y.D.; Hou, L.; Zhang, T.T.; Zhu, N.K.; Wang, H.; Yan, H. $(Na_{0.8}K_{0.2})_{0.5}Bi_{0.5}TiO_3$nanowires: Low-temperature sol-gel-hydrothermal synthesis and densification. *J. Am. Ceram. Soc.* **2007**, *90*, 1738–1743.

25. Lewis, J.A. Direct-write assembly of ceramics of colloidal inks. *Curr. Opin. Solid State Mater. Sci.* **2002**, *6*, 245–250.

26. Li, J.F.; Wang, K.; Zhu, F.Y.; Cheng, L.Q.; Yao, F.Z. (K, Na)NbO_3-based lead-free piezoceramics: Fundamental aspects,

processing technologies, and remaining challenges.*J. Am. Ceram. Soc.* **2013**, *96*, 3677–3696.

27. Li, Y.Y.; Li, B.; Li, L.T. Fabrication of 3D photocatalytic α-Fe_2O_3 structure using direct ink writing method. *Mod. Phys. Lett. B* **2014**, *28*, 1450051.

28. Simon, J.L.; Michna, S.; Lewis, J.A.; Rekow, E.D.; Thompson, V.P.; Smay, J.E.; Yampolsky, A.; Parsons, J.R.; Ricci, J.L. *In vivo* bone response to 3D periodic hydroxyapatite scaffolds assembled by direct ink writing. *J. Biomed. Mater. Res. A* **2006**, *83A*, 747–758.

29. San, C.M.; Kouzeli, M.; Rao, R.; Lewis, J.A.; Dunand, D.C. Alumina-aluminum interpenetrating-phase composites with three-dimensional periodic architecture. *Scr. Mater.* **2013**, *49*, 861–866.

30. Cai, K.P.; Li, Y.Y.; Sun, Z.X.; Sun, J.B.; Li, B.; Zhou, J. Preparation of 3D ceramic meshes by direct-write method and modulation of its photocatalytic properties by structure design. *J. Inorg. Mater.* **2012**, *27*, 102–106.

31. Li, Y.Y.; Li, B.; Sun, J.B.; Cai, K.P.; Zhou, J.; Li, L.T. Direct write assembly of 3-dimensional structures with aqueous-based piezoelectric inks. *Key Eng. Mater.* **2012**, *512*, 390–394.

Chapter 1

Recent Progress on PZT Based Piezoelectric Energy Harvesting Technologies

Min-Gyu Kang [1,2], Woo-Suk Jung [3], Chong-Yun Kang [1,4] and Seok-Jin Yoon [1,*]

[1]*Electronic Materials Research Center, Korea Institute of Science and Technology, 39-1 Hawolgok-Dong, Sungbuk-gu, Seoul 136-791, Korea*

[2]*Bio-inspired Materials and Devices Laboratory (BMDL), Center for Energy Harvesting Materials and Systems (CEHMS), Virginia Tech, Blacksburg, VA 24061, USA*

[3]*Department of Digital Electronics, Daelim University College, 29 Imgoklo Dongan-gu, Anyang-si, Gyunggi-do 431-715, Korea*

[4]*KU-KIST Graduate School of Converging Science and Technology, Korea University, Seoul 136-701, Korea*

ABSTRACT

Energy harvesting is the most effective way to respond to the energy shortage and to produce sustainable power sources from the surrounding environment. The energy harvesting technology enables scavenging electrical energy from wasted energy sources, which always exist everywhere, such as in heat, fluids, vibrations, *etc.* In particular, piezoelectric energy harvesting, which uses a direct energy conversion from vibrations and mechanical deformation to the electrical energy, is a promising technique to supply power sources in unattended electronic devices, wireless sensor nodes, micro-electronic devices,*etc.*, since it has higher energy conversion efficiency and a simple structure. Up to now, various technologies, such as advanced materials, micro- and macro-mechanics, and electric circuit design, have been investigated and emerged to improve performance and conversion efficiency of the piezoelectric energy harvesters. In this paper, we focus on recent progress of piezoelectric energy harvesting technologies based on $PbZr_xTi_{1-x}O_3$(PZT) materials, which have the most outstanding piezoelectric properties. The advanced piezoelectric energy harvesting technologies included materials, fabrications, unique designs, and properties are introduced to understand current technical levels and suggest the future directions of piezoelectric energy harvesting.

Keywords: PZT; energy harvesting; piezoelectric

1. INTRODUCTION

Energy harvesting, which collects useful energy from wasted energy sources, is the most promising technology to provide solutions for the shortage of a fossil fuel, various environmental problems, and the improvement of energy efficiency in smart grids. The basic concept of energy harvesting technology is to obtain the electrical energy by an energy conversion from a wasted heat, a vibration, a mechanical deformation, a potential energy, *etc.*The driving force for the piezoelectric energy harvesting is also the energy conversion effect using piezoelectricity. Up to now, the piezoelectric effect, which is an energy conversion phenomena leading to the electrical energy generated by an applied force and a mechanical deformation by an applied electric field, have been extensively utilized to demonstrate various devices, such as transducers [1,2], sensors [3,4,5], actuators [6,7], surface acoustic wave (SAW) devices [8,9], frequency control, *etc* [10]. In the past decade, piezoelectric energy harvesting technologies have been intensively investigated in various research fields, such as material science, electrical engineering, mechanical engineering, bio-medical science, *etc.*, since it has the remarkable energy conversion efficiency. The piezoelectric energy harvesting technologies can be utilized for the self-power sources in battery-less devices operated by small power at the mW-μW level, and power sources in portable systems, wireless sensor systems, health monitoring systems, and the military and vehicle industry [10,11].

Recently, the piezoelectric energy harvesters with unique structures and fabrication methods have been significantly developed, which can be classified with bulk type, microelectromechanical systems (MEMS), and flexible energy harvesters.

The piezoelectric materials, which contain dense polycrystalline and single crystalline ceramics have been investigated for achieving high piezoelectric charge constant (*d*), voltage constant (*g*), electromechanical coupling factor (*k*), and mechanical quality factor (Q_m), which are figures of merit (FOM) of the piezoelectric materials for the energy harvesting. Each factor is expressed as the following [12],

1. Piezoelectric strain (unit: pm/V) and charge (unit: pC/N) constant

$$x = dE\,(d = \text{piezoelectric strain and charge constant}) \quad (1)$$

where x and E are strain and external electric field.

2. Piezoelectric voltage constant (unit: Vm/N)

$$E = gX\,(g = d/\varepsilon_0\varepsilon_r) < \text{g} = \text{piezoelectric voltage constant} > \quad (2)$$

where ε_0 and ε_r are dielectric constant in a vacuum and relative dielectric constant.

3. Electromechanical coupling factor

$$k^2 = (\text{Stored mechanical energy/Input electrical energy}) \quad (3)$$

$$k^2 = (\text{Stored electrical energy/Input mechanical energy}) \quad (4)$$

Since the input electrical energy is $(1/2)\varepsilon_0\varepsilon_r E^2$ per unit volume and the stored mechanical energy per unit volume under zero external stress is given by $(1/2)x^2/s = (1/2)(dE)^2/s$, k^2can be calculated as

$$k^2 = [(1/2)(dE)^2/s]/[(1/2)\varepsilon_0\varepsilon_r E^2] = d^2/\varepsilon_0\varepsilon_r s \quad (5)$$

where *s* is stress.

4. Mechanical quality factor

$$Q_m = \omega_0/2\Delta\omega \quad (6)$$

where ω_0 is resonance frequency.

The piezoelectric material with higher $d \times g$ (transduction coefficient) and higher g value especially provides higher output electrical power. Hence, the d and g values are the essential criteria for the piezoelectric materials used for energy harvesting applications.

The piezoelectric thin films are also used in a variety of applications in the microelectronics industry for example in sensors and MEMS piezoelectric energy harvesting for a micro electricity generation. The effective piezoelectric constants d and e values are usually used as FOM factors for the thin film materials. Conventionally, the $d_{33,f}$, effective piezoelectric strain constant, has been determined by a piezoelectric force microscopy (PFM) technique. The electric field induced the surface deflection on the thin film piezoelectric material is measured by an atomic force microscopy (AFM) with a conductive cantilever tip and the $d_{33,f}$ value can be calculated from the displacement divided by a applied voltage (unit: pm/V). The effective piezoelectric constant $e_{31,f}$ is defined as $e_{31,f} = d_{31}/(s^E_{11} + s^E_{12})$ (unit: C/m^2) [13]. Where d_{31} is the in-plane piezoelectric constant and $s^E_{11} + s^E_{12}$ are the in-plane elastic compliances of the bulk material. The $e_{31,f}$ value has been evaluated by a wafer flexure method [14]. The wafer flexure method depends on the pressurization or evacuation of a cavity to which a simply supported wafer is sealed. During this stress oscillation, a charge integrator is connected to the sample, which measures the piezoelectric charge, and allows calculation of the stress-induced polarization.

Since large piezoelectric coefficients in $PbZr_xTi_{1-x}O_3$ (PZT) solid solution systems were discovered, the various composition of PZT based piezoelectric materials have been studied to obtain a higher piezoelectric coefficient (d_{33} = ~300–1000 pC/N) [15] and also alternate materials have been discovered to achieve the lead-free piezoelectric materials. $Na_xK_{1-x}NbO_3$ (NKN) piezoelectric ceramics are suggested for one of the promising alternate materials of the PZT, which includes non-toxic and environmental friendly materials with high piezoelectric coefficient (d_{33} = ~100–400 pC/N) [16,17]. However, the piezoelectric performance as the piezoelectric coefficient and the electromechanical coupling is still much lower than the PZT based materials. A piezoelectric AlN thin film is also attractive for the MEMS piezoelectric energy harvesting, because it can be obtained by a reactive sputtering method easily and has a low dielectric permittivity leading to high piezoelectric voltage coefficient. Nevertheless, the piezoelectric coefficient is still very small compared to the PZT based materials (d_{33} = ~3–5 pm/V) [18]. Since Wang *et al.* discovered a coupling of the piezoelectric and the semiconducting properties in single crystalline ZnO nanowires (d_{33} = ~14–26 pm/V) [19], the nano-structured ZnO materials have been studied as an attractive material for the piezoelectric energy harvesting [20]. The ZnO thin films and nanostructured materials are easily synthesized at low process temperature and can be integrated on flexible organic substrates for future flexible, stretchable, and portable electronics [21]. Moreover, the low dielectric permittivity of the ZnO materials improves the output performance of the piezoelectric energy harvesters. Despite of remarkable advantages in ZnO materials, the PZT based materials still have been considered as the most outstanding piezoelectric materials for the energy harvesting due to their much higher piezoelectric coefficients.

In this paper, we have reviewed recent progress on the PZT based piezoelectric materials and the devices for the energy harvesting. In recent years, a large amount of the considerable technical approaches regarding the PZT based piezoelectric energy harvesting have been reported. The piezoelectric material included bulks, thin films, and nanostructured materials with various composition and morphologies, and unique designs to improve the energy conversion efficiency of the piezoelectric energy harvesting, and desirable applications using bulk type, MEMS, and flexible piezoelectric harvesters have been suggested.

2. BULK TYPE PIEZOELECTRIC ENERGY HARVESTERS

Various structural designs and material compositions for the bulk type piezoelectric energy harvesters have been investigated to improve the conversion efficiency for the various applications. The operating performance of the energy harvesters is strongly affected by acceleration, applied force, vibration frequency, impedance matching, mass, and surrounding environment, structural design and materials. In this section, various design of bulk type piezoelectric energy harvesters are introduced including cantilever type, cymbal type and stack type energy harvesters for self-powered devices, water flow sensors, tire pressure monitoring system, *etc.*

2.1. Cantilever Type Piezoelectric Energy Harvesters

A cantilever type piezoelectric harvester is one of the promising structures to obtain high output power from the piezoelectric element, because it can apply larger strain to the piezoelectric elements under vibration condition. A general structure of the cantilever type piezoelectric energy harvester is illustrated in Figure 1a. The harvester consists of the piezoelectric ceramic, elastic body, and proof mass. This simple structure produces a large deformation under a vibration environment and can effectively collect the electrical power from the piezoelectric ceramic in 31 mode. Each physical factor in the devices (length, area, mass, thickness and position of the piezoelectric ceramics and elastic body, *etc.*) determines the operating performance of the harvester. To evaluate the performance of the cantilever type piezoelectric harvester, conventionally, such an experimental setup has been used as shown inFigure 1b. A function generator and an amplifier are used to operate an exciter with various frequencies and accelerations, and an electric circuit containing loads, capacitors, and a rectifier is connected with the harvester for an impedance matching and rectifying of the output signal. The volumetric output power (W/cm^3) is usually used to present the performance for the harvester and can be calculated from the generated voltage and the current value. The output power tends to be tuned by depending on the load resistance, the acceleration level, and the frequency of the vibration. Therefore, these factors should be optimized to obtain a maximum power output from the piezoelectric energy harvesters.

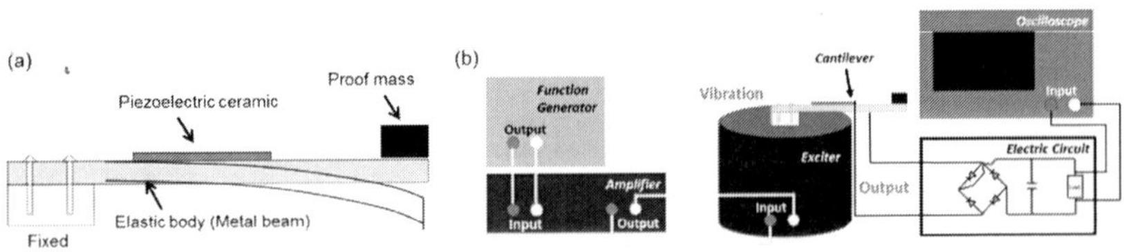

Figure 1. (**a**) Schematic of cantilever type piezoelectric energy harvester; (**b**) conventional experimental setup to evaluate cantilever type piezoelectric energy harvester.

Various research groups have investigated the simple cantilever and modified cantilever type piezoelectric harvesters. To enhance the output performance of the piezoelectric energy harvester, the material properties are the most important factor. Kim *et al.* demonstrated a unimorph cantilever type piezoelectric energy harvester using modified PZT thick films. Cr and Nb co-doping in the PZT ceramics improved $d_{33}{\cdot}g_{33}$ value up to 14,072 × 10^{-15} m^2/N, which is an FOM for the piezoelectric energy harvesting. From this, they obtained the maximum electrical power of 17.3 mW (2.08 mW/cm^3) at 20 Hz under ~4 g (g = 9.8 m/s^2) [22]. Choi *et al.* investigated a relationship between the piezoelectric properties and the output power of the piezoelectric energy harvester. They used the modified PZT ceramics with $Pb[(Zn_{0.4}Ni_{0.6})_{1/3}Nb_{2/3}]O_3$ (PZNN) in addition to exploring an optimum composition leading to the maximum piezoelectric properties. They revealed that the output power of the piezoelectric energy harvesters are linearly increased with increasing the $d_{33}{\cdot}g_{33}$ value and obtained the maximum electrical power of 231 mW/cm^3 from the piezoelectric cantilever at 84 Hz under 1 g [23]. Normally, the piezoelectric properties of the PZT based ceramics tend to be degraded at high temperature. Kim *et al.* studied the temperature dependent output power of the bimorph piezoelectric energy harvester with soft and hard PZT materials. Both cantilevers exhibited degraded output power in a high temperature range (50 °C < T < 150 °C) resulting from the sharply raised dielectric permittivity [24].

Generally, single crystal piezoelectric materials have significantly improved piezoelectric properties compared to the polycrystalline ceramics. Song *et al.* investigated the performance of the cantilever type piezoelectric energy harvesters using three different types of the piezoelectric materials including $0.7Pb(Mg_{1/2}Nb_{2/3})O_3$-$0.3PbTiO_3$ (PMN-PT) single crystals oriented along the <110> and <001> directions, and the PZT based polycrystalline ceramics [25]. The piezoelectric energy harvester with <110> oriented PMN-PT single crystal maximized the electrical power output among them. The maximum power was 1.07 mW at 89 Hz under 0.53 g. The cantilever type piezoelectric energy harvester was operated in 31 mode and the <110> oriented PMN-PT single crystal exhibited the maximized electromechanical coupling factor in the 31 mode. Karami *et al.* compared the piezoelectric performance of the unimorph cantilever type energy harvesters with single crystal PMN-PZT and the commercial PZT based materials, such as PZT-5A and PZT-5H. They revealed that the single crystal piezoelectric energy harvester produces superior power output compared to the polycrystalline piezoelectric ceramics. The maximum output from the cantilever was 226 μW/g^2 at 819 Hz [26].

To maximize the output power of the piezoelectric energy harvester, the cantilever type piezoelectric energy harvesters have been usually operated in a mechanical resonance frequency (natural frequency). Therefore, tuning of the natural frequency of the piezoelectric energy harvester is important to determine applications. The undammed natural frequency of the cantilever in a transverse vibration in given by

$$\omega_0 = \sqrt{\frac{k_{eq}}{m_{eq}}} = \sqrt{\frac{3YI/L^3}{(33/140)mL + M_t}} \quad (7)$$

where k_{eq} is the equivalent spring rate, m_{eq} is the equivalent mass, *YI* is the flexural rigidity of the beam, *L* is the length of the beam, *m* is the mass per unit length, and M_t is the tip mass, 33/140 mean that 33/140 (23.5%) of the beam mass participate in the motion. From Equation (7), it can be known that the natural frequency is tuned by changing the spring constant, mass, beam length, width, and thickness of the cantilever [27]. To modify natural frequency of the piezoelectric energy harvester, Berdy *et al.* demonstrated the piezoelectric energy harvester using a meandering structure. The meandering structure, which leads to increase of the body length, resulted in reducing the natural frequency of the cantilever, and the output power was 118 μW at 49.7 Hz of natural frequency under 0.2 g [28]. Shindo *et al.* also achieved the low natural frequency using a S-shape cantilever structure with the output power of 8.5 mW/cm^3 at 40 Hz under 3.2 g [29]. A wide bandwidth of the natural frequency enhances the capability of the piezoelectric energy harvesters for the various applications. Sobocinski *et al.* firstly introduced the fully embedded wideband piezoelectric energy harvester using a low temperature co-fired ceramic (LTCC) technique. They realized an integrated system including a reliable and hermetic package, a circuit board for electronics and three kinds of active piezoelectric harvester structures in the same process. Three cantilevers with different lengths had a variety of natural frequencies and exhibited 5.4% frequency bandwidth for 3 dB attenuation. The maximum power output was 32 μW at 1100~1165 Hz under 1 g acceleration [30]. From these results, it is clear that control of the natural frequency can be easily achieved by the tuning length of the cantilever. The low and wide resonance frequency of the cantilever expends the application field in the environmental vibration conditions. The output performance of the bulk cantilever energy harvesters is summarized in Table 1.

Table 1. Comparison of the output performance of the cantilever type bulk piezoelectric energy harvesters.

Structure	Power Density (mW/cm^3)	Normalized Power (mW/g^2)	Frequency (Hz)	Ref.
Cr and Nb doped PZT cantilever	2.1	1.1	20	[22]
PZNN cantilever	231	11.7	84	[23]
<110> oriented single crystalline PMN-PT cantilever	-	3.8	84	[25]
<001> oriented single crystalline PMN-PT cantilever	-	1.4	86	[25]
PMN-PZT single crystalline cantilever	-	0.2	819	[26]
Meandering structured cantilever	0.2	2.9	50	[28]
S-shape bulk cantilever	8.5	-	40	[29]
Wideband LTCC cantilever arrays	-	0.03	1100–1165	[30]

The cantilever type energy harvesters have been usually designated for operating under simple vertical vibration conditions. However, the environmental vibrations and the mechanical energy sources around it are actually not simple. Here, we introduce new challenges on the cantilever type energy harvesters for special applications without the simple vertical vibration. They determined the optimal structural design of the piezoelectric energy harvesters according to the various mechanical energy sources. Park *et al.* investigated bimorph piezoelectric energy harvesters with an asymmetric inertial mass for a multi-dimensional vibration ambient. The asymmetric inertial mass contributes to generating power from the vibration applied to the length and thickness directions of the PZT cantilever. They obtained the output power of 7.5 mW for the *z*-axis and 1.4 mW for the *x*-axis at 45 Hz under 1 g [31]. The multi-dimensional vibration energy harvester is an effective device to generate electrical power from the vibration of a vehicle. Hashimoto *et al.* demonstrated a multi-mode and multi-axis piezoelectric energy harvester using a two-mass device with two bending modes for the vehicle application. They performed a road test using the prototype energy harvester and acquired a high energy efficiency from the device [32]. For the multi-axis piezoelectric vibration energy harvesting, Erkan *et al.* fabricated a crab-leg like cantilever structure via individually assembling processed bulk-PZT leg on a patterns silicon die. They obtained 5–50 μW power at 387–389 Hz under 50 mg acceleration amplitude in the *x*, *y*, and *z* direction from the multi-axis piezoelectric energy harvester [33]. Khameneifar *et al.* proposed a rotary type piezoelectric energy harvester using four cantilever type harvesters. In this structure, mechanical vibration energy is applied on the PZT cantilevers due to a gravitational force while the hub is rotating. They obtained maximum power of 6.4 mW at a shaft speed of 138 rad/s [34]. This rotary type energy harvester is also useful for the wind energy harvesting using impact-induced resonance operation [35]. The off-resonance operation with the high power output of the piezoelectric energy harvester can achieve more practical ways to use the various applications around it. For this, Pozzi *et al.* demonstrated a prototype piezoelectric energy harvester with four PZT bimorph cantilevers that is wearable on knee joints. The energy harvester relies on the plucking technique, and the PZT cantilevers are deflected by a plectrum; when released due to loss of a contact, the cantilevers are free to vibrate at their resonant frequency. The average power output of the prototype harvester reached 2.06 mW [36]. Li *et al.*presented a low frequency acoustic energy harvester using a PZT cantilever. The acoustic piezoelectric energy harvester contained a quarter-wavelength straight tube resonator with the PZT piezoelectric cantilever placed inside the tube. The maximum output power of 12.697 mW was generated in an incident sound pressure of 110 dB [37].

2.2. Various Types of Bulk Piezoelectric Energy Harvesters

Unique structures of the bulk type piezoelectric harvesters have been proposed for the various applications. Palosaari *et al.* demonstrated cymbal type piezoelectric energy harvester. A piezoelectric disc was confined between two convex steel discs acting as a force amplifier delivering stress to the PZT discs and protecting the harvester. At 1.19 Hz compression frequency with 24.8 N force, a cymbal type harvester generated an average power of 0.66 mW and the maximum power densities of 0.31 mW/cm^3 were observed [38]. Chen *et al.* introduced a piezoelectric energy harvester with a clamped piezoelectric circular diaphragm structure, which is a common structure for pressure sensors. The electrical power of 12 mW was obtained when a pre-stress of 1.2 N was applied on the harvester at 113 Hz under 1 g [39]. Multilayer stack structure for an energy harvesting using 33 mode achieved enhancing of the mechanical to the electrical energy conversion efficiency [40,41]. Xu *et al.* obtained significantly higher electrical power than a similar size cantilever type piezoelectric energy harvester in both resonance and off-resonance modes. The generated electrical powers were 2 mW at 700 Hz (off-resonance) and 208 mW at 1714 Hz (resonance) [41]. The piezoelectric energy harvester can be utilized for battery- and wire-less tire pressure measurement systems (TPMS). Makki *et al.*embedded the stacked piezoelectric bender elements in a tire wheel for an in-wheel based power harvesting [42]. This utilizes the cyclic deformation of the tire during vehicle movement. The weight of the vehicle deforms the tire at the tire-road interface created on the piezoelectric sensors. As the tire rotates during vehicle movements, a new section of the tire's treadwall deforms within the piezoelectric sensors resulting in a cyclic deformation and relaxation pattern.

3. MEMS PIEZOELECTRIC ENERGY HARVESTERS

Decreasing flexural rigidity associated with the piezoelectric energy harvesters is very attractive to reduce the natural frequency and withstand large strain and deformation. The MEMS technology is one of the promising choices to utilize the low flexural rigidity of the piezoelectric energy harvesting devices. Moreover, tiny piezoelectric devices fabricated by the MEMS process can be easily integrated in the micro-electronics that require the power source of μW level [43,44,45]. Generally, the structure of the MEMS piezoelectric harvesters based on the cantilever structure have been mainly developed to obtain the high output power and easily enable tuning the resonance frequency [46,47,48,49]. The general MEM cantilever structure is similar to the bulk energy harvester [50]. The Si or metal is used for the elastic body, the piezoelectric film is the active piezoelectric materials, and the proof mass is formed to tune the resonance frequency and the applied force. Modified cantilever structure [51], diaphragm structure [52], *etc.* have also been suggested for the high efficiency and practical applications.

In this section, we introduce various structural designs and properties of the MEMS energy harvesters. Kim *et al.* demonstrated a bimorph PZT cantilever for energy harvesting application. They compared the piezoelectric properties under d_{31} unimorph, d_{31} bimorph, d_{33} unimorph, and d_{33} bimorph modes. At the resonance of 89.4 Hz, they obtained the maximum average power about 3.9 (d_{31} unimorph), 9.6 (d_{31} bimorph), 9.6 (d_{31} bimorph), 21.42 (d_{33}unimorph), and 22.47 (d_{33} bimorph) nW under 0.8 g, respectively [53]. To compare the output power of the d_{31} and d_{33} mode MEMS cantilevers, Kim *et al.* demonstrated both types of the energy harvesters with the same dimension in a cantilever structure leading to the same resonance frequency at 243 Hz. The results of modeling and experiment studies obtained almost the same output power. From the experiments, they obtained a slightly higher output performance from the d_{33} mode cantilever. Although the harvester with d_{33} mode showed higher output performances, however, they have mentioned that the electrode design of d_{33}mode should be conducted carefully, because the output power strongly depends on the dimensions of inter digital electrode (IDE) [54]. These results indicate that the output power of the MEMS cantilever is strongly affected by an electrode configuration and poling direction of the piezoelectric materials. To increase output power and energy conversion efficiency, Tang *et al.* fabricated a PZT microstructure using bulk ceramics [55]. Through the previous research, they obtained the optimized shape of the cantilever beam and mass, which was determined through simulations and experimental study, resulting in the output voltage of ~1 V at the resonance frequency around 1 kHz [56]. They used patterning on bulk ceramics by a dicing to demonstrate the rectangular micro-poles. The fabricated micro-cantilevers, which have different lengths and masses, and the resonance frequency around 520 Hz, were assembled on a printed circuit board (PCB). The output power and voltage were varied with the different frequency, acceleration, and load. The maximum output power was ~28.8 mW/cm^3 at the resonance frequency with load of 70 kΩ under 1 g of acceleration. Kuehne *et al.* presented a novel MEMS generator for fluid-actuated energy conversion [51]. They fabricated a narrow fluidic cantilever tip to increase an energy conversion efficiency. In the experimental study, the electrical quality factors and energy density of rectangular and triangular cantilevers were compared under the various ambient pressures and cantilever deflection of the free-end. The compared electrical energy density value of the rectangular shape was three times larger. The maximum energy density value for the triangular cantilever shape was measured to 35 nJ/mm^2.

A metal-based MEMS cantilever is a promising structure to get the flexibility to reduce the resonance frequency and enhance a strain. Wang *et al.* demonstrated the metal-based MEMS cantilever. The PZT thick film was deposited on the stainless steel beam by a sol-gel method, which has approximately 4 μm thickness. Around 90 Hz, the cantilever has the resonance frequency, and the maximum output power was 15 μW under 89 Hz and 1 g acceleration [57]. Lin *et al.* fabricated a piezoelectric micro energy harvester using a high-quality PZT thick film on stainless steel that was deposited by an aerosol deposition method. It generated the maximum output power of 200.28 μW with a 12.67 $V_{p\text{-}p}$ output voltage at 112.4 Hz under 1.5 g acceleration and the energy densities of the device in a vacuum and in atmosphere were calculated as 178.6 and 137.5 nJ/mm^3, respectively [58].

Normally, the conventional MEMS cantilever has been operated at a high frequency (>1 kHz), because of short length and small mass of the cantilever beam. Furthermore, the amplitude bandwidth at the resonance frequency is very narrow. However, ambient vibration sources around usually have a wide bandwidth and low frequency, normally less than 200 Hz. In the MEMS piezoelectric harvesting technology, achieving low resonance frequency and wide frequency range operation is the most important goal for the practical application. The tuning configuration of the mass on the cantilever is a general method to achieve low resonance frequency. Liu *et al.* demonstrated the piezoelectric MEMS energy harvester with a wideband operation range and steadily increased output power. The resonance frequency of the fabricated MEMS generator had around 36 Hz due to the Si mass configuration. To achieve the wide bandwidth, they also embedded the cantilever arrays with different PZT elements and limited spacers, which is a common technique to acquire the wide bandwidth. The whole chip was assembled onto a metal carrier with the limited spacer such that the operation frequency bandwidth can be widened to 17 Hz at the input acceleration of 1.0 g during frequency up-sweep. Load voltage and power generation for the different numbers of the PZT elements in series and in parallel connections were compared and discussed in this paper. Under the same input acceleration, the optimal power was the same. However, the impedance measured at the maximum output power was reduced by connecting PZT elements in parallel. The electrical power generated at 1.0 g with respect to the load resistance of 330 kΩ achieved a wideband output ranging from 32.3 nW to 85.5 nW within the operation bandwidth from 30 Hz to 47 Hz [59,60]. They also fabricated a new S-shaped MEMS PZT cantilever with a low frequency vibration. The long length of the S-shape cantilever beam effectively reduced the resonance frequency of the device. The S-shaped MEMS cantilever could be operated at 30 Hz with a low accelerations less than 0.4 g. The maximum voltage and normalized power are 42 mV and 0.31 $\mu W/g^2$, respectively. This cantilever was operated with the broad bandwidth from 3.4 to 11.1 Hz depending on the reduced stop distance from 1.7 to 0.7 mm under an acceleration of 0.3 g and the generated voltage and normalized power at the resonance frequency was reduced from 40 to 16 mV and from 17.8 to 2.8 nW/g^2, respectively [61]. Moreover, in order to use in low vibrational environments under 48 Hz, they fabricated a cantilever-type stopper beside a MEMS cantilever energy harvester, which also generates the output voltage. Thus, both the stopper and the generator could concurrently produce the output voltage of 20 V and output power of 140 nW, respectively [62,63]. The achieving of wide bandwidth of the cantilever basically causes the low power generation, because of vibration compensation in the limited spacer technique and a phase mismatch between the cantilever arrays. To overcome these challenges, Hajati and Kim *et al.* developed an ultra-wide bandwidth piezoelectric energy harvester by exploiting a nonlinear stiffness of a double clamped MEMS resonator [52]. In this structure, the stretching strain is always applied to the beam, which causes uniform tensile stress, while the bending strain in the conventional cantilever structure is not uniform along the length and thickness of the elastic body. The tensile strain with the nonlinear stiffness provides passive feedback and results in this amplitude-stiffened Duffing mode resonance. The nonlinear resonator showed wide bandwidth exceeding 20% of the arithmetic center frequency and generated maximum power of ~45 μW at ~1.3 kHz, which is much higher than other works that achieved wide bandwidth. The output performance of the MEMS piezoelectric vibration energy harvesters is summarized in Table 2.

Table 2. Comparison of the output performance of the microelectromechanical systems (MEMS) piezoelectric vibration energy harvesters.

Structure	Power Density (μW/cm³)	Normalized Power (μW/g²)	Frequency (Hz)	Ref.
Unimorph cantilever (d_{31} mode)	-	0.0061	89.4	[53]
Unimorph cantilever (d_{33} mode)	-	0.0335	89.4	[53]
Bimorph cantilever (d_{31} mode)	-	0.015	89.4	[53]
Bimorph cantilever (d_{33} mode)	-	0.035	89.4	[53]
Micro-cantilever using bulk PZT	28,856	11.6	520	[55]
Metal-based MEMS cantilever	98	15.4	89	[57]
Metal-based MEMS cantilever	15,453	89.0	112.4	[58]
Wideband cantilever	-	0.032–0.085	30–47	[59]
S-shape MEMS cantilever	-	0.31	27.4	[61]

4. FLEXIBLE PIEZOELECTRIC ENERGY HARVESTERS

The PZT based materials have an outstanding piezoelectric coefficient and have been utilized for actuators, transducers, energy harvesters, *etc.* However, they tend to be easily brittle under a large strain and an applied force leading to several restrictions to be installed in flexible, wearable, and biocompatible electronic devices. The flexibility extends the feature of the piezoelectric materials and the application field for the emerging energy harvesting technology. Moreover, the flexibility of the piezoelectric energy harvesters enables obtaining large output power under large strain by the off-resonance vibration and deformation due to a remarkable endurance under the large strain. To acquire the flexibility in the piezoelectric ceramics, nano- and micro-structured materials, and thin films have been used with a polymer composite and a freestanding structure. The idea of an energy harvester using nanostructures was firstly reported by Zhong Lin Wang in 2006. He introduced the piezoelectricity in the ZnO having enough feasibility for the flexible piezoelectric energy harvesting [3]. Since he demonstrated the piezoelectric energy harvester using the ZnO nanostructures, significant studies have been reported about the piezoelectric nanogenerators, which used ZnO nanostructures. The PZT based nanostructured materials have significant larger piezoelectric coefficients than ZnO [64,65]. Nevertheless, only a limited number of studies studies have been reported about the flexible piezoelectric energy harvesting using the PZT based material, because of critical limitations to synthesizing nanostructures due to a complex composition and a higher crystallization temperature, which makes it to be not applicable with the flexible polymer materials. Herein, we focused on flexible piezoelectric energy harvesting devices using the PZT based flexible materials including nanotubes, nanorods, nanowires, nanofibers, nanocomposites, thin films, *etc.*

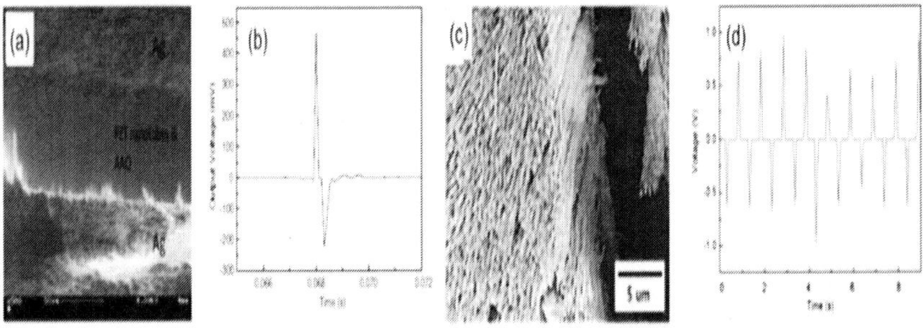

Figure 2. (**a**) Cross-sectional view of the device using PZT nanotubes [66]; (**b**) output voltage from the PZT nanotube arrays [66]; (**c**) PZT nanotubes [67]; (**d**) output performance of the PZT nanotube composite energy harvester [67].

4.1. Nanotubes

The PZT nanotubes have usually been synthesized by a template-assisted infiltration method using an anodic aluminum oxide (AAO). Xu *et al.* obtained electrical power from the PZT nanotubes, which have a high aspect ratio, by a template-assisted method. A pure perovskite phase with strong <110> preferred crystallographic orientation was obtained by a heat treatment at 650 °C. The diameters of the nanotubes ranging from 190 to 210 nm were obtained, corresponding to diameter and height of the nanopores in the AAO template as shown inFigure 2a. The fabricated nanotubes were sandwiched between two Ag electrodes and the energy harvester generated the electrical power up to 496 mV when a stainless-steel nugget was dropped on the electrode (Figure 2b) [66]. The Korea Institute of Science and Technology (KIST) presented a composite generator using the PZT nanotubes. An AAO template-assisted method was used to fabricate the PZT nanotubes. The fabricated nanotubes had the length of 60 μm and the diameter of 200 nm as shown in Figure 2c. For the flexibility of the device, polydimethylsiloxane (PDMS) elastomer was mixed with the fabricated nanotubes, and Pt coated polyimide (PI) films were attached on the bottom and top of the PZT–PDMS composite. The PZT–PDMS composite energy harvester generated about ~2 V and ~60 nA when the generator was bended and stretched (see Figure 2d) [67].

4.2. Nanorods and Nanowires

The PZT nanorods and nanowires have usually been synthesized by a hydrothermal method. Xu *et al.* demonstrated PZT nanowire arrays epitaxailly grown on conductive substrates using a hydrothermal method [68]. Pt electrodes were used on both the top and bottom of the nanowires by electron beam evaporation. The entire structure was packaged with a soft epoxy polymer to ensure the robustness of the device and also isolate a possible moisture erosion to the nanowires. They assessed the output voltage and current from the fabricated device, connecting a rectifier circuit and a light emitting diode (LED), under various frequencies. The output voltage of the device reached around 1 V and the output current is over 200 nA under pressing motion. Using a capacitor with 22 μF, they successfully tested the LED lighting. In addition, through the test with a range of stimulating frequencies from 1 to 50 Hz, they expected the energy harvester can be configured to work under a relatively high frequency of ~50 kHz [68]. Wang *et al.* demonstrated a PZT nanowire textile energy harvester with generating ~6 V output voltage and ~45 nA output current during bending motion, which are large enough to power a liquid crystal display (LCD). They fabricated PZT nanowires by an electrospinning method. For the flexibility of the device, they aligned the parallel PZT nanowires on PDMS template as shown in Figure 3 [69]. Zhou *et al.* synthesized high quality single crystal PZT nanowires by a two-step hydrothermal method. Measured piezoelectric strain coupling coefficient of the PZT nanowire was 80 pm/V. To fabricate the flexible piezoelectric energy harvester, a nanocomposite cantilever was demonstrated with the PDMS elastomer. The harvester exhibited 2.4 $\mu W/cm^3$ of the maximum power density when it is vibrating [70].

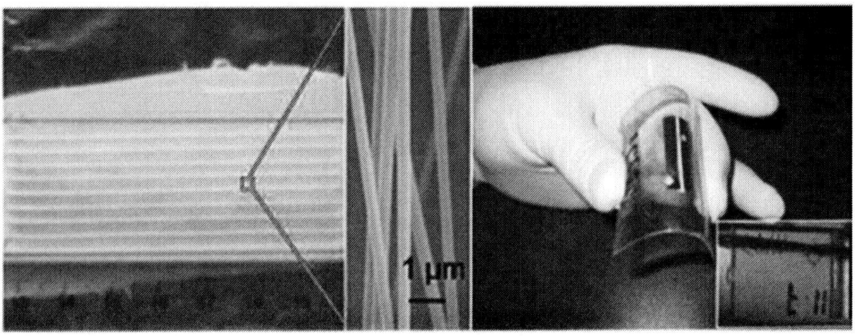

Figure 3. Nanocomposite generator using aligned PZT nanowires [69].

4.3. Nanofibers

Piezoelectric nanofibers are promising structures to obtain the large strain and the flexibility with the high energy density. Qi *et al.* in Princeton University fabricated piezoelectric nanoribbons printed onto rubber for the flexible energy conversion. The PZT ribbons have a micrometer-scale width and nanometer scale thickness. To fabricate PZT ribbons, PZT films were grown on a (100)-cleaved MgO crystal substrate and post-annealed to obtain the perovskite crystal structure. The fabricated films were patterned into nano-thick ribbons and printed onto the PDMS via a dry transfer, as shown in Figure 4a. By this process, they overcame the limitation of the high temperature heat treatment for the crystallization. In addition, they obtained the d_{33} value of 101 pm/V for the PZT ribbons after a poling. This value agrees well with the thin film data, which is understandable considering the relatively benign processing conditions for fabricating and transferring PZT ribbons [71]. They also represented a stretchable energy harvester using buckled PZT ribbons. The PZT ribbons were patterned on the MgO substrate and subsequently released from the substrate using the phosphoric acid. A slab of the PDMS was then elastically stretched and brought into a conformal contact with the ribbons. Finally, releasing the prestrain in the PDMS led to a compressive force on the PZT ribbons when the PDMS relaxed to zero strain, leading to a periodic de-adhesion and buckling as shown in Figure 4b. The measured d_{33} value was ~130 pm/V and the generated output current was about 100 pA [72]. Chen *et al.* utilized PZT nanofibers fabricated by an electrospinning method and deposited them onto the IDE of platinum fine wire arrays. The fabricated nanofibers have the diameter of 60 nm and were laterally aligned on electrodes for the longitudinal mode as shown in Figure 4c. The PZT nanofibers were polled by applying electric field across the electrodes, and then the silicon substrate was released for the flexibility of the device. In order to confirm the power generation from the device, the pressure was applied on the top surface. The output voltage was 0.6 V when applying a large load, and 1.63 V when applying the pressure with a finger [73].

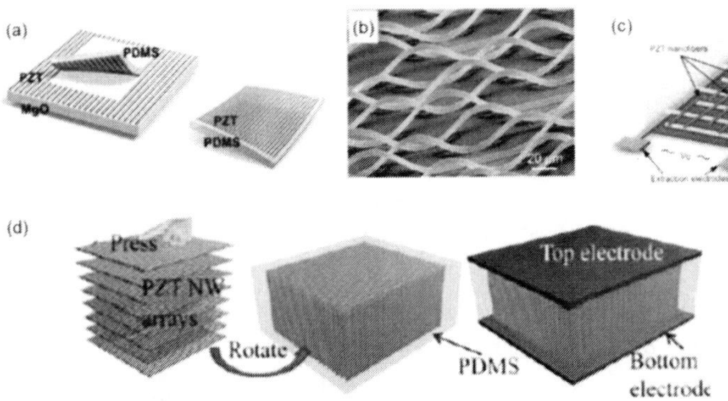

Figure 4. (**a**) Flexible PZT nanoribbons on the PDMS substrate [71]; (**b**) buckled PZT nanoribbons [72]; (**c**) energy harvester using aligned PZT nanofiber [73]; (**d**) nanogenerator using PZT nanofiber composite [75].

They also demonstrated a flexible acoustic emission sensor based on the PZT nanofiber composite membrane with a similar structure. The electrospinning method was used for the fabrication of the PZT nanofibers. The plastic substrate was used for the flexibility of the device. It was attached on the carbon fiber composite to measure an acoustic sensing performance. The device produced the output voltage of ~10 mV when the steel ball was dropped down to the carbon composite and up to ~100 mV when a steel bar hit the composite directly. They showed flexibility and self-powered sensing, and high sensitivity, which allows the possibility of monitoring structures by the curved surfaces or integrating into composite [74]. Gu *et al.* presented a flexible fiber nanogenerator, which enables turning on the light-emitting diodes. To fabricate vertically aligned PZT nanofiber arrays, the PZT nanofiber film was obtained first and it was stacked layer by layer to form multilayer films (see Figure 4d). PDMS was filled into the interspace of the films and nanofibers using a capillary force, and the multilayer film was gently pressed without damage of its structure. The electrodes were attached on the top and bottom surfaces of the fabricated device. In experimental results, the maximum output voltage was 209 V and the peak current was 53 μA. The output power was available for the self-powered system, such as sensor network [75].

4.4. Nanoparticles

The PZT nanoparticles have been used as active materials in the composite piezoelectric energy harvesters. Park *et al.* reported the flexible and large-area nanocomposite generators (NCG) based on the PZT nanoparticles and carbon nanotubes. They used the PZT nanoparticles as an energy generation source to achieve the excellent output signal. The nanoparticles were mixed with the PDMS and CNTs in order to improve the flexibility and stretchability of the device. The CNTs, as dispersing agents, stress reinforcement, and conduction path, improved the output performance of the nanocomposite-based piezoelectric device. It produced the output voltage and current up to 10 V and 1.3 μA, respectively. From a large-area NCG device with 30 × 30 cm^2, they obtained the output voltage up to 100 V and the current signal up to 10 μA. Using these devices, they turned on the LEDs and LCD screen [76]. Zhang *et al.*demonstrated a self-charging Li battery using a polyvinylidene fluoride (PVDF)-PZT nanoparticle composite separator (Figure 5). The piezoelectric potential from the separator could effectively drive the diffusion of Li ions. This device directly converted the mechanical energy into the electrochemical energy and stored it. The piezoelectric separator generated ~1 V under periodic compressive stress of 10 N. The storage capacity of the PVDF-PZT based Li battery was ~0.010 μAh in 240 s [77].

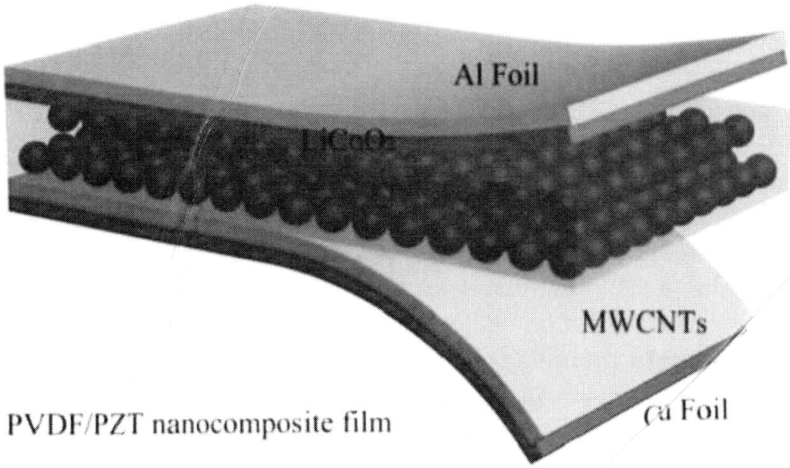

Figure 5. Self-charging Li battery using PVDF-PZT nanoparticle composite separator [77].

4.5. Thin Films

Thin film materials also have remarkable potential for the flexible piezoelectric energy harvesting technology. However, the PZT thin film has many limitations to demonstrate the flexible devices, which need to be installed on flexible substrates because of their high crystallization temperature (<600 °C). To overcome these challenges, Do *et al.* firstly reported a film transfer method using laser lift-off (LLO) process. They fabricated a transparent flexible piezoelectric energy harvester based on the PZT films by the LLO process as shown in Figure 6a. They synthesized the fully crystallized PZT thin film on the sapphire substrate. By the LLO process, the PZT thin film was transferred from the sapphire substrate to the flexible receptor substrate. In this process, an additional heat treatment was unnecessary to form the crystallized PZT film on the flexible substrate. Indium-tin-oxide (ITO) coated polyethylene terephthalate (PET) film was used as the electrodes and flexible substrate. Finally, they obtained the output signal generated from the device, resulting in the maximum output voltage of ~0.28 V and output current of ~30 nA. The average power density of the device with a total active area about 1 cm × 1 cm was approximately 8.4 nW/cm^2 [78]. Park *et al.* also reported the flexible PZT energy harvester fabricated by the LLO process. They transferred the PZT thin films from the sapphire substrate to a plastic substrate and formed IDE patterns on the PZT thin film to fabricate the energy harvester. The energy harvester generated ~200 V of output voltage and 150 μA/cm^2 of current, and turned on over 100 blue LEDs without any external electric source [79]. Using this technique, they also fabricated a self-powered flexible embedded LED system [80] and a nanosensor for biomedical applications [81]. Chun *et al.* demonstrated a highly-stretchable composite nanogenerator using a hemisphere PZT thin film [82]. The hemisphere PZT thin film was produced using a polystyrene (PS) sphere template. To fabricate the composite energy harvester, PDMS was filled into the hemisphere PZT thin film and the substrate was removed by the lift-off method as shown in Figure 6b. The energy harvester generated 4 V of the output voltage and 0.13 μA/cm^2 of the current during a periodic bending motion. The output performance of the flexible piezoelectric energy harvesters is summarized in Table 3.

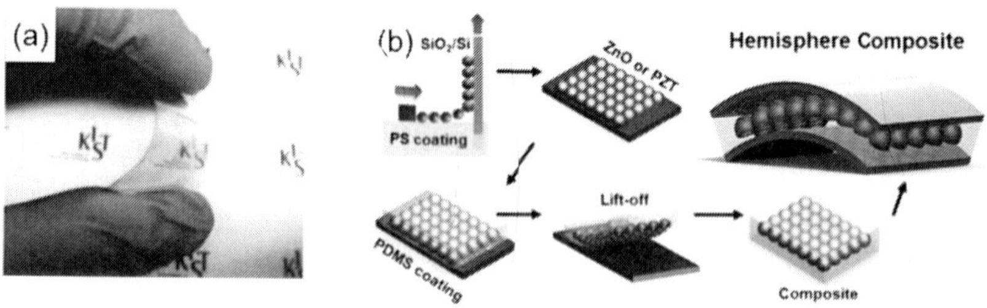

Figure 6. (**a**) Transparent flexible piezoelectric energy harvester based on PZT films by LLO process [78]; (**b**) highly-stretchable composite nanogenerator using hemisphere PZT thin film [82].

Table 3. Comparison of the output performance of flexible piezoelectric energy harvesters.

Structure	Max. Voltage	Max. Current	Max. Power	Ref.
Freestanding PZT nanotube	0.495 V	-	-	[66]
PZT nanotube composite	2 V	60 nA	-	[67]
PZT nanorod arrays	1 V	200 nA	-	[68]
PZT nanowire textile	6 V	45 nA	-	[69]
PZT nanowire composite	4 V	88 nA	2.4 μW/cm^3	[70]
Buckled PZT nano-ribbons	-	100 pA	-	[72]
PZT nano fibers	1.63 V	-	0.03 μW	[73]
Vertically aligned PZT nanofiber arrays	209 V	53 μA	4.9 mW/cm^2	[75]
PZT nanoparticle composite	10 V	1.3 μA	13 μW	[76]
Transparent PZT thin film harvester (sandwich structure)	0.28 V	30 nA	8.4 nW/cm^2	[78]
PZT thin film harvester (planar structure)	200 V	150 μA/cm^2	30 mW/cm^2	[79]
Hemisphere PZT thin film composite	4 V	0.13 μA/cm^2	0.52 μW/cm^2	[82]

5. CONCLUSIONS

PZT based energy harvesting technologies reported in the last few years are summarized in this paper. The following three focuses were derived from this study. First, the PZT based bulk piezoelectric energy harvesters have been suggested for self-powered systems. For the practical application, the resonance frequency has been tuned to the low frequency in the cantilever type piezoelectric energy harvesters by controlling the structural parameters. Unique designs of the bulk type energy harvesters provided a higher efficiency in certain environmental conditions including resonance vibration, off-resonance, impact, *etc.* Second, various designs of the MEMS piezoelectric energy harvesters have usually been based on the cantilever structure, and have been studied to control the resonance frequencies and achieve wide bandwidth for practical applications. Finally, flexible piezoelectric energy harvesters using nano-materials, which allow flexibility for the piezoelectric materials, enabled the off-resonance operation of the piezoelectric energy harvester with the remarkable strength under the large strain. However, although the larger strain is induced on the flexible energy harvesters, the output power is still lower than the bulk and MEMS piezoelectric energy harvesters that operated at resonance frequency. The off-resonance operation and very lower packing density of the active piezoelectric materials of the flexible energy harvesters resulted in a low output power. Nevertheless, from a different point of view, the flexible piezoelectric energy harvesters are very attractive in a random vibration environment, such as the wind, water flow, human movement, *etc.*

From this paper, here, we anticipate that the bulk and MEMS energy harvesters should be designed to determine the resonance frequency for an exact application because the vibration sources that are vibrating in a certain frequency regularly are the only optimum sources for these two techniques. Therefore, achieving wide bandwidth is the most reasonable way for them to be used in practical applications. For enhancing the performance of the flexible piezoelectric energy harvesting technology, the packing density of the active piezoelectric materials should be increased while maintaining flexibility. Moreover, the specific units of the output voltage, current, and power need to be uniform and standardized for the piezoelectric energy harvesting technology except for cantilever type harvesters.

ACKNOWLEDGMENTS

This work was supported by the Converging Research Center Program by the Ministry of Education, Science and Technology (2010K000979), Korea.

AUTHOR CONTRIBUTIONS

Min Gyu Kang and Woo-Suk Jung built and wrote the majority of the manuscript text. This review paper was directed by Chong-Yun Kang and Seok-Jin Yoon. All authors reviewed the manuscript.

REFERENCES

1. Park, S.E.; Shrout, T.R. Characteristics of relaxor-based piezoelectric single crystals for ultrasonic transducers. *IEEE Trans. Ultrason. Ferroelectr. Freq. Control* **1997**, *44*, 1140–1147.
2. Dubois, M.A.; Muralt, P. Properties of aluminum nitride thin films for piezoelectric transducers and microwave filter applications. *Appl. Phys. Lett.* **1999**, *74*, 3032–3034.
3. Wang, X.D.; Zhou, J.; Song, J.H.; Liu, J.; Xu, N.S.; Wang, Z.L. Piezoelectric field effect transistor and nanoforce sensor based on a single ZnO nanowire. *Nano Lett.* **2006**, *6*, 2768–2772.
4. Giurgiutiu, V. *Structural Health Monitoring with Piezoelectric Wafer Active Sensors*, 2nd ed.; Academic Press: Cambridge, MA, USA, 2014; pp. 1–1012.
5. Ihn, J.B.; Chang, F.K. Detection and monitoring of hidden fatigue crack growth using a built-in piezoelectric sensor/actuator network: I. Diagnostics. *Smart Mater. Struct.* **2004**,*13*, 609–620.
6. Uchino, K. Materials issues in design and performance of piezoelectric actuators: An overview. *Acta Mater.* **1998**, *46*, 3745–3753.

7. Wang, Q.M.; Zhang, Q.M.; Xu, B.M.; Liu, R.B.; Cross, L.E. Nonlinear piezoelectric behavior of ceramic bending mode actuators under strong electric fields. *J. Appl. Phys.* **1999**, *86*, 3352–3360.

8. Chu, S.Y.; Chen, T.Y.; Tsai, I.T.; Water, W. Doping effects of Nb additives on the piezoelectric and dielectric properties of PZT ceramics and its application on saw device. *Sens. Actuators A Phys.* **2004**, *113*, 198–203.

9. Tang, I.T.; Chen, H.J.; Hwang, W.C.; Wang, Y.C.; Houng, M.P.; Wang, Y.H. Applications of piezoelectric ZnO film deposited on diamond-like carbon coated onto Si substrate under fabricated diamond saw filter. *J. Cryst. Growth* **2004**, *262*, 461–466.

10. Kim, H.S.; Kim, J.H.; Kim, J. A review of piezoelectric energy harvesting based on vibration. *Int. J. Precis. Eng. Man.* **2011**, *12*, 1129–1141.

11. Beeby, S.P.; Tudor, M.J.; White, N.M. Energy harvesting vibration sources for microsystems applications. *Meas. Sci. Technol.* **2006**, *17*, R175–R195.

12. Uchino, K. *Ferroelectric Devices*, 2nd ed.; CRC Press: Boca Raton, FL, USA, 2010; pp. 161–169.

13. Ouyang, J.; Ramesh, R.; Roytburd, A.L. Intrinsic effective piezoelectric coefficient $e_{31,f}$ for ferroelectric thin films. *Appl. Phys. Lett.* **2005**, *86*, 152901.

14. Maria, J.P.; Shepard, J.F.; Trolier-McKinstry, S.; Watkins, T.R.; Payzant, A.E. Characterization of the piezoelectric properties of $Pb_{0.98}Ba_{0.02}(Mg_{1/3}Nb_{2/3})O_3$-$PbTiO_3$epitaxial thin films. *Int. J. Appl. Ceram. Technol.* **2005**, *2*, 51–58.

15. Shrout, T.R.; Zhang, S.J. Lead-free piezoelectric ceramics: Alternatives for PZT? *J. Electroceram.* **2007**, *19*, 113–126.

16. Tao, H.; Wu, J.G.; Xiao, D.Q.; Zhu, J.G.; Wang, X.J.; Lou, X.J. High strain in (K,Na)NbO_3-based lead-free piezoceramics. *ACS Appl. Mater. Inter.* **2014**, *6*, 20358–20364.

17. Rubio-Marcos, F.; Ochoa, P.; Fernandez, J.F. Sintering and properties of lead-free (K,Na,Li)(Nb,Ta,Sb)O_3 ceramics. *J. Eur. Ceram. Soc.* **2007**, *27*, 4125–4129.

18. Zhang, M.; Yang, J.; Si, C.W.; Han, G.W.; Zhao, Y.M.; Ning, J. Research on the piezoelectric properties of AlN thin films for MEMS applications. *Micromachines* **2015**, *6*, 1236–1248.

19. Zhao, M.H.; Wang, Z.L.; Mao, S.X. Piezoelectric characterization of individual zinc oxide nanobelt probed by piezoresponse force microscope. *Nano Lett.* **2004**, *4*, 587–590.

20. Wang, Z.L.; Song, J.H. Piezoelectric nanogenerators based on zinc oxide nanowire arrays.*Science* **2006**, *312*, 242–246.

21. Kumar, B.; Kim, S.W. Energy harvesting based on semiconducting piezoelectric ZnO nanostructures. *Nano Energy* **2012**, *1*, 342–355.

22. Kim, K.B.; Kim, C.I.; Jeong, Y.H.; Lee, Y.J.; Cho, J.H.; Paik, J.H.; Nahm, S. Performance of unimorph cantilever generator using Cr/Nb doped $Pb(Zr_{0.54}Ti_{0.46})O_3$ thick film for energy harvesting device applications. *J. Eur. Ceram. Soc.* **2013**, *33*, 305–311.

23. Choi, C.H.; Seo, I.T.; Song, D.; Jang, M.S.; Kim, B.Y.; Nahm, S.; Sung, T.H.; Song, H.C. Relation between piezoelectric properties of ceramics and output power density of energy harvester. *J. Eur. Ceram. Soc.* **2013**, *33*, 1343–1347.

24. Kim, S.B.; Park, J.H.; Ahn, H.; Liu, D.; Kim, D.J. Temperature effects on output power of piezoelectric vibration energy harvesters. *Microelectron. J.* **2011**, *42*, 988–991.

25. Song, H.C.; Kang, C.Y.; Yoon, S.J.; Jeong, D.Y. Engineered domain configuration and piezoelectric energy harvesting in $0.7Pb(Mg_{1/3}Nb_{2/3})O_3$–$0.3PbTiO_3$ single crystals. *Metals Mater. Int.* **2012**, *18*, 499–503.

26. Karami, M.A.; Bilgen, O.; Inman, D.J.; Friswell, M.I. Experimental and analytical parametric study of single-crystal unimorph beams for vibration energy harvesting.*IEEE Trans. Ultrason. Ferroelectr. Freq. Control* **2011**, *58*, 1508–1520.

27. Priya, S.; Inman, D.J. *Energy Harvesting Technologies*; Springer: New York, NY, USA, 2009; p. 517.

28. Berdy, D.F.; Srisungsitthisunti, P.; Jung, B.; Xu, X.F.; Rhoads, J.F.; Peroulis, D. Low-frequency meandering piezoelectric vibration energy harvester. *IEEE Trans. Ultrason. Ferroelectr. Freq. Control* **2012**, *59*, 846–858.

29. Shindo, Y.; Narita, F. Dynamic bending/torsion and output power of s-shaped piezoelectric energy harvesters. *Int. J. Mech. Mater. Des.* **2014**, *10*, 305–311.

30. Sobocinski, M.; Leinonen, M.; Juuti, J.; Jantunen, H. Monomorph piezoelectric wideband energy harvester integrated into LTCC. *J. Eur. Ceram. Soc.* **2011**, *31*, 789–794.

31. Park, J.C.; Park, J.Y. Asymmetric PZT bimorph cantilever for multi-dimensional ambient vibration harvesting. *Ceram. Int.* **2013**, *39*, S653–S657.

32. Hashimoto, S.; Zhang, Y.; Nagai, N.; Fujikura, Y.; Takahashi, J.; Kumagai, S.; Kasai, M.; Suto, K.; Okada, H.; Jiang, W. Multi-mode and multi-axis vibration power generation effective for vehicles. In Proceedings of the 2013 IEEE International Symposium on Industrial Electronics, Taipei, Taiwan, 28–31 May 2013.

33. Ethem Erkan Aktakka, K.N. Three-axis piezoelectric vibration energy harvester. In Proceedings of 28th IEEE International Conference on Micro Electro Mechanical Systems, Estoril, Portugal, 18–22 January 2015.

34. Khameneifar, F.; Arzanpour, S.; Moallem, M. A piezoelectric energy harvester for rotary motion applications: Design and experiments. *IEEE ASME Trans. Mechatron.* **2013**, *18*, 1527–1534.

35. Yang, Y.; Shen, Q.L.; Jin, J.M.; Wang, Y.P.; Qian, W.J.; Yuan, D.W. Rotational piezoelectric wind energy harvesting using impact-induced resonance. *Appl. Phys. Lett.* **2014**, *105*, 053901.

36. Pozzi, M.; Aung, M.S.H.; Zhu, M.L.; Jones, R.K.; Goulermas, J.Y. The pizzicato knee-joint energy harvester: Characterization with biomechanical data and the effect of backpack load. *Smart Mater. Struct.* **2012**, *21*, 076023.

37. Li, B.; You, J.H.; Kim, Y.J. Low frequency acoustic energy harvesting using PZT piezoelectric plates in a straight tube resonator. *Smart Mater. Struct.* **2013**, *22*, 055013.

38. Palosaari, J.; Leinonen, M.; Hannu, J.; Juuti, J.; Jantunen, H. Energy harvesting with a cymbal type piezoelectric transducer from low frequency compression. *J. Electroceram.* **2012**, *28*, 214–219.

39. Chen, X.R.; Yang, T.Q.; Wang, W.; Yao, X. Vibration energy harvesting with a clamped piezoelectric circular diaphragm. *Ceram. Int.* **2012**, *38*, S271–S274.

40. Zhao, S.; Erturk, A. Deterministic and band-limited stochastic energy harvesting from uniaxial excitation of a multilayer piezoelectric stack. *Sens. Actuators. A Phys.* **2014**, *214*, 58–65.

41. Xu, T.B.; Siochi, E.J.; Kang, J.H.; Zuo, L.; Zhou, W.L.; Tang, X.D.; Jiang, X.N. Energy harvesting using a PZT ceramic multilayer stack. *Smart Mater. Struct.* **2013**, *22*, 065015.

42. Makki, N.; Pop-Iliev, R. Battery- and wire-less tire pressure measurement systems (TPMS) sensor. *Microsyst. Technol.* **2012**, *18*, 1201–1212.

43. Roundy, S.; Wright, P.K. A piezoelectric vibration based generator for wireless electronics. *Smart Mater. Struct.* **2004**, *13*, 1131–1142.

44. Renaud, M.; Karakaya, K.; Sterken, T.; Fiorini, P.; Van Hoof, C.; Puers, R. Fabrication, modelling and characterization of mems piezoelectric vibration harvesters. *Sens. Actuators A Phys.* **2008**, *145*, 380–386.

45. Shen, D.; Park, J.H.; Ajitsaria, J.; Choe, S.Y.; Wikle, H.C.; Kim, D.J. The design, fabrication and evaluation of a MEMS PZT cantilever with an integrated Si proof mass for vibration energy harvesting. *J. Micromech. Microeng.* **2008**, *18*, 055017.

46. Janphuang, P.; Lockhart, R.; Uffer, N.; Briand, D.; de Rooij, N.F. Vibrational piezoelectric energy harvesters based on thinned bulk PZT sheets fabricated at the wafer level. *Sens. Actuators A Phys.* **2014**, *210*, 1–9.

47. Cui, Y.; Zhang, Q.Y.; Yao, M.L.; Dong, W.J.; Gao, S.Q. Vibration piezoelectric energy harvester with multi-beam. *AIP Adv.* **2015**, *5*, 041332.

48. Deng, L.C.; Wen, Z.Y.; Zhao, X.Q.; Yuan, C.W.; Luo, G.X.; Mo, J.K. High voltage output MEMS vibration energy harvester in d_{31} mode with PZT thin film. *J. Microelectromech. Syst.* **2014**, *23*, 855–861.

49. Liu, H.C.; Zhang, S.S.; Kobayashi, T.; Chen, T.; Lee, C. Flow sensing and energy harvesting characteristics of a wind-driven piezoelectric $Pb(Zr_{0.52}, Ti_{0.48})O_3$microcantilever. *Micro Nano Lett.* **2014**, *9*, 286–289.

50. Kim, M.; Hwang, B.; Min, N.K.; Jeong, J.; Kwon, K.H.; Park, K.B. Design and fabrication of a PZT cantilever for low frequency vibration energy harvesting. *J. Nanosci. Nanotechnol.* **2011**, *11*, 6510–6513.

51. Kuehne, I.; Schreiter, M.; Seidel, J.; Seidel, H.; Frey, A. A novel MEMS design of a piezoelectric generator for fluid-actuated energy conversion. *Proc. SPIE* **2011**, *8066*, 806617.

52. Hajati, A.; Kim, S.G. Ultra-wide bandwidth piezoelectric energy harvesting. *Appl. Phys. Lett.* **2011**, *99*, 083105.

53. Kim, M.; Hwang, B.; Jeong, J.; Min, N.K.; Kwon, K.H. Micromachining of a bimorph $Pb(Zr,Ti)O_3$ (PZT) cantilever using a micro-electromechanical systems (MEMS) process for energy harvesting application. *J. Nanosci. Nanotechnol.* **2012**, *12*, 6011–6015.

54. Kim, S.B.; Park, H.; Kim, S.H.; Wikle, H.C.; Park, J.H.; Kim, D.J. Comparison of MEMS PZT cantilevers based on d(31) and d(33) modes for vibration energy harvesting. *J. Microelectromech. Syst.* **2013**, *22*, 26–33.

55. Tang, G.; Liu, J.Q.; Yang, B.; Luo, J.B.; Liu, H.S.; Li, Y.G.; Yang, C.S.; He, D.N.; Dao, V.D.; Tanaka, K.; *et al.* Fabrication and analysis of high-performance piezoelectric MEMS generators. *J. Micromech. Microeng.* **2012**, *22*, 065017.

56. Tang, G.; Liu, J.Q.; Liu, H.S.; Li, Y.G.; Yang, C.S.; He, D.N.; DzungDao, V.; Tanaka, K.; Sugiyama, S. Piezoelectric MEMS generator based on the bulk PZT/silicon wafer bonding technique. *Phys. Status Solidi A* **2011**, *208*, 2913–2919.

57. Wang, Q.; Cao, Z.P.; Kuwano, H. Metal-based piezoelectric energy harvesters by direct deposition of PZT thick films on stainless steel. *Micro Nano Lett.* **2012**, *7*, 1158–1161.

58. Lin, S.C.; Wu, W.J. Piezoelectric micro energy harvesters based on stainless-steel substrates. *Smart Mater. Struct.* **2013**, *22*, 045016.

59. Liu, H.C.; Tay, C.J.; Quan, C.G.; Kobayashi, T.; Lee, C. Piezoelectric MEMS energy harvester for low-frequency vibrations with wideband operation range and steadily increased output power. *J. Microelectromech. Syst.* **2011**, *20,* 1131–1142.

60. Liu, H.C.; Quan, C.G.; Tay, C.J.; Kobayashi, T.; Lee, C. A MEMS-based piezoelectric cantilever patterned with PZT thin film array for harvesting energy from low frequency vibrations. *Phys. Proc.* **2011**, *19*, 129–133.

61. Liu, H.C.; Lee, C.; Kobayashi, T.; Tay, C.J.; Quan, C.G. A new S-shaped MEMS PZT cantilever for energy harvesting from low frequency vibrations below 30 Hz. *Microsyst. Technol.* **2012**, *18*, 497–506.

62. Liu, H.C.; Zhang, S.S.; Kathiresan, R.; Kobayashi, T.; Lee, C. Development of piezoelectric microcantilever flow sensor with wind-driven energy harvesting capability.*Appl. Phys. Lett.* **2012**, *100*, 223905.

63. Liu, H.C.; Lee, C.; Kobayashi, T.; Tay, C.J.; Quan, C.G. A MEMS-based wideband piezoelectric energy harvester system using mechanical stoppers. In Proceedings of the 2011 IEEE International Electron Devices Meeting, Washington, DC, USA, 5–7 December 2011.

64. Xu, G.; Ren, Z.H.; Du, P.Y.; Weng, W.J.; Shen, G.; Han, G.R. Polymer-assisted hydrothermal synthesis of single-crystalline tetragonal perovskite $PbZr_{0.52}Ti_{0.48}O_3$nanowires. *Adv. Mater.* **2005**, *17*, 907–910.

65. Zhang, X.Y.; Zhao, X.; Lai, C.W.; Wang, J.; Tang, X.G.; Dai, J.Y. Synthesis and piezoresponse of highly ordered $Pb(Zr_{0.53}Ti_{0.47})O_3$ nanowire arrays. *Appl. Phys. Lett.***2004**, *85*, 4190–4192.

66. Xu, S.Y.; Shi, Y. Power generation from piezoelectric lead zirconate titanate nanotubes. *J. Phys. D Appl. Phys.* **2009**, *42*, 085301.

67. Jung, W.S.; Do, Y.H.; Kang, M.G.; Kang, C.Y. Energy harvester using PZT nanotubes fabricated by template-assisted method. *Curr. Appl. Phys.* **2013**, *13*, S131–S134.

68. Xu, S.; Hansen, B.J.; Wang, Z.L. Piezoelectric-nanowire-enabled power source for driving wireless microelectronics. *Nat. Commun.* **2010**, *1*, 1098.

69. Wu, W.W.; Bai, S.; Yuan, M.M.; Qin, Y.; Wang, Z.L.; Jing, T. Lead zirconate titanate nanowire textile nanogenerator for wearable energy-harvesting and self-powered devices. *ACS Nano* **2012**, *6*, 6231–6235.

70. Zhou, Z.; Tang, H.X.; Sodano, H.A. Scalable synthesis of morphotropic phase boundary lead zirconium titanate nanowires for energy harvesting. *Adv. Mater.* **2014**, *26*, 7547–7554.

71. Qi, Y.; Jafferis, N.T.; Lyons, K.; Lee, C.M.; Ahmad, H.; McAlpine, M.C. Piezoelectric ribbons printed onto rubber for flexible energy conversion. *Nano Lett.* **2010**, *10*, 524–528.

72. Qi, Y.; Kim, J.; Nguyen, T.D.; Lisko, B.; Purohit, P.K.; McAlpine, M.C. Enhanced piezoelectricity and stretchability in energy harvesting devices fabricated from buckled PZT ribbons. *Nano Lett.* **2011**, *11*, 1331–1336.

73. Chen, X.; Xu, S.Y.; Yao, N.; Shi, Y. 1.6 V nanogenerator for mechanical energy harvesting using PZT nanofibers. *Nano Lett.* **2010**, *10*, 2133–2137.

74. Chen, X.; Guo, S.; Li, J.; Zhang, G.T.; Lu, M.; Shi, Y. Flexible piezoelectric nanofiber composite membranes as high performance acoustic emission sensors. *Sens. Actuators A Phys.* **2013**, *199*, 372–378.

75. Gu, L.; Cui, N.Y.; Cheng, L.; Xu, Q.; Bai, S.; Yuan, M.M.; Wu, W.W.; Liu, J.M.; Zhao, Y.; Ma, F.; *et al.* Flexible fiber nanogenerator with 209 V output voltage directly powers a light-emitting diode. *Nano Lett.* **2013**, *13*, 91–94.

76. Park, K.I.; Jeong, C.K.; Ryu, J.; Hwang, G.T.; Lee, K.J. Flexible and large-area nanocomposite generators based on lead zirconate titanate particles and carbon nanotubes. *Adv. Energy Mater.* **2013**, *3*, 1539–1544.

77. Zhang, Y.; Zhang, Y.J.; Xue, X.Y.; Cui, C.X.; He, B.; Nie, Y.X.; Deng, P.; Wang, Z.L. PVDF-PZT nanocomposite film based self-charging power cell. *Nanotechnology* **2014**, *25*, 105401.

78. Do, Y.H.; Jung, W.S.; Kang, M.G.; Kang, C.Y.; Yoon, S.J. Preparation on transparent flexible piezoelectric energy harvester based on PZT films by laser lift-off process. *Sens. Actuators A Phys.* **2013**, *200*, 51–55.

79. Park, K.I.; Son, J.H.; Hwang, G.T.; Jeong, C.K.; Ryu, J.; Koo, M.; Choi, I.; Lee, S.H.; Byun, M.; Wang, Z.L.; *et al.* Highly-efficient, flexible piezoelectric PZT thin film nanogenerator on plastic substrates. *Adv. Mater.* **2014**, *26*, 2514–2520.

80. Jeong, C.K.; Park, K.I.; Son, J.H.; Hwang, G.T.; Lee, S.H.; Park, D.Y.; Lee, H.E.; Lee, H.K.; Byun, M.; Lee, K.J. Self-powered fully-flexible light-emitting system enabled by flexible energy harvester. *Energ. Environ. Sci.* **2014**, *7*, 4035–4043.

81. Hwang, G.T.; Byun, M.; Jeong, C.K.; Lee, K.J. Flexible piezoelectric thin-film energy harvesters and nanosensors for biomedical applications. *Adv. Healthc. Mater.* **2015**, *4*, 646–658.

82. Chun, J.; Kang, N.R.; Kim, J.Y.; Noh, M.S.; Kang, C.Y.; Choi, D.; Kim, S.W.; Wang, Z.L.; Baik, J.M. Highly anisotropic power generation in piezoelectric hemispheres composed stretchable composite film for self-powered motion sensor. *Nano Energy* **2015**, *11*, 1–10.

Chapter 8

Acoustic Wave Velocity Measurement on Piezoelectric Ceramics

Toshio Ogawa[1]

[1] *Department of Electrical and Electronic Engineering, Shizuoka Institute of Science and Technology, Toyosawa, Fukuroi, Shizuoka, Japan*

1. INTRODUCTION

1.1. Why Is "Acoustic Wave Velocity Measurement" Needed?

1.1.1. Relationships between young's modulus and electromechanical coupling factor in piezoelectric ceramics and single crystals

Material research on lead-free piezoelectric ceramics has received much attention because of global environmental considerations. The key practical issue is the difficulty in realizing excellent piezoelectricity, such as electromechanical coupling factors and piezoelectric strain constants, by comparison with lead zirconate titanate [$Pb(Zr,Ti)O_3$] ceramics. The coupling factor is closely related to the degree of ferroelectric-domain alignment in the DC poling process. We have already shown the mechanism of domain alignment in $Pb(Zr,Ti)O_3$ (PZT) [1-5], $PbTiO_3$ (PT) [6], and $BaTiO_3$ (BT) [7] ceramics and in relaxor single crystals of $Pb[(Zn_{1/3}Nb_{2/3})_{0.91}Ti_{0.09}]O_3$ (PZNT91/09) [8] and $Pb[(Mg_{1/3}Nb_{2/3})_{0.74}Ti_{0.26}]O_3$ (PMNT74/26) [9] by measuring the piezoelectricity vs DC poling field characteristics. Throughout the studies, we have discovered relatively high coupling factor of transverse vibration mode k_{31} over 80 % (it called giant k_{31}) in the relaxor single-crystal plates [10,11]. Furthermore, we reported that the origin of giant k_{31} is due to the lowest Young's modulus of the single-crystal plates [11,12]. The results indicate the important relationships between Young's modulus and electromechanical coupling factor to realize high piezoelectricity.

Concrete experimental results are mentioned as follows. The Young's modulus (Y^E) of PZNT91/09 single crystals with giant k_{31} (Y^E=0.89x10^{10} N/m^2) is one order of magnitude smaller than the Y^E(6~9x10^{10} N/m^2) of PZT ceramics, and roughly speaking, one order of magnitude larger than the Y^E(0.05x10^{10} N/m^2) of rubber [12]. It was thought that the origin of giant k_{31} in PZNT91/09 and PMNT74/26 single crystals was due to the mechanical softness of the materials. Namely, the most important factor to realize high piezoelectricity is easy deformation by the DC poling field. Figure 1shows the relationships with k_{31} and k_{33} (a coupling factor of longitudinal vibration mode) vs Y^E on various kinds of single crystals and ceramics reported [13]. There is a linear relationship with k_{31} vs Y^E and k_{33} vs Y^E. Furthermore, there is a blank space between Y^E of PZNT91/09 and PMNT74/26 single crystals and Y^E of ordinary piezoelectric materials. Therefore, the important viewpoint is that new piezoelectric materials including lead-free compositions with higher coupling factor should be investigated to clarify the blank space, such as the research for elements softened the materials in lead-free ceramics.

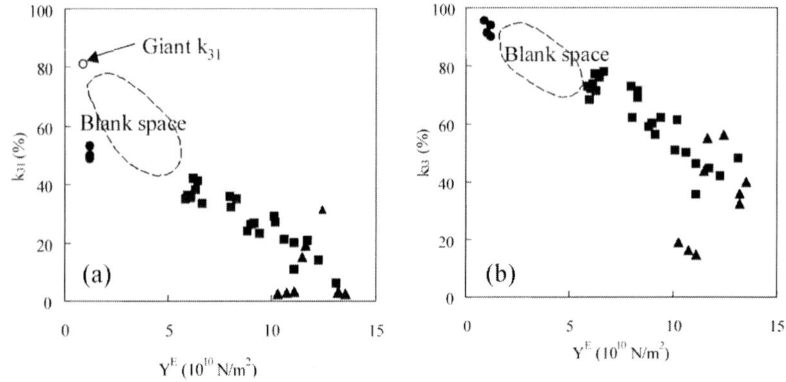

Figure 1. Relationships with coupling factors of (a) k_{31} and (b) k_{33} vs Young's modulus (Y^E) in piezoelectric materials.

1.1.2. Newly developed measurement method for acoustic wave velocities

Elastic constants of piezoelectric ceramics such as Young's modulus and Passion's ratio were basically evaluated by impedance responses on frequency of piezoelectric resonators with various relationships between DC poling directions and vibration modes [14-17]. Therefore, it needs all kinds of resonators with different shapes [14], and furthermore, with uniform poling-degree in spite of different poling-thicknesses such as 1.0 mm for plate resonators (for example, dimensions of 12 mm length, 3.0 mm width and 1.0 mm poling-thickness) and 15 mm for rod resonators (dimensions of 6.0 mm diameter and 15 mm poling-thickness). While applying a DC poling field of 3.0 kV/mm, 3.0 kV DC voltage must be applied to the former sample and 45 kV for the later sample. There was no guarantee to obtain uniform poling-degree between two samples even through the same poling field of 3.0 kV/mm. In addition, it is difficult to measure in the cases of as-fired ceramics, namely ceramics before poling, or weak polarized ceramics because of none or weak impedance responses on frequency.

On the other hand, there was a well-known method to measure the elastic constants by pulse echo measurement [18,19]. Acoustic wave velocities in ceramics were directly measured by this method. However, since the frequency of pulse oscillation for measurement was generally below 5 MHZ, it needs to prepare the rod samples with thickness of 10-20 mm in order to guarantee accuracy of the velocities. Moreover, it is unsuitable for measuring samples such as disks with ordinary dimensions (10-15 mm diameter and 1.0-1.5 mm thickness) and with many different compositions for piezoelectric materials R & D.

We developed a method to be easy to measure acoustic wave velocities suitable for the above mentioned disk samples by an ultrasonic thickness gauge with high-frequency pulse oscillation. This method was applied to hard and soft PZT [3], lead titanate [6] (both lead-containing ceramics) and lead-free ceramics composed of alkali niobate [20,21] and alkali bismuth titanate [22]. In this chapter, first of all, we report on the measurement of acoustic wave velocities in PZT ceramics, and the following, the calculation results of Young's modulus and Poisson's ratio, especially to obtain high piezoelectricity in lead-free ceramics.

For the measurement, the ultrasonic precision thickness gauge (Olympus Co., Model 35DL) has PZT transducers with 30 MHz for longitudinal-wave generation and 20 MHz for transverse-wave generation. The acoustic wave velocities were evaluated by the propagation time between second pulse echoes in thickness of ceramic disks with dimensions of 14 mm in diameter and 0.5-1.5 mm in thickness. The sample thickness was measured by a precision micrometer (Mitutoyo Co., Model MDE-25PJ). Piezoelectric ceramic compositions measured were as follows: $0.05Pb(Sn_{0.5}Sb_{0.5})O_3$-(0.95-x)$PbTiO_3$-x$PbZrO_3$ (x=0.33, 0.45, 0.48, 0.66, 0.75) with (hard PZT) and without 0.4 wt% MnO_2(soft PZT) [3]; (1-x)(Na,K,Li,Ba)$(Nb_{0.9}Ta_{0.1})O_3$-x$SrZrO_3$(SZ) (x=0, 0.02, 0.04, 0.05, 0.06, 0.07) [20,21]; (1-x)$(Na_{0.5}Bi_{0.5})TiO_3$(NBT)-x$(K_{0.5}Bi_{0.5})TiO_3$ (KBT) (x=0.08, 0.11, 0.15, 0.18, 0.21, 0.28) and 0.79NBT-0.20KBT-0.01$Bi(Fe_{0.5}Ti_{10.5})O_3$(BFT) [22]; and (1-x)NBT-x$BaTiO_3$(BT) (x=0.03, 0.07, 0.11) [22]. In addition to evaluate ceramic compositions, we investigated ceramics with different ceramic manufacturing processes such as firing processes of normal firing in air atmosphere and oxygen atmosphere firing to realize pore-free ceramics [23], and DC poling processes of as-fired (before poling), weakly and fully polarized ceramics.

2. YOUNG'S MODULUS AND POISSON'S RATIO IN PIEZOELECTRIC CERAMICS

2.1. Longitudinal and Transverse Wave Velocities by Pulse Echo Measurement

Figure 2 shows pulse echoes of longitudinal acoustic wave in hard PZT ceramics at a composition of x=0.48. The longitudinal wave velocity was calculated from the propagation time between second-pulse echoes (␣) and the thickness of ceramic disks. The dependences of the longitudinal and transverse wave velocities on the composition of x in hard and soft PZT ceramics and their fluctuation of the individual ceramic disks were shown in Figs. 3(a)-(d). The fluctuation of the velocities in transverse wave was smaller than the ones in longitudinal wave when the samples of n=16-21 pieces were measured at each composition. In addition, it was clarified that the fluctuation in soft PZT was smaller than the one in hard PZT ceramics because of easy alignment of ferroelectric domains by poling field in soft PZT, namely lower coercive fields than the ones of hard PZT.

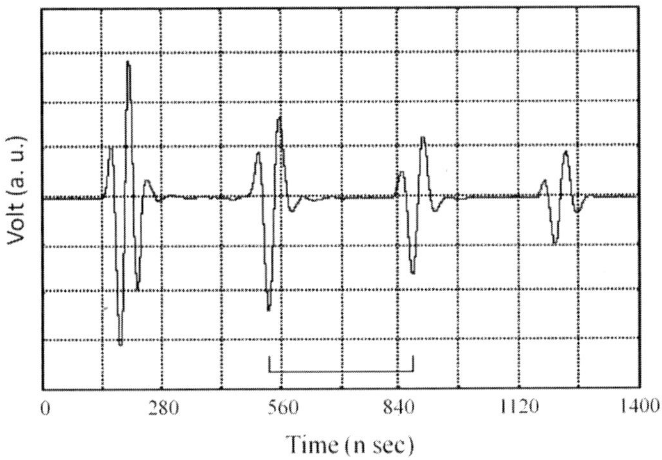

Figure 2. Pulse echoes of longitudinal acoustic wave in hard PZT ceramics (disk dimensions: 13.66 mm diameter and 0.735 mm thickness) before poling at x=0.48; the calculated wave velocity is 4,319 m/s.

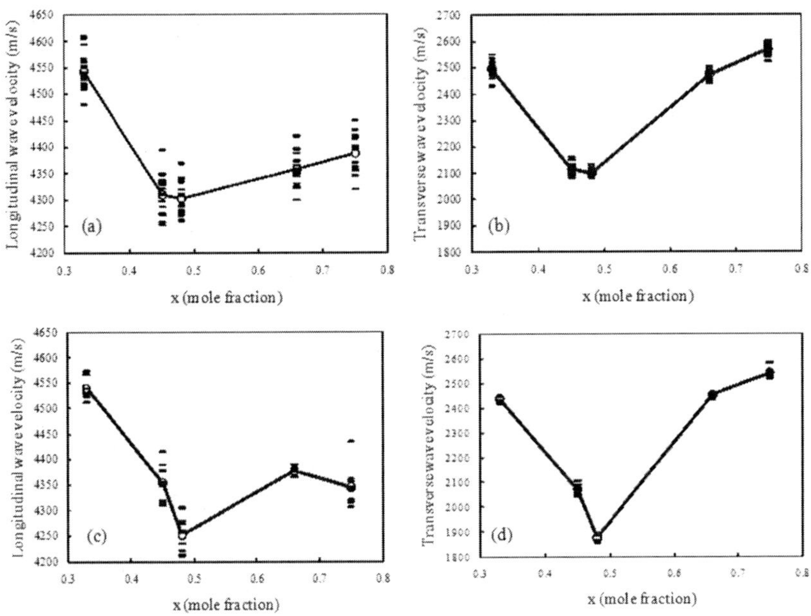

Figure 3. Composition x dependence of longitudinal and transverse wave velocities in hard [(a) and (b)] and soft PZT [(c) and (d)] ceramics before poling; all the samples of n=16-21 are shown in the figure to evaluate their fluctuation.

2.2. Young's modulus and poisson's ratio in pzt ceramics

Figures 4(a)-(d) show the dependences of the longitudinal and transverse wave velocities, Young's modulus and Poisson's ratio on the composition of x in hard and soft PZT ceramics, respectively. Here, the elastic constants were calculated from equations in the following session 3.1. From Fig. 4, it was found that the large differences in these values between hard and soft PZT ceramics only occurred around a composition of x=0.48, which corresponds to a morphotropic phase boundary (MPB) [23].

In the case of the evaluation on ceramic manufacturing processes, the values of pore-free ceramics fired in oxygen atmosphere also show in the figures at compositions of x=0.66 and 0.75 (in dotted circles, the samples of n=2). The effect of reducing pores in ceramics on the values works like the increase of the longitudinal wave velocity and Poisson's ratio, and the decrease of the transverse wave velocity and Young's modulus.

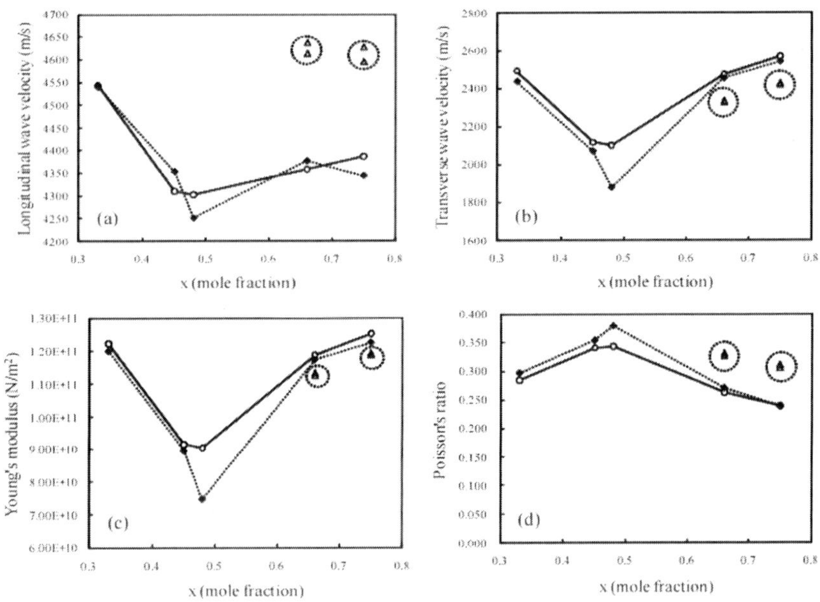

Figure 4. Composition x dependence of (a) longitudinal and (b) transverse wave velocities, (c) Young's modulus and (d) Poisson's ratio in hard (—) and soft (---) PZT ceramics before poling; the values of pore-free hard PZT ceramics (x=0.66, 0.75 and the samples of n=2) were shown in dotted circles.

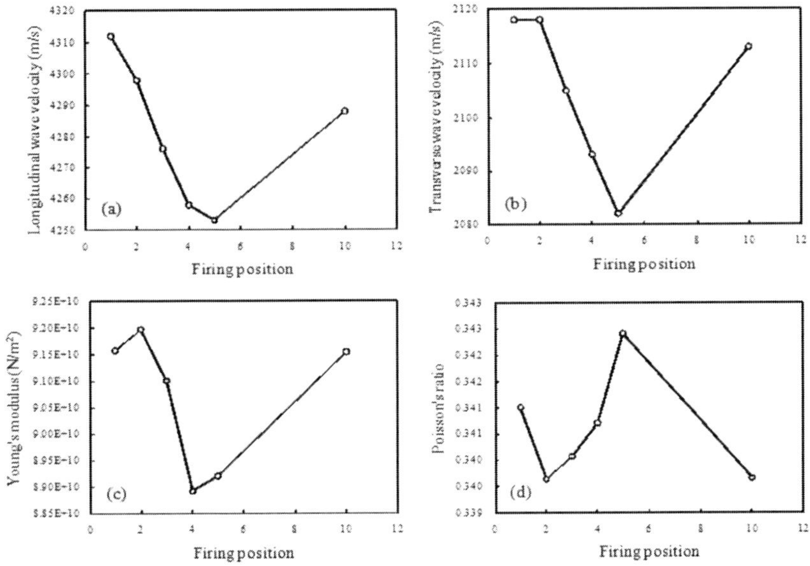

Figure 5. Firing position (Nos. 1 and 10 correspond to ceramic disks of top piled up and bottom piled up in sagger) dependence of (a) longitudinal and (b) transverse wave velocities, (c) Young's modulus and (d) Poisson's ratio in hard PZT ceramics at x=0.45.

As mentioned later in the session 4, the decrease in Young's modulus and the increase in Poisson's ratio correspond to improve the piezoelectricity. Therefore, it was confirmed by using the pulse echo measurement that oxygen atmosphere firing is an effective tool to improve piezoelectric properties as well as production of pore-free ceramics. Figures 5(a)-(d) show the dependences of the longitudinal and transverse wave velocities, Young's modulus and Poisson's ratio on the firing position in sagger at a composition of x=0.45 in hard PZT ceramics, respectively. Nos. 1 and 10 in the figure correspond to the top position piled up and the bottom position piled up in sagger (Fig. 6). It became clear that there was fluctuation of the velocities and elastic constants in the case of as-fired (before poling) samples. In addition, the sample of the middle position piled up of No. 5 possesses low wave velocities, low Young's modulus and high Poisson's ratio because of the firing under high lead oxide (PbO) atmosphere. On the other hand, the samples of the top and bottom positions of Nos. 1, 2 and 10 possess high wave velocities, high Young's modulus and low Poisson's ratio because of the firing under low PbO atmosphere due to vaporization of PbO during firing (see Fig. 6). Young's modulus and Poisson's ratio in PZT ceramics of No. 5 could be improved the piezoelectricity (see the session of 4) by high PbO atmosphere firing. However, there is a different tendency to decrease the longitudinal wave velocity [No.5 position in Fig. 5(a)] in comparison with the case of oxygen atmosphere firing [dotted circles at x=0.66, 0.75 in Fig. 4(a)], which also can be improved the elastic constants.

We believe that the increase in longitudinal wave velocity [dotted circles at x=0.66, 0.75 in Fig. 4(a)] comes from pore-free microstructure, and further, higher dense and higher Passion's ratio ceramics compared with in the case of normal firing in air atmosphere. Furthermore, it is thought that the fluctuation of as-fired ceramics connected to the fluctuation of ceramics after DC poling.

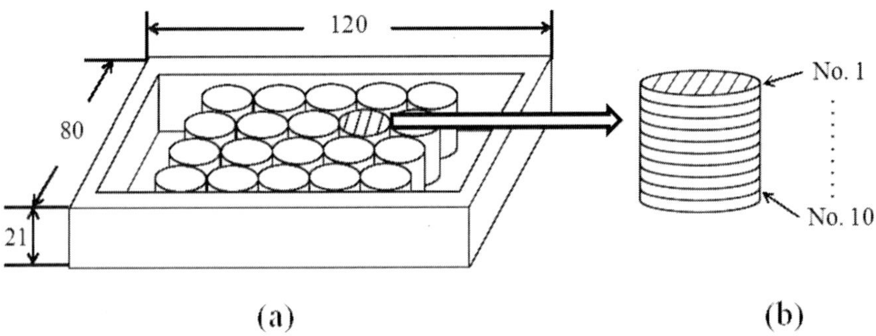

Figure 6. Schematic pictures of (a) PZT green (as-pressed) disks piled up in dense Al_2O_3 sagger (container dimensions: 120 mm length, 80 mm width and 21 mm height) and (b) firing positions (Nos. 1-10) of ceramic green disks (16 mm diameter and 1.2 mm thickness); the ceramic green disk of No. 5 is surrounded with high PbO atmosphere by the other disks piled up, and it is understood that PbO vaporization is easy to occur at the disks of Nos.1, 2 (top) and No.10 (bottom) even through a cover (dense Al_2O_3 ceramic plate) is put on the container while firing [23].

2.3. Young's Modulus and Poisson's Ratio In Lead-Free Ceramics

Figures 7(a)-(d) show the dependences of the longitudinal and transverse wave velocities, Young's modulus and Poisson's ratio on the composition of lead-free ceramics after poling; the compositions were (1-x)(Na,K,Li,Ba)($Nb_{0.9}Ta_{0.1}$)O_3-xSZ (x=0, 0.02, 0.04, 0.05, 0.06, 0.07), (1-x)NBT-xKBT (x=0.08, 0.11, 0.15, 0.18, 0.21, 0.28), 0.79NBT-0.20KBT-0.01BFT and (1-x)NBT-xBT (x=0.03, 0.07, 0.11), respectively. While the compositions to obtain high electro-mechanical coupling factor in (Na,K,Li,Ba)($Nb_{0.9}Ta_{0.1}$)O_3-SZ (x=0.05-0.06) and NBT-KBT (x=0.18) [21,22] correspond to the compositions with low Young's modulus and high Poisson's ratio as well as PZT (x=0.48) [3]. Although a MPB existed around x=0.18 in the phase diagram of NBT-KBT [24,25], a MPB did not exist in the phase diagram of (Na,K,Li,Ba)($Nb_{0.9}Ta_{0.1}$)O_3-SZ [20,21]. On the other hand, the compositions with high coupling factor in NBT-BT (x=0.07) [22] possessed low Young's modulus and low Poisson's ratio such as $PbTiO_3$ [6]. In the phase diagram of NBT-BT, it is confirmed that there is a MPB near x=0.07 [26]. Furthermore, there was no significant difference in Young's modulus and Poisson's ratio between as-fired (△) (before poling) and after poling (▲) in (Na,K,Li,Ba)($Nb_{0.9}Ta_{0.1}$)O_3-SZ.

From the above mentioned viewpoint of the study on the elastic constants in lead-free ceramic, it was clarified that there are compositions with high coupling factor in the cases of (1) low Young's modulus and high Poisson's ratio (PZT type) and (2) low Young's modulus and low Poisson's ratio ($PbTiO_3$type). Namely, it was found that there are two kinds of MPB compositions, (1) and (2) in lead-free ceramics with high coupling factor. Therefore, lead-free ceramic compositions with high piezoelectricity must be primarily focused on to realize the compositions with low Young's modulus such as MPB compositions [27]. Secondarily, the importance to measure Poisson's ratio was understood for the research to recognize the compositions of (1) and (2) in lead-free ceramics with high piezoelectricity.

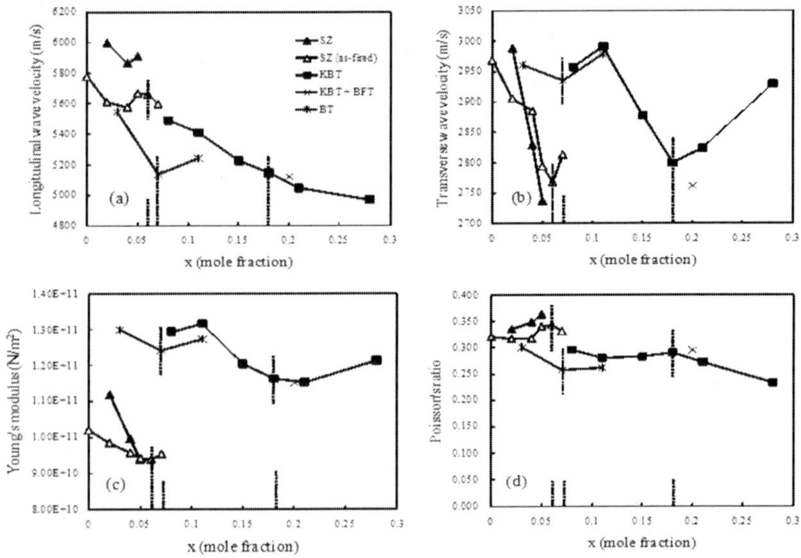

Figure 7. Composition x dependence of (a) longitudinal and (b) transverse wave velocities, (c) Young's modulus and (d) Poisson's ratio in lead-free ceramics of $(1\text{-}x)(Na,K,Li,Ba)(Nb_{0.9}Ta_{0.1})O_3\text{-}xSZ$ (x=0, 0.02, 0.04, 0.05, 0.06, 0.07), (1-x)NBT-xKBT (x=0.08, 0.11, 0.15, 0.18, 0.21, 0.28), 0.79NBT-0.20KBT-0.01BFT and (1-x)NBT-xBT (x=0.03, 0.07, 0.11).

3. POLING FIELD DEPENDENCE OF LONGITUDINAL AND TRANSVERSE WAVE VELOCITIES, YOUNG'S MODULUS AND POISSON'S RATIO IN PIEZOELECTRIC CERAMICS

3.1. Dc Poling Field Dependence of Elastic Constants

As mentioned previously in the session 1, the behavior of ferroelectric domains toward a DC field was investigated on the basis of the DC poling field dependence of dielectric and piezoelectric properties in various types of piezoelectric ceramics and single crystals. In addition, we have developed a method to evaluate elastic constants, such as Young's modulus and Passion's ratio, by measuring longitudinal and transverse wave velocities using an ultrasonic thickness gauge with high-frequency pulse oscillation in comparison with a conventional method as described in the session 2. The acoustic wave velocities can be measured by this method in the cases of ceramics and single crystals despite the strength of the DC poling field, including as-fired ceramics and as-grown single crystals.

Therefore, in order to clarify the relationships between elastic constants and electrical properties vs DC poling fields, we studied the poling field dependence of acoustic wave velocities and dielectric and piezoelectric properties in ceramics. Here, we report the relationships between DC poling fields, acoustic wave velocities, Young's modulus, and Passion's ratio to realize a high piezoelectricity, especially in lead-free ceramics.

The piezoelectric ceramic compositions measured were as follows: $0.05Pb(Sn_{0.5}Sb_{0.5})O_3$-$(0.95-x)PbTiO_3$-$xPbZrO_3$ (x=0.33, 0.45, 0.66) with (hard PZT) and without 0.4 wt% MnO_2 (soft PZT); $0.90PbTiO_3$-$0.10La_{2/3}TiO_3$(PLT) and $0.975PbTiO_3$-$0.025La_{2/3}TiO_3$(PT); (1-x)(Na,K,Li,Ba)$(Nb_{0.9}Ta_{0.1})$ O_3-x$SrZrO_3$(SZ) (x=0.00, 0.05, 0.07); (1-x)$(Na_{0.5}Bi_{0.5})TiO_3$(NBT)-x$(K_{0.5}Bi_{0.5})TiO_3$(KBT) (x=0.08, 0.18) and 0.79NBT-0.20KBT-0.01$Bi(Fe_{0.5}Ti_{10.5})O_3$(BFT) (x= 0.20); and (1-x)NBT-x$BaTiO_3$(BT) (x=0.03, 0.07, 0.11).

DC poling was conducted for 30 min at the most suitable poling temperature depending on the Curie points of the ceramic materials when the poling field (E) was varied from 0→+0.25→+0.5→+0.75→ +1.0→ --- $+E_{max}$→0→$-E_{max}$→0 to $+E_{max}$ kV/mm. $\pm E_{max}$ depended on the coercive fields of the piezoelectric ceramics. After each poling process, the dielectric and piezoelectric properties were measured at room temperature using an LCR meter (HP4263A), a precision impedance analyzer (Agilent 4294A), and a piezo d_{33} meter (Academia Sinica ZJ-3D). Furthermore, the acoustic wave velocities were measured using an ultrasonic precision thickness gauge (Olympus 35DL), which has PZT transducers with 30 MHz for longitudinal-wave generation and 20 MHz for transverse-wave generation. The acoustic wave velocities were evaluated on the basis of the propagation time between the second-pulse echoes in the thickness of ceramic disks parallel to the poling field with dimensions of 14 mm diameter and 0.5-1.5 mm thickness. The sample thickness was measured using a precision micrometer (Mitutoyo MDE-25PJ). Moreover, Young's modulus in the thickness direction of ceramic disks (Y_{33}^E) and Passion's ratio (σ) were calculated on the basis of the longitudinal (V_L) and transverse (V_S) wave velocities as shown in the following equations:

$$Y_{33}{}^{E}=3\rho V_S{}^2\frac{V_L{}^2-\frac{4}{3}V_S{}^2}{V_L{}^2-V_S{}^2} \text{ and } \sigma=\frac{1}{2}\left\{1-\frac{1}{\left(\frac{V_L}{V_S}\right)^2-1}\right\},$$

where ρ is the density of ceramic disks.

3.2. Poling Field Dependence In Pzt Ceramics

Figures 8-10 show the poling field dependence of longitudinal (V_L) and transverse (V_S) wave velocities, Young's modulus ($Y_{33}{}^{E}$), and Passion's ratio (σ) in $0.05Pb(Sn_{0.5}Sb_{0.5})O_3$-$(0.95\text{-}x)PbTiO_3$-$xPbZrO_3$ (x=0.33, 0.45, 0.66) with (hard PZT) and without 0.4 wt% MnO_2 (soft PZT) ceramics at a poling temperature (T_P) of 80 °C (hard PZT in Fig. 8 and soft PZT in Fig. 9), $0.90PbTiO_3$-$0.10La_{2/3}TiO_3$ (abbreviate to PLT/T_p=80 °C) and $0.975PbTiO_3$-$0.025La_{2/3}TiO_3$(abbreviate to PLT /T_p=200 °C) ceramics (Fig. 10), respectively. While the poling field dependence of V_L has almost same tendency in spite of hard and soft PZT, the one of V_S at x=0.45, which corresponds to MPB, abruptly decreases in the both cases of hard and soft PZT (Figs. 8, 9). Since the highest coupling factor in PZT is obtained at the MPB (x=0.45), the origin of the highest piezoelectricity is due to the decrease in V_S with increasing the domain alignment by DC poling field. Furthermore, the lowest $Y_{33}{}^{E}$ and the highest σ are realized at the MPB. The change in V_L, V_S, $Y_{33}{}^{E}$and σ vs E of soft PZT is smaller than the one of hard PZT because of the softness of the materials. As mentioned details to the next session 3.3, it was indicated that minimum V_L, σ and maximum V_S, $Y_{33}{}^{E}$ were obtained at the domain clamping such as the domain alignment canceled each other (↑↓), at which the lowest piezoelectricity is realized. On the other hand, the values of V_L, V_S, $Y_{33}{}^{E}$ and σ vs E in PT are smaller than the ones of PLT. Moreover, at the domain clamping fields, minimum V_L, σ, $Y_{33}{}^{E}$ and maximum V_S were obtained in both the PLT and PT. The reason of minimum $Y_{33}{}^{E}$ at the DC field of the domain clamping will be discuss in the next session. In addition, the change in $Y_{33}{}^{E}$ of PLT and PT while applying ±E is smaller than the one of PZT, and higher $Y_{33}{}^{E}$ and lower σ appear in comparison with the ones of PZT, because these come from the hardness of PLT and PT ceramics.

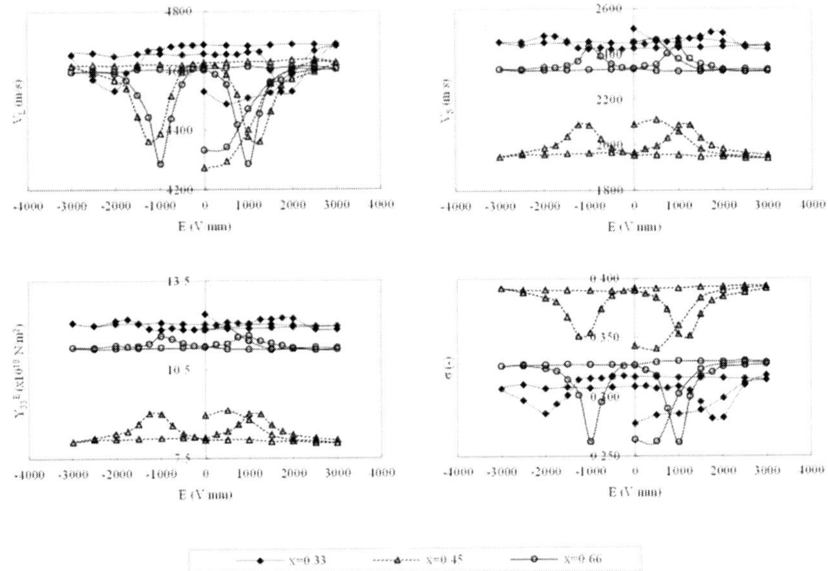

Figure 8. Poling field dependence of longitudinal (V_L) and transverse (V_S) wave velocities, Young's modulus (Y_{33}^E), and Poisson's ratio (σ) in $0.05Pb(Sn_{0.5}Sb_{0.5})O_3$-$(0.95-x)PbTiO_3$-$xPbZrO_3$ (x=0.33, 0.45, 0.66) with 0.4 wt% MnO_2 (hard PZT) ceramics.

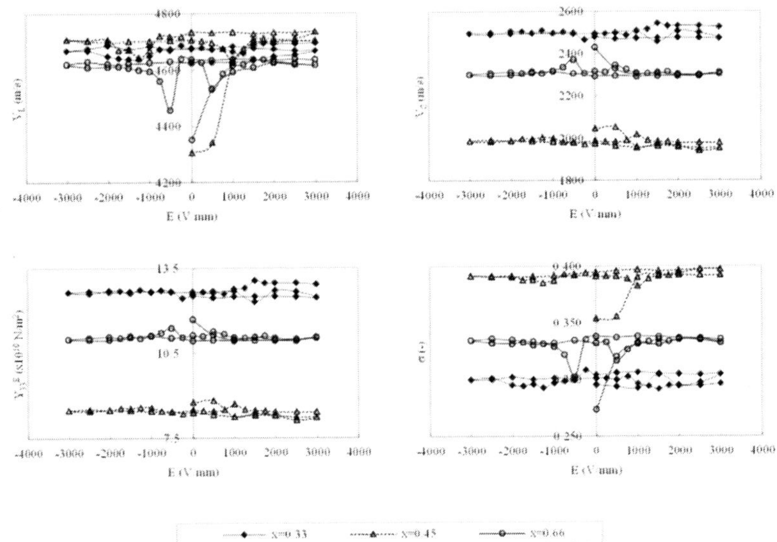

Figure 9. Poling field dependence of longitudinal (V_L) and transverse (V_S) wave velocities, Young's modulus (Y_{33}^E), and Poisson's ratio (σ) in $0.05Pb(Sn_{0.5}Sb_{0.5})O_3$-$(0.95-x)PbTiO_3$-$xPbZrO_3$ (x=0.33, 0.45, 0.66) without 0.4 wt% MnO_2 (soft PZT) ceramics.

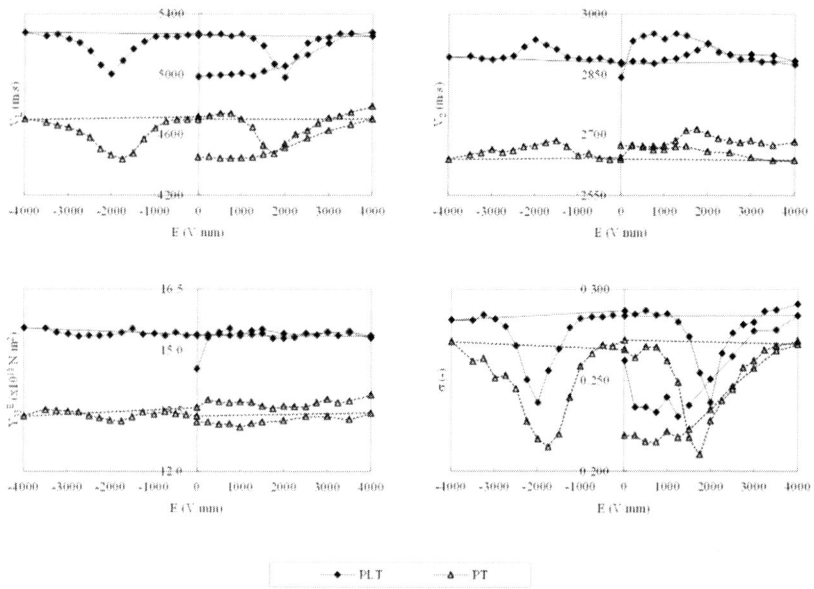

Figure 10. Poling field dependence of longitudinal (V_L) and transverse (V_S) wave velocities, Young's modulus (Y_{33}^E), and Poisson's ratio (σ) in $0.90PbTiO_3$-$0.10La_{2/3}TiO_3$ (PLT) and $0.975PbTiO_3$-$0.025La_{2/3}TiO_3$ (PLT) ceramics.

3.3. Poling field dependence in lead-free ceramics

Figures 11-13 show the poling field dependence of longitudinal (V_L) and transverse (V_S) wave velocities, Young's modulus (Y_{33}^E), and Passion's ratio (σ) in (1-x)(Na,K,Li,Ba)($Nb_{0.9}Ta_{0.1}$)O_3-x$SrZrO_3$(SZ) (x=0.00, 0.05, 0.07) ceramics at a poling temperature (T_P) of 150 °C (Fig. 11), in (1-x)($Na_{0.5}Bi_{0.5}$)TiO_3(NBT)-x($K_{0.5}Bi_{0.5}$)TiO_3(KBT) (x=0.08, 0.18/ T_p=70 °C) and 0.79NBT-0.20KBT-0.01Bi($Fe_{0.5}Ti_{10.5}$)O_3(BFT) (x=0.20/ T_p=70 °C) ceramics (Fig. 12), and in (1-x)NBT-x$BaTiO_3$(BT) (x=0.03, 0.07, 0.11/ T_p=70 °C) ceramics (Fig. 13), respectively. With the enhancement of domain alignment with an increase in poling field from E=0 to +E_{max}, V_L increased and V_S decreased independently of the ceramic composition. From the composition dependence of Y_{33}^E and σ, high piezoelectricity [high planar coupling factor (k_p)] compositions show lower Y_{33}^E and higher σ values at 0.95(Na,K,Li,Ba)($Nb_{0.9}Ta_{0.1}$)O_3-0.05SZ (k_p=46%) (Fig. 11) and 0.82NBT-0.18KBT (k_p=27%) (Fig. 12) as well as the ones at 0.05Pb($Sn_{0.5}Sb_{0.5}$)O_3-0.47$PbTiO_3$-0.48$PbZrO_3$ (k_p=65% in soft ceramics and k_p=52% in hard ceramics) than the other compositions. Although morphotropic phase boundaries (MPBs) were observed in the NBT-KBT [24,25] and PZT [3,23,28] ceramics, there was no evidence of the existence of MPBs in the $(Na,K,Li,Ba)(Nb_{0.9}Ta_{0.1})O_3$-SZ ceramics [20,21]. The effects of 0.01BFT modification in NBT-KBT on Y_{33}^E and σ were as follows: the composition of 0.79NBT-0.20KBT-0.01BFT (k_p=22%; x=0.20 in Fig. 12) showed the highest piezoelectric d_{33}constant of 150 pC/N in NBT-KBT and higher Y_{33}^E and lower σ values than that of 0.82NBT-0.18KBT (k_p=27%; x=0.18 in Fig. 12) because the relative dielectric constant increased from 800 (0.82NBT-0.18KBT) to 1250 (0.79NBT-0.20KBT-0.01BFT) [22]. Moreover, a high-coupling-factor composition at 0.93NBT-0.07BT (k_p=16%) existed in a MPB [26] with a low Y_{33}^E (Fig. 13).

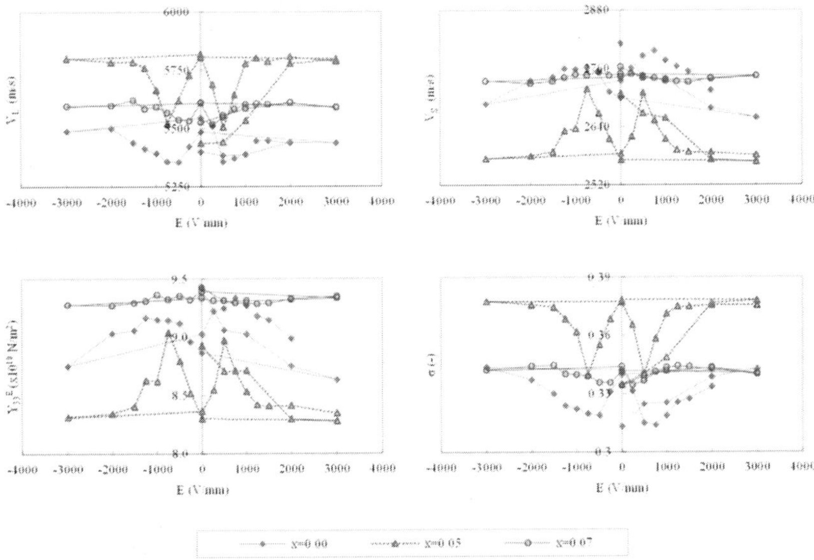

Figure 11. Poling field dependence of longitudinal (V_L) and transverse (V_S) wave velocities, Young's modulus (Y_{33}^E), and Poisson's ratio (σ) in $(1-x)(Na,K,Li,Ba)(Nb_{0.9}Ta_{0.1})O_3$-$xSrZrO_3$ (x=0.00, 0.05, 0.07) ceramics.

Basically, research on piezoelectric ceramics with high planar coupling factors, especially in lead-free ceramics, has been focused on determining the MPB composition because many different polarization axes are generated in MPB.

Through our study on acoustic wave measurement, it was found that there was an important factor for obtaining a high piezoelectricity regarding Y_{33}^{E} and σ; lower Y_{33}^{E} and higher σ values in the cases of PZT, alkali bismuth titanate [NBT-KBT], and alkali bismuth barium titanate [NBT-BT] with MPB, and in the case of alkali niobate [(Na,K,Li,Ba)($Nb_{0.9}Ta_{0.1}$)O_3-SZ] without MPB. Therefore, as mentioned later, we need a new concept in addition to conventional research on MPB compositions for developing piezoelectric ceramics with high coupling factors.

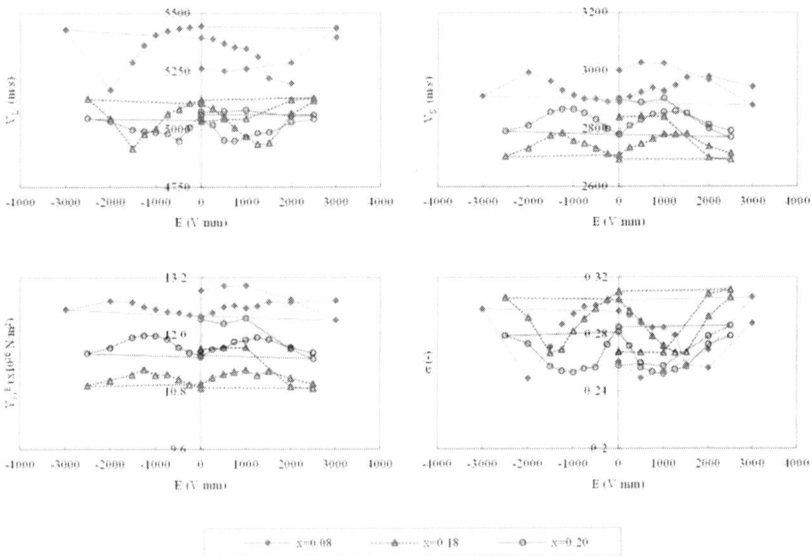

Figure 12. Poling field dependence of longitudinal (V_L) and transverse (V_S) wave velocities, Young's modulus (Y_{33}^{E}), and Poisson's ratio (σ) in (1-x)NBT-xKBT (x=0.08, 0.18) and 0.79NBT-0.2KBT-0.01BFT (x=0.20) ceramics.

The DC poling fields under domain clamping (E_d) can be estimated on the basis of E values to realize the minimum V_L and maximum V_S in Figs. 11-13. Although the maximum Y_{33}^{E} and minimum σ were obtained at E_d, which corresponds to the E of the minimum coupling factor in (Na,K,Li,Ba)($Nb_{0.9}Ta_{0.1}$)O_3-SZ (Fig. 11), NBT-KBT, and 0.79NBT-0.20KBT-0.01BFT (Fig. 12), the minimum Y_{33}^{E} and minimum σ were obtained at

E_d in NBT-BT (Fig. 13). It was clarified that the minimum Y_{33}^E in NBT-BT was realized in the cases of $\Delta V_S/\Delta V_L<1/4$ at E_d, where ΔV_S and ΔV_L denote the variations in V_S and V_L at E_d, respectively. This may be due to the poor domain alignment (lower Poisson's ratio) perpendicular to the poling field (radial direction in disk ceramics) in comparison with the domain alignment parallel to the poling field (thickness direction in disk ceramic). Furthermore, the increase in coupling factor corresponds to the increase in σ at all compositions including lead-containing and lead-free ceramic compositions.

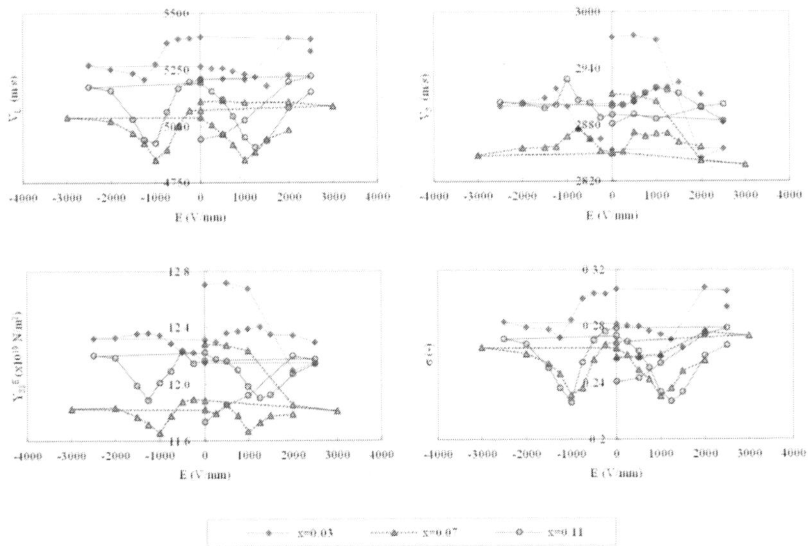

Figure 13. Poling field dependence of longitudinal (V_L) and transverse (V_S) wave velocities, Young's modulus (Y_{33}^E), and Poisson's ratio (σ) in (1-x)NBT-xBT (x=0.03, 0.07, 0.11) ceramics.

4. MATERIALS ROAD MAPS IN PIEZOELECTRIC CERAMICS ON ELASTIC CONSTANTS

Figure 14 shows the relationships between longitudinal (V_L) and transverse (V_S) wave velocities, Young's modulus (Y_{33}^E), and Passion's (σ) ratio vs planar coupling factors (k_p) in (1-x)(Na,K,Li,Ba)($Nb_{0.9}Ta_{0.1}$)O_3-xSZ (abbreviated to "SZ"), (1-x)NBT-xKBT ("KBT"), 0.79NBT- 0.20KBT-0.01BFT ("KBT"), and (1-x)NBT-xBT ("BT") lead-free ceramics compared with 0.05Pb($Sn_{0.5}Sb_{0.5}$)O_3-(0.95-x)$PbTiO_3$-x$PbZrO_3$ ceramics with ("hard PZT")

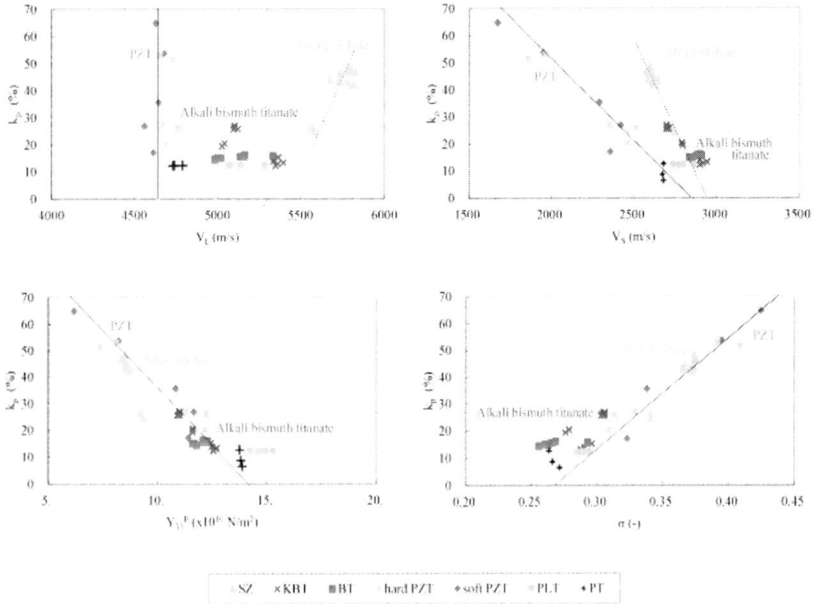

Figure 14. Relationships between longitudinal (V_L) and transverse (V_S) wave velocities, Young's modulus (Y_{33}^E), and Passion ratio (σ) vs planar coupling factors (k_p) in (1-x)(Na,K,Li,Ba) $(Nb_{0.9}Ta_{0.1})O_3$-xSZ (abbreviated to "SZ"), (1-x)NBT-xKBT ("KBT"), 0.79NBT-0.20KBT-0.01BFT ("KBT"), and (1-x)NBT-xBT ("BT") lead-free ceramics compared with $0.05Pb(Sn_{0.5}Sb_{0.5})O_3$-(0.95-x) $PbTiO_3$-$xPbZrO_3$ ceramics with ("hard PZT") and without 0.4 wt% MnO_2 ("soft PZT"), and with $0.90PbTiO_3$-$0.10La_{2/3}TiO_3$ ("PLT") and $0.975PbTiO_3$-$0.025La_{2/3}TiO_3$ ("PT") lead-containing ceramics after full DC poling.

and without 0.4 wt% MnO_2 ("soft PZT"), and with $0.90PbTiO_3$-$0.10La_{2/3}TiO_3$ ("PLT") and $0.975PbTiO_3$-0.025 $La_{2/3}TiO_3$ ("PT") lead-containing ceramics after full DC poling. Although the V_L values of the PZT ceramics were almost constant at approximately 4,600-4,800 m/s independently of the composition x, their V_S values linearly decreased from 2,500 to 1,600 m/s with increasing k_p from 20 to 65% (solid lines). In addition, the V_L and $V_S$values of the PZT ceramics were smaller than those of the lead-free ceramics (V_L=5,000-5,800 m/s and V_S=2,600-3,000 m/s; dashed and dotted lines). Although the V_L values of the PT ceramics were almost the same (4,800 m/s) as those of the PZT ceramics, the V_S values of the PT ceramics were approximately 2,700 m/s. On the other hand, the V_L values of the SZ ceramics were relatively high (5,500-5,800 m/s); furthermore, the V_S values of the SZ ceramics also increased (2,600-2,700 m/s) and linearly decreased with increasing k_p from 25 to 50% (dashed lines), the behavior of which was almost the same as that of the V_S values of the PZT ceramics. The V_L values of the KBT, BT, and PLT ceramics (5,000-5,400 m/s) were between those of the PZT, PT, and SZ ceramics. However, the V_S values of the KBT, BT, and PLT ceramics (2,800-3,000 m/s) were the highest. Therefore, it was possible to divide V_L and V_S into three material groups, namely, PZT and PT/ KBT, BT (alkali bismuth titanate), and PLT/ SZ (alkali niobate). In addition, k_p lineally increased from 4 to 65% with decreasing Y_{33}^E from 15×10^{10} to 6×10^{10} N/m^2 and lineally increased with increasing σ from 0.25 to 0.43. It was clarified that higher k_p values can be realized at lower Y_{33}^E and higher σ values.

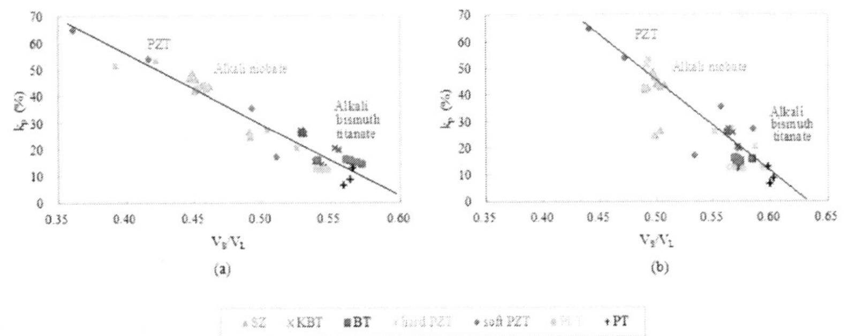

Figure 15. Ratio of longitudinal wave velocity (V_L) to transverse wave velocity (V_S): V_S/V_L vs k_p in lead-containing ceramics [$0.05Pb(Sn_{0.5}Sb_{0.5})O_3$-(0.95-x)$PbTiO_3$-x$PbZrO_3$ with ("hard PZT") and without 0.4 wt% MnO_2 ("soft PZT"), $0.90PbTiO_3$-$0.10La_{2/3}TiO_3$ ("PLT"), and $0.975PbTiO_3$-$0.025La_{2/3}TiO_3$ ("PT")] and in lead-free ceramics [(1-x)(Na,K,Li,Ba)$(Nb_{0.9}Ta_{0.1})O_3$-xSZ (abbreviated to "SZ"), (1-x)NBT-xKBT ("KBT"), 0.79NBT-0.20KBT-0.01BFT ("KBT"), and (1-x)NBT-xBT ("BT")] (a) after and (b) before full DC poling.

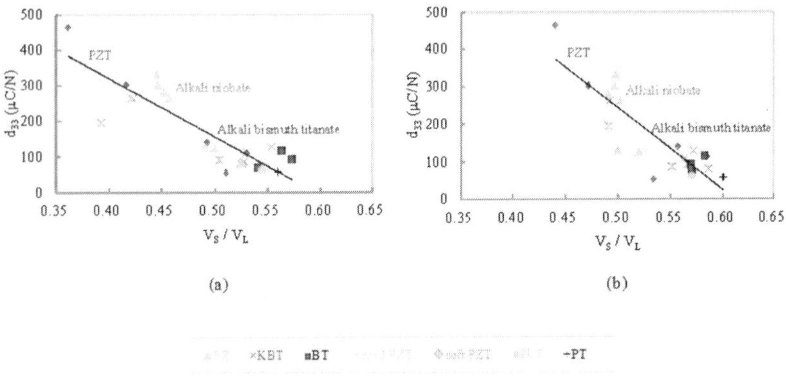

Figure 16. Ratio of longitudinal wave velocity (V_L) to transverse wave velocity (V_S): V_S/V_L vs piezoelectric strain constant d_{33} in lead-containing ceramics ["hard PZT", "soft PZT", "PLT" and "PT"] and in lead-free ceramics ["SZ", "KBT" and "BT"] (a) after and (b) before full DC poling.

Figures 15 and 16 show the ratio of V_L to V_S vs k_p and V_S/V_L vs piezoelectric strain constant d_{33} in lead-containing and lead-free ceramics (a) after and (b) before full DC poling, respectively. There are linear relationships between V_S/V_L vs k_p and d_{33} in spite of DC poling treatment; with decreasing V_S/V_L from 0.58 to 0.36, k_p increased from 4 to 65% (correlation coefficient r=-0.96 after poling and r=-0.89 before poling) and d_{33} also increased from 52 to 463 μC/N (r=-0.87 in both after and before poling). This means that it is possible to estimate the degree of piezoelectricity (the values of k_p and d_{33}) from V_S/V_L, even though the ceramics were without DC poling (as-fired).

Furthermore, we can mention that it is significant to evaluate the both values of k_p and d_{33}, which are not directly corresponding to each other because of $d=k(\varepsilon \cdot s)^{1/2}$ (d: piezoelectric strain constant, k: coupling factor, ε: dielectric constant, s: compliance). Therefore, our developed method for acoustic wave velocity measurement, especially V_S/V_L, is useful for materials research on piezoelectric ceramics including lead-free ceramics with higher coupling factor and higher piezoelectric strain constant d_{33} in the cases of as-fired [Figs. 15(b) and 16(b)] as well as DC fully polarized [Figs. 15(a) and 16(a)] ceramics.

This session was described the road maps in piezoelectric ceramics between longitudinal and transvers wave velocity, Young's modulus and Poisson's ratio to research and develop new piezoelectric ceramic materials, especially lead-free ceramics with high piezoelectricity.

5. CONCLUSIONS

Longitudinal and transverse wave velocities of PZT, lead titanate and lead-free ceramics were measured by an ultrasonic precision thickness gauge with high-frequency pulse oscillation to calculate elastic constants such as Young's modulus and Poisson's ratio. The dependences of compositions and manufacturing processes of the ceramics on the elastic constants were clarified in PZT ceramics. Moreover, while the compositions around MPB in lead-free ceramics were investigated by this method, it was found significant relationships between Young's modulus and Poisson's ratio in lead-free ceramics with high piezoelectricity. The DC poling field dependence of the longitudinal and transverse wave velocities of PZT, lead titanate, and lead-free ceramics was also investigated by this method. The effect of the poling field on domain alignment, such as in the cases of full DC poling and domain clamping, could be explained by the relationships between acoustic wave velocities, Young's modulus, and Poisson's ratio vs poling field. The directions of the research and development of piezoelectric ceramics including lead-free ceramics with high coupling factors could be proposed on the basis of the findings of this study.

ACKNOWLEDGMENTS

This work was partially supported by a Grant-in-Aid for Scientific Research C (No. 21560340) and a Grant of Strategic Research Foundation Grant-aided Project for Private Universities 2010-2014 (No. S1001032) from the Ministry of Education, Culture, Sports, Science and Technology, Japan (MEXT).

REFERENCES

1. Ogawa T, Yamada A, Chung YK, Chun DI. Effect of Domain Structures on Electrical Properties in Tetragonal PZT Ceramics. J. Korean Phys. Soc. 1998;32: S724-S726.

2. Ogawa T, Nakamura K. Poling Field Dependence of Ferroelectric Properties and Crystal Orientation in Rhombohedral Lead Zirconate Titanate Ceramics. Jpn. J. Appl. Phys. 1998;37: 5241-5245.

3. Ogawa T, Nakamura K. Effect of Domain Switching and Rotation on Dielectric and Piezoelectric Properties in Lead Zirconate Titanate Ceramics. Jpn. J. Appl. Phys. 1999;38: 5465-5469.

4. Ogawa T. Domain Switching and Rotation in Lead Zirconate Titanate Ceramics by Poling Fields. Ferroelectrics 2000;240: 75-82.

5. Ogawa T. Domain Structure of Ferroelectric Ceramics. Ceram. Int. 2000;25: 383-390.

6. Ogawa T. Poling Field Dependence of Crystal Orientation and Ferroelectric Properties in Lead Titanate Ceramics. Jpn. J. Appl. Phys. 2000;39: 5538-5541.

7. Ogawa T. Poling Field Dependence of Ferroelectric Properties in Barium Titanate Ceramics. Jpn. J. Appl. Phys. 2001;40: 5630-5633.

8. Ogawa T. Poling Field Dependence of Ferroelectric Properties in Piezoelectric Ceramics and Single Crystal. Ferroelectrics 2002;273: 371-376.

9. Kato R, Ogawa T. Chemical Composition Dependence of Giant Piezoelectricity on k31 Mode in Pb(Mg1/3Nb2/3)O3–PbTiO3 Single Crystals. Jpn. J. Appl. Phys. 2006; 45: 7418-7421.

10. Ogawa T, Yamauchi Y, Numamoto Y, Matsushita M, Tachi Y. Giant Electromechanical Coupling Factor of k31 Mode and Piezoelectric d31 Constant in Pb[(Zn1/3Nb2/3)0.91Ti0.09]O3 Piezoelectric Single Crystal. Jpn. J. Appl. Phys. 2002;41: L55-L57.

11. Ogawa T. Giant k31 Relaxor Single-Crystal Plate and Their Applications. In: Lallart M. (ed.) Ferroelectrics-Applications. Rijeka: InTech; 2011. P.3-34.

12. Ogawa T, Numamoto Y. Origin of Giant Electromechanical Coupling Factor of k31 Mode and Piezoelectric d31 Constant in Pb[(Zn1/3Nb2/3)0.91Ti0.09]O3 Single Crystal. Jpn. J. Appl. Phys. 2002;41: 7108-7112.

13. Technical catalogues 2004 for piezoelectric materials of TDK Corporation, Murata Manufacturing Co., Ltd., Fuji Ceramics Corporation, NEC Tokin Corporation, etc.

14. Kadota M, Saito Y, Yoshida T, Ozeki H, Yamaguchi T, Hasegawa M, Sato H. EMAS-6100, Standard of Electronic Materials Manufacturers Association of Japan. Tokyo: Piezoelectric Ceramics Technical Committee; 1993 [in Japanese].

15. Bechmann R. IRE Standards on Piezoelectric Crystals. Proc. IRE, 1958.

16. Shibayama K. Danseiha Soshi Gijyutsu Handbook (Technical Handbook of Acoustic Wave Devices). Tokyo: Ohmsha; 1991 [in Japanese].

17. Mason WP. Physical Acoustics I, Part A. New York: Academic Press; 1964.

18. Takagi K. Cyouonpa Binran (Ultrasonic Handbook). Tokyo: Maruzen; 1999 [in Japanese].

19. Negishi K, Takagi K. Cyouonpa Gijutsu (Ultrasonic Technology). Tokyo: Univ. of Tokyo Press; 1984 [in Japanese].

20. Furukawa M, Tsukada T, Tanaka D, Sakamoto N. Alkaline Niobate-based Lead-Free Piezoelectric Ceramics. Proc. 24th Int. Japan-Korea Semin. Ceramics, 2007.

21. Ogawa T, Furukawa M, Tsukada T. Poling Field Dependence of Piezoelectric Properties and Hysteresis Loops of Polarization versus Electric Field in Alkali Niobate Ceramics. Jpn. J. Appl. Phys. 2009;48, 709KD07-1-5.

22. Ogawa T, Nishina T, Furukawa M, Tsukada T. Poling Field Dependence of Ferroelectric Properties in Alkali Bismuth Titanate Lead-Free Ceramics. Jpn. J. Appl. Phys. 2010; 49: 09MD07-1-4.

23. Ogawa T. Highly Functional and High-Performance Piezoelectric Ceramics. Ceramic Bulletin 1991;70: 1042-1049.

24. Zhao W, Zhou H, Yan Y, Liu D. Morphotropic Phase Boundary Study of the BNT-BKT Lead-Free Piezoelectric Ceramics. Key Eng. Mater. 2008;368-372: 1908-1910.

25. Yang Z, Liu B, Wei L, Hou Y. Structure and Electrical Properties of (1-x)Bi0.5Na0.5TiO3-xBi0.5K0.5TiO3 Ceramics near Morphotropic Phase Boundary. Mater. Res. Bull. 2008;43: 81-89.

26. Dai Y, Pan J,Zhang X. Composition Range of Morphotropic Phase Boundary and Electrical Properties of NBT-BT System. Key Eng. Mater. 2007; 336-338: 206-209.

27. Saito Y, Takao H, Tani T, Nonoyama T, Takatori K, Homma T, Nagaya T, Nakamura M. Lead-Free Piezoceramics. Nature 2004;432: 84-87.

28. Jaffe B, Cook WR, and Jaffe H. Piezoelectric Ceramics. New York: Academic Press; 1971.

Chapter 9

Development of a Flexible Lead-Free Piezoelectric Transducer for Health Monitoring in the Space Environment

Marco Laurenti, Denis Perrone, Alessio Verna, Candido F. Pirri and Alessandro Chiolerio*

Center for Space Human Robotics, Istituto Italiano di Tecnologia, C.so Trento 21, Turin 10129, Italy

ABSTRACT

In this work we report on the fabrication process for the development of a flexible piezopolymeric transducer for health monitoring applications, based on lead-free, piezoelectric zinc oxide (ZnO) thin films. All the selected materials are compatible with the space environment and were deposited by the RF magnetron sputtering technique at room temperature, in view of preserving the total flexibility of the structures, which is an important requirement to guarantee coupling with cylindrical fuel tanks whose integrity we want to monitor. The overall transducer architecture was made of a *c*-axis-oriented ZnO thin film coupled to a pair of flexible Polyimide foils coated with gold (Au) electrodes. The fabrication process started with the deposition of the bottom electrode on Polyimide foils. The ZnO thin film and the top electrode were then deposited onto the Au/Polyimide substrates. Both the electrodes and ZnO layer were properly patterned by wet-chemical etching and optical lithography.

The assembly of the final structure was then obtained by gluing the upper and lower Polyimide foils with an epoxy resin capable of guaranteeing low outgassing levels, as well as adequate thermal and electrical insulation of the transducers. The piezoelectric behavior of the prototypes was confirmed and evaluated by measuring the mechanical displacement induced from the application of an external voltage.

Keywords: piezopolymeric transducer; zinc oxide; sputtering; health monitoring systems; flexible devices; space environment

1. INTRODUCTION

In parallel to the development of strong approaches to face reliability problems, two solutions are normally proposed: redundancy, with an obvious drawback for space exploration purposes in terms of weight and complexity, and fault detection/recovery. The last approach, smarter in principle, relies almost entirely on Health Monitoring Systems (HMS), which are becoming one of the most intriguing challenges faced from the world of applied science and engineering. Their development involves a huge number of sectors, from materials scientists, to physicists and engineers, and their application is strongly required in different fields; civil infrastructures, aerial, and space vehicles [1,2] are just some examples where on-site, real-time structural monitoring and identification systems are mandatory for safety reasons and for preserving the integrity of the overall structure, as well. The main factors affecting the mechanical integrity of such systems are numerous: damages due to fatigue, corrosion, impact, cracks, and extreme loading are some examples of events which can lead to catastrophic structural failures, for ground-based applications. Regarding space, we may add huge temperature oscillations, vibrations, cosmic rays, gamma rays, charged particles (solar wind), atomic oxygen, micrometeorites, and space debris. Their effects on the structural properties of the considered systems must be properly contextualized to the particular case. For example, when considering satellites or spacecrafts, the mechanical impacts acting on the structure shells can be caused by different conditions with respect to the terrestrial environment. One example is the action of corrosive chemicals, like atomic oxygen, which strikes on the external bodies of aerospace vehicles at a velocity of about 8 km/s [3]. These impact forces affect the materials and strongly reduce their overall lifetime. Another important issue is represented by meteorite impacts, which can induce severe structural damage like cracks. In order to face such problems, active/passive elements sensitive to the aforementioned events are required to monitor the structural integrity and perform failure analysis [3,4]. Hence, HMSs are based on sensing elements which must be able to provide real-time information on the mechanical status of the analyzed systems. For these reasons, a lot of efforts must be devoted to the development of sensing platforms capable of being sensitive to the mechanical events affecting the considered systems, but also of periodically analyzing the external surface of aerospace apparatus, in order to determine, in advance, the eventual presence of structural damages. To this purpose, meaningful work has been carried out in recent years with the final aim of developing sensors, damage detection algorithms, and structural health monitoring methods [1].

The most diffused HMS technology is based on the use of strain gauges [5] or transducers [6] as active sensing elements. These can be classified as optical sensors, resistance-based sensors, or piezoelectric (PE) sensors. In the case of optical ones, the Fiber Optical Sensor (FOS) technology is widely diffused, since FOS-based HMS are lightweight, with reduced dimensions, insensitive to electromagnetic interference, and show good resistance to many chemical agents [7]. Nevertheless, these can only be used as sensors or actuators but not as a combination of both. Furthermore, FOS technologies are based on quite complex systems, requiring a light source and light detection component with the final aim of illuminating and acquiring the sensor output, respectively [8]. On the other hand, strain gauges represent a well-established and widely-commercialized technology [5], which is normally based on metal and metal alloys, or better on nanocomposite (NC) materials, that allow reaching higher sensitivities [9]. One of the major limitations of the aforementioned technologies is their impossibility of working simultaneously as both active and passive elements; that is they can operate either as diagnostics or as passive elements sensitive to structural damages. For these reasons, PE transducers found large applications for the development of single active/passive elements to be integrated in HMS. The main advantage of this class of materials is their possibility of generating an electric field, *i.e.*, an electrical output voltage, as a consequence of mechanical strains (direct PE effect). On the other hand, a mechanical strain can be induced on the PE material when subjected to an electric field (converse PE effect). Consequently, these represent an intriguing technological solution, since in this way a single device can operate serially in sensing/actuating mode, thus allowing not only to analyze the investigated system (passive mode), but also to scan the overall system itself and detect in advance the presence of eventual failures (active mode). PE-based HMSs are, thus, considered potential candidates for mechanical and structural damage monitoring and diagnostic systems [10]. In particular, a discussion of the use of many piezoelectric patches for structural-damage detection was proposed by dell'Isola, who described an electro-mechanical system based on electric measurements [11]. The same group also discussed the dynamics of this kind of electro-mechanical systems in a wide range of frequencies and in case of non-linear evolution equations [12].

The most diffused working principle of PE transducers is the Lamb wave-based technique [13]. This is quite convenient, since Lamb waves can propagate on relatively large areas without suffering too high of an attenuating phenomenon [2,8]. In this way, the number of patch transducers required to cover and scan a large area can be properly minimized. As an alternative to the conventional piezoceramic patch transducers, the new class of so-called smart fasteners, *i.e.*, a piezoceramic stack transducer embedded in the body of a conventional fastener, has been proposed [4]. The main difference among the conventional patch-type piezoceramic transducers and piezoceramic stack transducers is the direction of application of the mechanical stress. In the case of stack transducers, normal stress is applied to the structure while the patch-type transducers apply shear stress [2]. A successful method for the analysis and detection of mechanical damages using piezotransducers is the electromechanical impedance method (EMI) [14]. The EMI method analyzes and compares the mechanical impedance value of the structure to establish if damages are present or not [15]. Structural damage detection can be also carried out by coupling the electromechanical impedance (EMI) technique with the pencil lead break (PLB) method [16]. The working principle of PLB is quite simple: a pencil lead is carefully broken against the block or rod on which an acoustic emission (AE) sensor is installed [16,17]. The breaking of the pencil, thus, generates an impulsive stress,*i.e.*, an elastic wave. Additionally, in this case PE materials play a fundamental role, since the AE sensor generally consists of a shielded PE element mounted in a metal shell [16].

Table 1 summarizes some selected properties of different PE materials commonly used for the development of PE-based transducers. Each one of the considered materials can potentially satisfy the functional properties required for the fabrication of reusable, flexible, and conformable PE transducers for HMS. However, some differences can be appreciated, and help in the selection of the PE material to be used. The first distinctive element is by far the PE behavior, quantitatively represented by the PE coefficients. This is found to be at the maximum in the case of lead zirconate titanate (PZT), and different works reported about the development of HMS based on PZT PE transducers, able to detect damage in different kind of structures. For example, Sohn *et al.*, showed that PZT transducers are capable of generating Lamb waves to detect the presence of delamination phenomena in a fiber-reinforced polymer (FRP) composite test structure [18]. Wang *et al.*, measured debonding between steel reinforcement bars and concrete by embedding PE transducers within the structure [19]. Lamb wave generation in PZT transducers was also successfully exploited for sensing damage in metallic structural members [20]. Despite its promising PE response, PZT shows several disadvantages. It is a non-biocompatible, toxic material which requires high deposition temperatures or post-deposition thermal treatments (generally higher than 500 °C) in order to get the right crystalline structure and avoid the formation of the pyrochlore phase which suppresses piezo/ferroelectricity [21,22]. Another important issue of PZT is its superior brittleness [23], representing a strong limitation in view of the development of mechanical flexible transducers conformable to structural surfaces. Moreover, PZT transducers require an additional poling treatment in order to induce the alignment of ferroelectric domains and to promote the emergence of piezoelectricity [24]. In order to overcome some of these limitations the use of flexible, low-cost PE polymers took place and different works demonstrated that poly(vinyledenefluoride) (PVDF) and its copolymer poly(vinylidene fluoride-trifluoroethylene) (PVDF-TrFE) could be good alternatives to brittle piezoceramics. Indeed, it has been demonstrated that PVDF can be used to generate and receive guided waves in a way very similar to PZT [25]. Monkhouse *et al.*, produced interdigitated transducers (IDT) from PVDF-TrFE [26]. These IDTs structures successfully generated Lamb waves propagating in an aluminum substrate, and demonstrated their potential applications for HMS. Despite the advantages of using these flexible structures, also in this case a poling treatment is required. Moreover, the lower PE response of PVDF and PVDF-TrFE, together with their lower sensitivity, has limited their applications for HMS despite their favorable mechanical properties. In order to take advantage of both the high PE coefficient of PZT and the promising mechanical characteristics of polymeric materials, novel PE composite materials mainly based on PZT and PVDF/PVDF-TrFE started to be investigated [27,28,29]. Parallel to this, lots of efforts have been devoted to the study of lead-free biocompatible PE materials as alternatives to PZT. Among them, zinc oxide (ZnO), barium titanate (BT), and aluminum nitride (AlN) are the most investigated. Barium titanate could be a non-toxic valid alternative, showing a good d_{33} piezocoefficient and high dielectric constant [30]. However, the relatively low Curie temperature of BT strictly limits the operating temperature range of this ceramic. Moreover, low-temperature depositions of BT result in amorphous nanostructures, thus strongly suppressing the promotion of PE phenomena. Polycrystalline PE BT films thus require high deposition temperatures (often not compatible with the use of polymeric substrates) and also require quite extreme post-deposition poling treatments in order to align the ferroelectric domains and induce PE phenomena [31]. Other concepts were explored, such as the use of BT nanoparticles embedded in polymeric PVDF matrices, producing encouraging results, but still with a poor conformability [32]. On the contrary, AlN and ZnO are non-ferroelectric materials showing piezoelectricity in a wide range of operating temperatures and the desired crystal orientation at room temperature [33,34]. Thus, their synthetic process can be merely carried out at room temperature without preventing the use of polymeric substrates. AlN and ZnO are lead-free biocompatible materials, and so represent a good alternative for replacing the most performing, but toxic, PZT one. AlN and ZnO PE thin films are widely used for the preparation of surface acoustic wave devices [35]. In particular, AlN is suitable for Micro Electro-Mechanical Systems (MEMS) application, with Complementary Metal-Oxide Semiconductor (CMOS)-compatible processes applied even on flexible substrates, such as Polyimide [36]. Despite both these materials showing several advantages and similar properties, AlN is characterized by a lower d_{33} value. This aspect, together with the high crystalline quality of room-temperature deposited ZnO, drove our choice of ZnO thin films as PE materials for HMS transducers.

Zinc oxide is a wide band-gap *n*-type semiconductor (3.37 eV), showing a large exciton binding energy at room temperature (60 meV) [37]. ZnO is of great interest thanks to the co-presence of both semiconducting and PE properties, together with the possibility of being easily synthesized in a lot of different micro/nano structures, like porous [38] and compact [39] thin films, micro/nanowires [40,41], nanorods [42], or nanobelts [43]. Among all of the aforementioned structures, the thin film one is of particular interest since many deposition techniques, like magnetron sputtering [44], pulsed-laser deposition [45], sol-gel [46], chemical vapor deposition [47], and atomic layer deposition [48] can be exploited for growing high-quality ZnO layers. Thin-film technologies have the main advantage of being fully compatible with semiconductor processing, and allow the deposition of ZnO on large-area substrates through repeatable and controllable processes, fully extendable to large-scale industrial production. In particular, the sputtering technology shows several advantages. First of all, the scalability of the process is promoted, since allowing the deposition of PE materials on large-area substrates. Moreover, a strict control on the deposition rate, chemical composition, stoichiometry, and morphology of the final sputtered thin films can be simply achieved by properly changing one or more deposition parameters, such as gas flows and partial pressure, substrate temperature, and/or the supplied direct current (DC)/RF signal power. Different works reported about the successful use of ZnO thin films and nanostructures as flexible PE devices [39], nanogenerators [49], anodes for Li-ion batteries [50], photoanodes for dye-sensitized solar cells [51], and for wetting controllable systems [52]. The possibility of growing ZnO micro/nanostructures also promoted the study and development of ZnO-based composite materials, which were successfully tested as PE elements for HMS applications or for other applications [40]. Dodds *et al.* [53] reported about the preparation of PVDF-TrFE nanocomposites (NCs) using ZnO nanoparticles. The resulting nanocomposite was flexible and characterized by a higher PE response than pristine PVDF-TrFE. Moreover, it showed sensitivity to impact vibrations and strains [54]. The piezoelectricity of ZnO-based NCs has been further demonstrated [1,55] and used as a dynamic strain sensor [56]. Recently, negative capacitance in ZnO thin films was also shown to be enhanced thanks to piezoelectricity, showing a new potential application of the material as programmable dynamic device [57]. While ZnO-based NCs seem to be promising candidates for HMS, such types of materials are not easily processed with the standard microfabrication techniques required in view of scaling their production to an industrial level. Moreover, strong poling processes are still required in order to induce piezoelectricity and make the transducers working. On the other hand, the well-known thin-film technology represents a valid alternative for the fabrication of ZnO PE transducers [3]. The possibility of growing high (002)-oriented ZnO thin films on flexible polymeric substrates have been already highlighted. For example, Prepelita *et al.*, reported about the successful deposition of transparent, uniform, and adherent ZnO thin films sputtered on Polyimide substrates [58]. They showed that post-deposition annealing treatments can lead to superior electrical and optical properties of the deposited material. Moreover, the annealing process could stabilize the ZnO thin film and improve the mechanical adhesion to the underlying polymer. The improved electro-optical properties of the annealed samples thus suggested their promising application to solar cell technology. Changji *et al.* also discussed the room temperature deposition of ZnO thin films by cathodic vacuum arc on Polyimide foils [59]. The effect of the deposition pressure on the mechanical adhesion of the resulting (002)-oriented ZnO films was investigated, showing that low deposition pressures were useful to enhance the adhesion between ZnO and the Polyimide foils. Despite both these works succeeding in the fabrication of ZnO thin films on flexible Polyimide substrates, their investigation was still limited in view of exploiting ZnO piezoelectricity for flexible transducer applications. Indeed, the analyzed samples merely consisted of ZnO thin films directly deposited on Polyimide foils, and no functional characterization regarding their piezoelectric properties was provided. Nevertheless, the presence of a high (002) orientation could be inferred for ZnO thin films grown on Polyimide either at room temperature, or after annealing. This aspect represents an additional advantage in using piezoelectric ZnO thin films, since it prevents requiring any additional poling treatment.

Table 1. Properties of most common PE materials.

Properties	Symbol	Unit	Material				PVDF β Phase *	P(VDF-TrFE) β Phase *
			PZT	ZnO	$BaTiO_3$	AlN		
Cost	P	€/g	15.2	5.9	15.4	15.2	1–2	7
Toxicity	LD	mg/kg	–	No	–	–	–	–
Reusable in laboratory	–	–	Yes	Yes	Yes	Yes	Yes	Yes
Reproducible in laboratory	–	–	Yes	Yes	Yes	Yes	Yes	Yes
Conformable	–	–	Yes	Yes	Yes	Yes	Yes	Yes
Metallization on both sides	–	–	Yes	Yes	Yes	Yes	Yes	Yes
Actuator/sensor function	–	–	Yes	Yes	Yes	Yes	Yes	Yes
Adhesion to metal and CFRP substrate	–	–	Yes	Yes	Yes	Yes	Yes	Yes
PE coefficient	d_{33}	pm/V	280–380	12	140–250	4.5–6.4	13–22	−20
	d_{31}	pm/V	−123	−5	−30	−3.2	6–10	6
Curie temperature	T_C	°C	300	N/A	120	>2000	N/A	112
relative permittivity	ε_r	–	>300	8.3	190–700	9.5–10.5	10–12	8–12
Density	ρ	g/cm^3	7.5	5.6	5.9	3.3	1.8	3
Tangent loss angle	tan δ	–	0.08–0.06	0.053	0.012–0.015	0.02–0.06	0.010–0.035	0.016
Young modulus	E	GPa	60–70	80–140	67–200	300–350	2–3.4	1

* data refer to commercial products by PIÉZOTECH.

The final aim of this work consists in the development of flexible and conformable piezopolymeric transducers (PPT) for HMS applications, based on the use of a piezoceramic ZnO thin film deposited by the sputtering technique. The final PPT structure is made of a PE ZnO thin film embedded between a couple of gold (Au) metal electrodes, featuring the so-called cross-point electrode (CPE) configuration. A commercial flexible polyimide, namely Polyimide, is considered as the substrate material. Suitable deposition conditions are selected for growing high-quality ZnO thin films. The ZnO properties are analyzed by using Field Emission Scanning Electron Microscopy (FESEM) to determine the film morphology and thickness, while the X-Ray Diffraction (XRD) technique is used to study the crystallographic orientation of the material. Energy-Dispersive X-ray spectroscopy is carried out to investigate the chemical composition while the ZnO electrical behavior, its resistivity, and carrier concentration are evaluated by IV characteristics and Hall effect measurements. The assembly of the different parts composing the transducer is pursued by a bonding procedure. This was performed with a special adhesive fully compatible with aerospace environment. The fabricated PPT prototypes were characterized by measuring their mechanical displacement under the application of an electric voltage, which resulted in the PE behavior of the fabricated transducers, but also provided an estimation of the ZnO PE coefficient d_{33}.

2. EXPERIMENTAL SECTION

The deposition process of ZnO thin films and gold metal electrodes was made by a dual chamber co-sputtering system (KS 300 CONFOCAL Dual, Kenosistec, Binasco, Italy). Each process takes place in a cylindrical stainless steel chamber (AISI 304, 40 cm diameter, 45 cm height approximately), equipped with three circular cathodes (3 inch and 2 inch diameter), in a confocal geometry. Targets are clamped on the cathodes, in the inner part of the chamber, while the substrates are placed in the upper part, facing down towards the cathodes. In order to get homogeneity and uniformity of the as-deposited materials, substrates are rotating during the deposition, and coatings with a maximum area of about 180 cm^2 can be achieved. A radio-frequency (RF) signal generator, operating at a working frequency of 13.56 MHz, and a matching network needed to adjust the reflected power down to zero are used, respectively, to supply and maximize the signal power provided to the cathodes. Before starting the sputter deposition of materials, suitable vacuum conditions are obtained by a two-stage pumping system, made by a rotary and a turbomolecular pumps connected in series, allowing to reach pressures as low as 2×10^{-7} Torr. Gas flows are properly controlled by a set of mass flow controllers while the substrate temperature is monitored by a thermocouple. All of the depositions were performed at room temperature, without heating the substrates.

The material selected for the preparation of the metal electrodes was gold (Au). According to the deposition conditions summarized in Table 2, Au electrodes for the CPE geometry were grown starting from a 2 inch diameter Au target (Kurt J. Lesker, Hastings, UK, purity 99.99%), without providing any substrate heating. The average deposition rate was 1.38 Å/s.

ZnO thin films were grown at room temperature with the set of parameters summarized inTable 2. The final average thickness for ZnO was set to 700 nm in order to get, at the same time, the maximization of the thickness-dependent piezoelectric response, without inducing too pronounced a stress release phenomenon of the polymeric substrate once that thickest films are grown. Despite no intentional heating being provided to the substrates, the temperature locally increased up to 50 °C because of the energy transfer during the ion bombardment of the sample. The depositions were performed starting from a 3 inch diameter ZnO target (Kurt J. Lesker, purity 99.999%). To prevent a lack of oxygen typical of ZnO films, depositions were carried out in a mixed atmosphere of Ar and O_2. The average deposition rate was 0.94 Å/s.

Au and ZnO depositions were performed on silicon (simplifying cross-sectional imaging of the films during electron microscopy measurements) and flexible Polyimide substrates (DuPont, Wilmington DE, USA), *i.e.*, a commercial polyimide resistant up to temperatures around 400 °C.

The morphology and thickness of ZnO thin films and metal layers were investigated through Field Emission Scanning Electron Microscope (FESEM) measurements (Auriga Dual Beam, from Carl Zeiss, Jena, Germany). The presence of a suitable electrical conductivity of the Au electrodes was verified with a tester, checking the presence of a short circuit over the whole metal layer. The ZnO crystal structure and orientation was analyzed by X-Ray Diffraction

(XRD) measurements (PANalytical, Almelo, Netherlands), using a Bragg-Brentano diffractometer (X'Pert Diffractometer, Cu Kα radiation, λ = 1.54). Energy Dispersive X-Ray (EDX) spectrometry was carried out to evaluate the chemical composition of the investigated ZnO thin films. The electrical behavior of ZnO was also investigated at room temperature by Hall effect measurements using an MMR K2500-3RSLP instrument. Electrical and PE measurements were carried out on the fabricated PPT structures through a Piezo Evaluation System (PES, TF Analyzer 2000 HS, Aixacct, Aachen, Germany) coupled to a single-point laser vibrometer (Polytec OVF-505, Dexter, MI, USA).

Table 2. Deposition conditions and average thickness of sputtered Au and ZnO thin films.

Target	RF Power Density ($W{\cdot}cm^{-2}$)	Gas Pressure (bar)	Ar Flow (sccm)	O_2 Flow (sccm)	Thickness (nm)
Au	4.9	4×10^{-6}	20	–	100
ZnO	1.8	4×10^{-6}	19	1	700

3. RESULTS AND DISCUSSION

3.1. Gold Electrodes—Preparation and Characterization

The morphology and average thickness of Au metal layers were analyzed by means of FESEM analyses for the samples deposited on Polyimide foils and Si substrates, respectively. In the case of the polymeric substrate, particular attention was paid to the evaluation of the surface characteristics of the Au layer, investigating the eventual presence of cracks, voids or detachments as well. Figure 1 shows FESEM images of a 100 nm-thick Au layer deposited on Polyimide foil (Figure 1a) and Si substrate (Figure 1b). A smooth surface, characterized by nanometer-sized, closely-packed grains, was observed. Even if no adhesion promoter was used, the Au layer showed a continuous and uniform structure on the whole area without evidence of cracks. All of these factors witnessed the good mechanical adhesion of the Au film to the polymeric substrate. This is further visible in the inset of Figure 1a, which shows a picture representative of the flexible Au/Polyimide structure. The good mechanical quality of the metal layer directly deposited on the flexible Polyimide foil is further noticed after bending the flexible specimen, indeed no mechanical degradation affecting the metal was observed, confirming the good adhesion of the Au layer to the polymeric substrate, despite no adhesion layers being previously deposited on the uncoated polymer.

Figure 1. FESEM images of Au layers deposited on (**a**) Polyimide and (**b**) Si wafer. Inset of panel (**a**) shows a picture of the Au-coated Polyimide specimen.

Figure 2. FESEM images showing (**a**) the surface morphology and (**b**) the cross-section structure of ZnO thin films grown on Au-coated Si substrates.

3.2. ZnO Thin Films—Preparation and Characterization

The set of deposition parameters used for growing ZnO thin films are summarized in Table 2. Figure 2 shows the surface morphology and cross-sectional nanostructure representative of a 700 nm-thick ZnO film deposited on Si/Au substrate. The surface morphology was first analyzed (Figure 2a), highlighting the presence of a fine-grained nanostructure, with an average grain size ranging between 20 and

100 nm. The cross-section of the ZnO film is shown in Figure 2b and reveals the presence of highly densely-packed, vertically-aligned nanocrystals. This aspect is representative of the columnar growth regime which took place during the thin film growth, and is typical of *c*-axis-(002)-oriented polycrystalline ZnO thin films [60].

X-Ray diffraction measurements were performed in order to analyze the crystal structure and orientation of the ZnO material. This aspect is really important. Indeed, the (002) crystallographic orientation is the one along with ZnO PE is maximized [61]. For this reason the deposition of highly *c*-axis-(002)-oriented films with a hexagonal wurtzite-like structure is essential for the PE response of ZnO to be promoted. The diffraction pattern collected from 700 nm-thick ZnO film grown on Si/Au substrate is shown in Figure 3a. A single diffraction peak positioned at 34.35° was detected. This can be ascribed to the family of (002)-oriented crystal planes (Joint Committee on Powder Diffraction Standards-International Centre for Diffraction Data, JCPDS-ICDD card no. 89–1397) and is representative of the presence of the hexagonal wurtzite-like crystal structure in the analyzed material. The presence of such preferential orientation also witnesses that approximately all of the crystal grains are already aligned along the desired PE (002) orientation. Therefore, no additional poling treatment is required for PE to occur.

The chemical composition of ZnO thin film was investigated by Energy-Dispersive X-ray (EDX) spectroscopy. From the EDX spectrum reported in Figure 3b it can be observed that zinc (21.93 at.%) and oxygen (41.06 at.%) are detected, as expected. However, carbon (12.15 at.%) and silicon (24.68 at.%) are visible as well, because of contaminants absorbed on the sample surface and of the Si substrate contribution, respectively. A small amount of Al (0.17 at.%) coming from the sample holder is also detected. Since, from the EDX measurement the contributions of silicon, carbon, and aluminum cannot be excluded, a precise estimation of the ZnO stoichiometry could not be carried out.

Hall effect measurements were carried out on the 700 nm-thick ZnO film in order to examine the nature of majority carriers, film resistivity, and carriers mobility and concentration as well. ZnO is a well-known *n*-type semiconductor. This aspect is further confirmed in the present work, since electrons were found as majority charge carriers. The measured ZnO resistivity was $4.41 \cdot 10^{-3}$ Ω·cm, while the carriers mobility and concentration values were 80 $cm^2 \cdot V^{-1} \cdot s^{-1}$ and 2.29×10^{17} cm^{-3}, respectively. Generally oxygen vacancies, together with the presence of unintentional hydrogen impurities, are believed to be at the basis of the intrinsic *n*-type nature of ZnO-based materials [62].

The current-voltage (IV) characteristic describing the electrical behavior of 700 nm-thick ZnO thin film is shown in Figure 4. The IV response was representative of the semiconductor nature of ZnO, showing currents ranging in the order of tens of μA and a nonlinear Schottky behavior. When the applied voltage was ±40 V or more, short circuits started to be present, witnessing that breakdown phenomena occurred in the investigated ZnO sample.

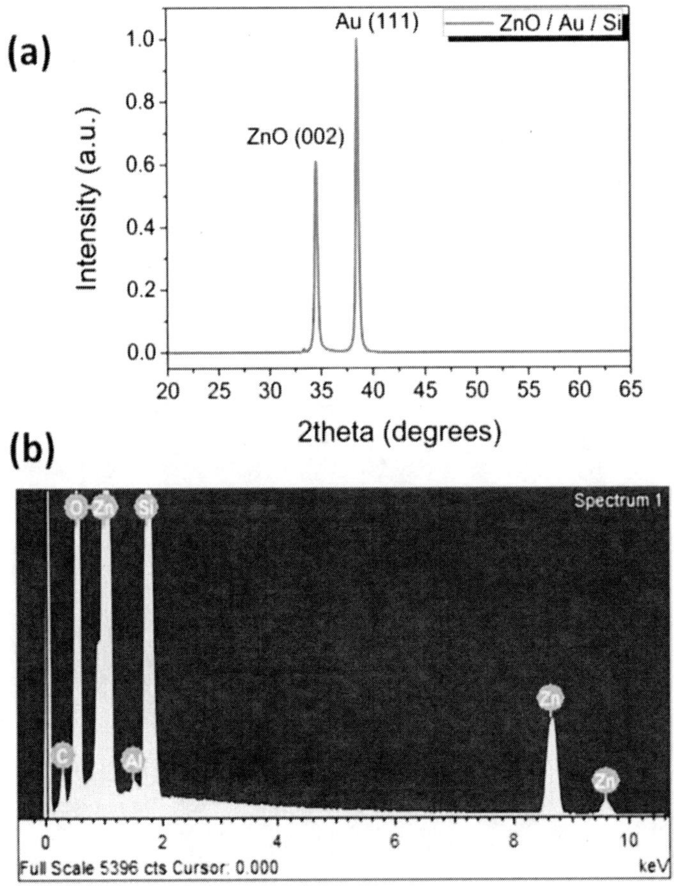

Figure 3. (**a**) XRD pattern and (**b**) EDX spectrum of sputtered ZnO thin films.

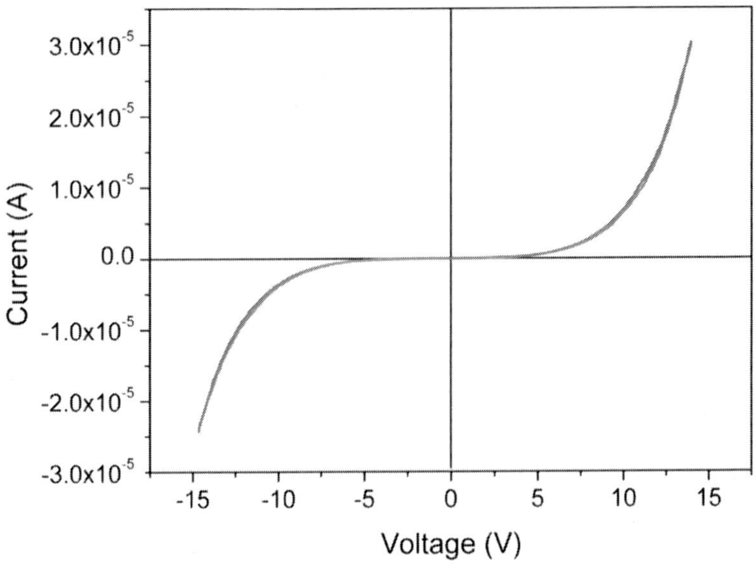

Figure 4. IV curve measured on a 700 nm-thick ZnO thin film.

Figure 5. Sketch of the main parts composing the PPT structure in the CPE configuration—Top view.

3.3. Device Fabrication and Characterization

The design of the final PPT structure in its CPE configuration is qualitatively presented inFigure 5.

Each transducer was composed of:

1. Lower and upper Polyimide supports (total patch dimensions: 10 mm × 10 mm);
2. 100 nm-thick Au bottom electrode (5 mm × 5 mm);
3. 700 nm-thick ZnO piezoceramic layer (8 mm × 8 mm);
4. 100 nm-thick Au top electrode (4 mm × 4 mm).

The process steps followed for the fabrication of the transducers are represented in Figure 6. The first steps (a–c) consisted in the preparation of the patterned Au bottom electrode deposited on Polyimide foils by optical lithography and wet etching processes. A 1 μm–thick positive photoresist (HPR504) was deposited onto the samples. The desired geometry was then obtained by aligning a Mylar® mask with the UV radiation, exposing to UV radiation, and developing the photoresist. The Au layer was then removed by wet etching with a solution of $KI:I_2:H_2O$ (4 g:1 g:40 mL). Finally, the patterned samples were rinsed in deionized water and then with acetone and isopropanol to remove the unexposed photoresist. A 700 nm-thick ZnO film was then deposited onto the patterned Au-coated Polyimide substrates (d–f). The same patterning procedure described so far was performed to define the geometry of the ZnO layer and top electrode (g–h).

The last step of the fabrication process (Figure 6i,j) consisted in the assembly of the final structure. This was obtained by gluing the upper and lower Polyimide foils with an epoxy resin suitable for space applications. The assembly of the final transducer requires the use of adhesives in order to fix the different parts together, guaranteeing insulation between facing electrodes, electrical continuity between metallic electrodes, and the active layer and the sealing of the structure from the external environment, without worsening the flexibility of the device. The adhesive used for the assembly was, thus, selected among those commercially available, by considering thermal and electrical conductivity, mechanical, and out-gassing characteristics of the commercial products. On the basis of these considerations, an epoxy resin was purchased from ELANTAS (Epoxylite TSA 220, Wesel, Germany) and selected for being an electrically-insulating epoxide resistant to high strains, bondable to the required materials, curable from room to high temperature, stable during the service temperature range, and characterized by low outgassing values. The curing procedure consisted in a thermal treatment performed in air at 165 °C for 12 h.

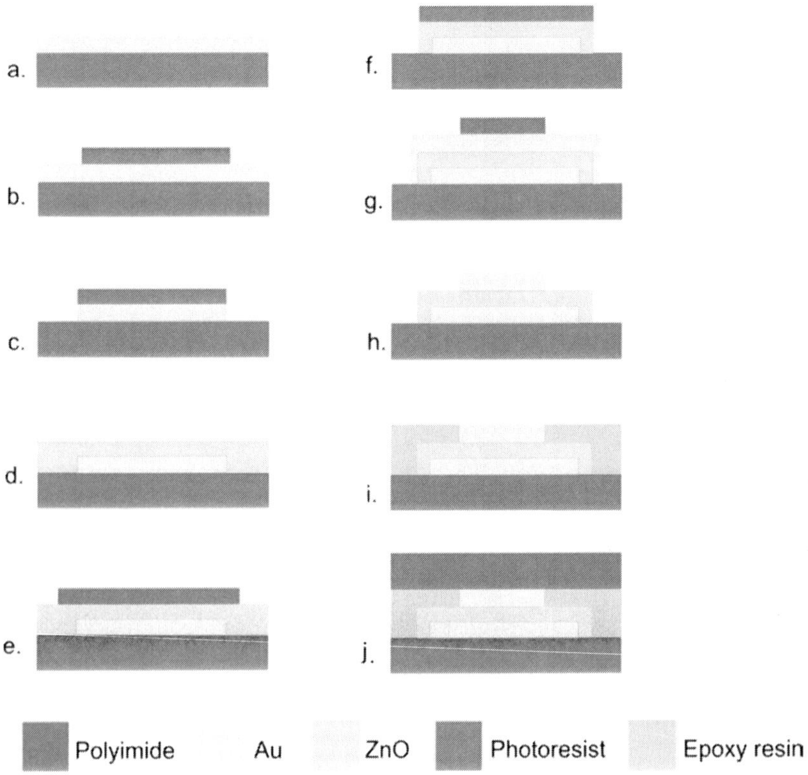

Figure 6. Flow-chart describing the process steps of the PPT in CPE configuration fabrication process: (**a**–**h**) deposition and patterning of the electrodes and ZnO thin film; (**i**,**j**) assembly of the final PPT structure.

Table 3 summarizes some technical characteristics of the prototypes prepared by following the fabrication process reported in Figure 6 and the curing process described above.

Table 3. Technical characteristics of the PPT prototypes.

Parameter	Value
Active Area	4 mm × 4 mm
Active layer thickness	900 nm
Minimum curvature radius	6 mm
Operating temperature	−40 °C up to 80 °C
Breakdown voltage *	±40 V
d_{33}	12.3 pm/V
Typical operating voltage	±15 V
Typical operating frequency	100 Hz

* Tested in the frequency range between 1 and 100 Hz.

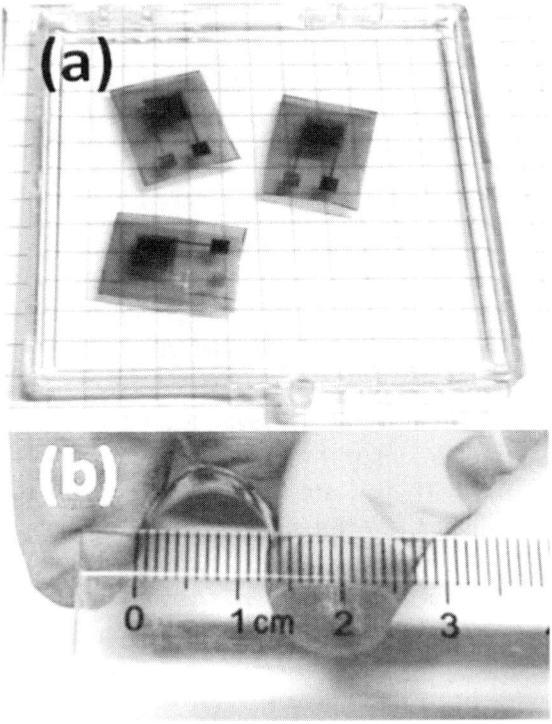

Figure 7. Image representative of the different PPT prototypes (**a**) as prepared and (**b**) under bending conditions.

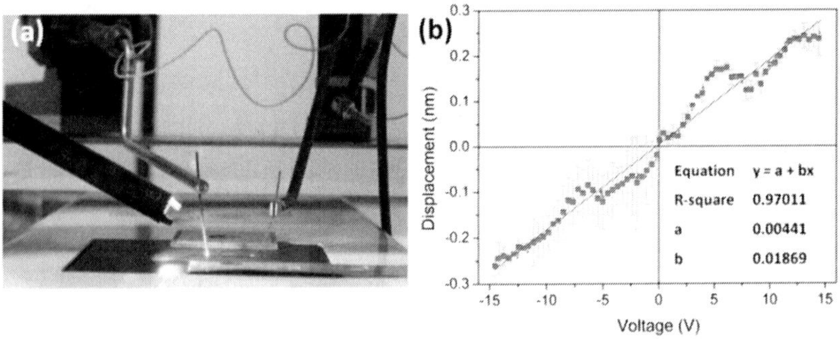

Figure 8. (**a**) Picture of the PES setup used for piezoelectric characterization; (**b**) Mechanical displacement *vs.* applied voltage measured on the PPT prototypes.

In order to qualify the operation of the prepared prototypes as piezopolymeric transducers, the converse PE effect was evaluated on the samples, as shown in Figure 7. Measurements were performed by using the PES setup described in the Experimental Section. An electric voltage (±15 V) was applied on the PPT and the PE-induced mechanical displacement recorded by a single-point laser vibrometer (see Figure 8a). The displacement *vs.* voltage curve representative of the converse PE behavior of the investigated PPT structures is shown in Figure 8b together with the corresponding linear fit.

An almost-linear behavior was obtained and no hysteresis phenomena were observed. This last aspect was expected, since ZnO does not show the presence of ferroelectric domains, being a non-perovskite PE material. Thus, the PE characteristic is expected to be quite linear, as for other non-centrosymmetric materials like AlN [63]. However, some deviations from a perfect linear behavior are visible (*R*-square ~0.97). These are mainly ascribed to additional contributions coming from the substrate bending, which cannot be excluded during the measurements, with our setup. The measured peak-to-peak mechanical displacement was around 530 nm. From the slope of the linear characteristic an estimation of the d_{33} PE coefficient was also possible, with an average value of 12.3 pm/V that well agrees with the 12.84 pm/V theoretically determined by direct first-principles density functional calculations [64]. The direct PE effect of flexible ZnO-based structures very similar to the ones reported in this work was already investigated and reported in a previously published paper [39]. In that case a PE output voltage generation of 0.361 V was obtained for a 700 nm-thick ZnO film under the application of a periodical mechanical stimulus of 30 N. The sensitiveness of our sputtered ZnO thin films towards input mechanical forces was thus already pointed out, and further supported the investigation of such materials as PE transducers for HMS.

The device described in this paper was conceived to be coupled to the cylindrical fuel tank of a system launcher. Normally those vessels are fabricated by means of filament winding using carbon fibers, an extremely tough material, in which the formation and propagation of a crack, due to accidental impacts or mechanical stresses could eventually result in mission failure and payload loss. The real-time monitoring of eventual crack formation/propagation is thought to be done by injecting a train of Lamb waves through our PPT and using the same device as a receiver to correlate eventual echoes to the cracks. The material choice and trade-off, the device design, and fabrication, were all conceived to provide an engineering solution to the onboard requirements. As an example, the breakdown voltage of 40 V well fits with the standard onboard power bus (28 V).

Both read-out and control circuitry fabrication and testing, as well as PPT operation in the relevant environment, are outside of the scope of the present paper.

4. CONCLUSIONS

This work reports about the fabrication of conformable, reusable, and reproducible piezopolymeric transducers (PPT) for health monitoring systems in space applications. Suitable materials and design for the fabrication of the PPT structures were defined, on the basis of fundamental requirements for HMS applications. The design of the PPT consisted of a piezoceramic layer coupled to a pair of flexible polymeric patches covered with metal layers acting as electrodes. Sputtering was selected as the synthesis technique for growing all the involved materials, while Au and ZnO were selected as the electrode material and the piezoceramic, respectively. In particular, ZnO was considered a promising candidate thanks to its biocompatibility and compatibility to the space environment, thus representing a valid alternative to the conventional lead-based, toxic PZT. Suitable deposition conditions were selected for the sputter deposition of high-quality Au electrodes and PE ZnO thin films on flexible Polyimide substrates. The fabrication process for the PPT structure consisted in the following steps. Au metal layers were deposited on Polyimide foils and patterned in order to get the desired bottom electrode geometry. After that, ZnO thin films were deposited onto the Au/Polyimide substrates and patterned. Finally, the Au top electrode was deposited on the ZnO/Au/Polyimide structure, and patterned again. Wet-chemical etching and optical lithography were performed for patterning the different parts of the transducer. An epoxy resin suitable for space applications was selected among those commercially available and used for the final assembly of the PPT structure, which consisted in a thermal bonding procedure needed to fix the upper Polyimide foil to the underlying Au/ZnO/Au/Polyimide structure. Following this fabrication process, different PPT prototypes were successfully prepared and their converse PE behavior evaluated by applying an electric voltage across the ZnO layer, and measuring the induced mechanical displacement. In this way an estimation of the d_{33} PE coefficient was also possible, with a measured value of 12.3 pm/V, in agreement with the one theoretically predicted for ZnO. Our results encourage the use of PE ZnO thin films for the fabrication of flexible and conformable PE transducers for HMS applications.

ACKNOWLEDGMENTS

Research was funded by Regione Piemonte, Piattaforma Aerospazio, under the STEPS2 project (Sistemi e TEcnologie Per lo Spazio 2).

AUTHOR CONTRIBUTIONS

This work was conceived and scientifically coordinated by Alessandro Chiolerio. Marco Laurenti performed ZnO depositions and their characterization. Denis Perrone and Alessio Verna designed and carried out all the technological processes for patterning the different parts composing the transducers. The final assembly of the transducers and their piezoelectric characterization was carried out by Marco Laurenti. This paper was written by Marco Laurenti with the contribution of Alessandro Chiolerio. This work was carried out under the further supervision of Candido F. Pirri.

REFERENCES

1. Dodds, J.S.; Meyers, F.N.; Loh, K.J. Piezoelectric nanocomposite sensors assembled using zinc oxide nanoparticles and poly(vinylidene fluoride). *Smart Struct. Syst.* **2013**, *12*, 055–071.

2. Yoon, H.S.; Jung, D.; Kim, J.H. Lamb wave generation and detection using piezoceramic stack transducers for structural health monitoring applications. *Smart Mater. Struct.* **2012**,*21*, 055019.

3. Joshi, S.; Hegde, G.M.; Nayak, M.M.; Rajanna, K. A novel piezoelectric thin film impact sensor: Application in non-destructive material discrimination. *Sens. Actuators A* **2013**,*199*, 272–282.

4. Yan, S.; Wu, J.; Sun, W.; Ma, H.; Yan, H. Development and application of structural health monitoring system based on piezoelectric sensors. *Int. J. Distrib. Sens. Netw.* **2013**,*270927*, 12.

5. Ko, J.M.; Ni, Y.Q. Technology developments in structural health monitoring of large-scale bridges. *Eng. Struct.* **2005**, *27*, 1715–1725.

6. Lu, Y.; Wang, X.; Tang, J.; Ding, Y. Damage detection using piezoelectric transducers and the Lamb wave approach: II. Robust and quantitative decision making. *Smart Mater. Struct.* **2008**, *17*, 025034.

7. Betz, D.C.; Thursby, G.; Culshaw, B.; Staszewski, W.J. Structural damage location with fiber bragg grating rosettes and lamb waves. *Struct. Health Monit.* **2007**, *6*, 299–308.

8. Barazanchy, D.; Martinez, M.; Rocha, B.; Yanishevsky, M. A hybrid structural health monitoring system for the detection and localization of damage in composite structures.*J. Sens.* **2014**, *109403*.

9. Chiolerio, A.; Roppolo, I.; Sangermano, M. Radical diffusion engineering: Tailored nanocomposite materials for piezoresistive inkjet printed strain measurement. *RSC Adv.***2013**, *3*, 3446–3452.

10. Giorgio, I.; Galantucci, L.; Della Corte, A.; del Vescovo, D. Piezo-electromechanical smart materials with distributed arrays of piezoelectric transducers: Current and upcoming applications. *Int. J. Appl. Electromagn. Mech.* **2015**, *47*, 1051–1084.

11. Dell'Isola, F.; Vestroni, F.; Vidoli, S. Structural-damage detection by distributed piezoelectric transducers and tuned electric circuits. *Res. Nondestruct. Eval.* **2005**, *16*, 101–118.

12. Dell'Isola, F.; Vestroni, F.; Vidoli, S. A class of electro-mechanical systems: Linear and nonlinear dynamics. *J. Theor. Appl. Mech.* **2002**, *40*, 47–71.

13. Su, Z.; Ye, L.; Ye, L. Guided lamb waves for identification of damage in composite structures: A review. *J. Sound Vib.* **2006**, *295*, 753–780.

14. Giurgiutiu, V.; Zagrai, A. Damage detection in thin plates and aerospace structures with the electro-mechanical impedance method. *Struct. Health Monit.* **2005**, *4*, 99–118.

15. Ciang, C.C.; Lee, J.R.; Bang, H.J. Structural health monitoring for a wind turbine system: A review of damage detection methods. *Meas. Sci. Technol.* **2008**, *19*, 122001.

16. Daré de Almeida, V.A.; Guimarães Baptista, F.; de Aguiar, P.R. Piezoelectric transducers assessed by the pencil lead break for impedance-based structural health monitoring.*IEEE Sens. J.* **2015**, *15*, 693–702. (In Portuguese).

17. Gary, J.; Hamstad, M.A. On the far-field structure of waves generated by a pencil lead break on a thin plate. *J. Acoust. Emiss.* **1994**, *12*, 157–170.

18. Sohn, H.; Park, G.; Wait, J.R.; Limback, N.P.; Farrar, C.R. Wavelet-based active sensing for delamination detection in composite structure. *Smart Mater. Struct.* **2004**, *13*, 153–160.

19. Wang, C.S.; Wu, F.; Chang, F.K. Structural health monitoring from fiber-reinforced composites to steel-reinforced concrete. *Smart Mater. Struct.* **2001**, *10*, 548.

20. Park, S.; Yun, C.B.; Roh, Y.; Lee, J.J. PZT-based active damage detection techniques for steel bridge components. *Smart Mater. Struct.* **2006**, *15*, 957.

21. Law, C.W.; Tong, K.Y.; Li, J.H.; Li, K. Effect of pyrolysis temperature on the characteristics of PZT films deposited by the sol-gel method. *Thin Solid Films* **1998**, *335*, 220–224.

22. Chiolerio, A.; Quaglio, M.; Lamberti, A.; Celegato, F.; Balma, D.; Allia, P. Magnetoelastic coupling in multilayered ferroelectric/ferromagnetic thin films: A quantitative evaluation. *Appl. Surface Sci.* **2012**, *258*, 8072–8077.

23. Makino, H.; Kamiya, N. Effects of DC electric fields on mechanical properties of piezoelectric ceramics. *Jpn. J. Appl. Phys.* **1994**, *33*, 5323–5327.

24. Watanabe, S.; Fujiu, T.; Fujii, T. Effect of poling on piezoelectric properties of lead zirconate titanate thin films formed by sputtering. *Appl. Phys. Lett.* **1995**, *66*, 1481–1483.

25. Lin, B.; Giurgiutiu, V. Modeling and testing of PZT and PVDF piezoelectric wafer active sensors. *Smart Mater. Struct.* **2006**, *15*, 1085–1093.

26. Monkhouse, R.; Wilcox, P.; Cawley, P. Flexible interdigital PVDF transducers for the generation of Lamb waves in structures. *Ultrasonics* **1997**, *35*, 489–498.

27. Furukawa, T.; Ishida, K.; Fukada, E. Piezoelectric properties in the composite systems of polymers and PZT ceramics. *J. Appl. Phys.* **1979**, *50*, 4904–4912.

28. Venkatragavaraj, E.; Satish, B.; Vinod, P.R.; Vijaya, M.S. Piezoelectric properties of ferroelectric PZT-polymer composites. *J. Phys. D: Applied Physics* **2001**, *34*, 487.

29. Dietze, M.; Es-Souni, M. Structural and functional properties of screen-printed PZT–PVDF-TrFE composites. *Sens. Actuators A Phys.* **2008**, *143*, 329–334.

30. Panda, P.K. Review: Environmental friendly lead-free piezoelectric materials. *J. Mater. Sci.* **2009**, *44*, 5049–5062.

31. Park, K.I.; Xu, S.; Liu, Y.; Hwang, G.T.; Kang, S.J.L.; Wang, Z.L.; Lee, K.J. Piezoelectric $BaTiO_3$ thin film nanogenerator on plastic substrates. *Nano Lett.* **2010**, *10*, 4939–4943.

32. Chiolerio, A.; Lombardi, M.; Guerriero, A.; Canavese, G.; Stassi, S.; Gazia, R.; Cauda, V.; Manfredi, D.; Chiodoni, A.;

Verna, A.; *et al.* Effect of the fabrication method on the functional properties of $BaTiO_3$: PVDF nanocomposites. *J. Mater. Sci.* **2013**, *48*, 6943–6951.

33. Iriarte, G.F.; Rodriguez, J.G.; Calle, F. Synthesis of c-axis oriented AlN thin films on different substrates: A review. *Mater. Res. Bull.* **2010**, *45*, 1039–1045.

34. Youssef, S.; Combette, P.; Podlecki, J.; Asmar, R.A.; Foucaran, A. Structural and optical characterization of ZnO thin films deposited by reactive RF magnetron sputtering. *Cryst. Growth Des.* **2009**, *9*, 1088–1094.

35. Gualtieri, J.G.; Kosinski, J.; Ballato, A. Piezoelectric materials for acoustic wave applications. *IEEE Trans. Ultrason. Ferroelectr. Freq. Control* **1994**, *41*, 53–59.

36. Jackson, N.; Keeney, L.; Mathewson, A. Flexible-CMOS and biocompatible piezoelectric AlN material for MEMS applications. *Smart Mater. Struct.* **2013**, *22*, 115033.

37. Ottone, C.; Laurenti, M.; Motto, P.; Stassi, S.; Demarchi, D.; Cauda, V.A. ZnO nanowires: Synthesis approaches and electrical properties. In *Nanowires. Synthesis, Electrical Properties and Uses in Biological Systems*; Nova Publisher: New York, NY, USA, 2014; pp. 1–57.

38. Gazia, R.; Motto, P.; Stassi, S.; Sacco, A.; Virga, A.; Lamberti, A.; Canavese, G. Photodetection and piezoelectric response from hard and flexible sponge-like ZnO-based structures. *Nano Energy* **2013**, *2*, 1294–1302.

39. Laurenti, M.; Stassi, S.; Lorenzoni, M.; Fontana, M.; Canavese, G.; Cauda, V.; Pirri, C.F. Evaluation of the piezoelectric properties and voltage generation of flexible zinc oxide thin films. *Nanotechnology* **2015**, *26*, 215704.

40. Chiolerio, A.; Roppolo, I.; Cauda, V.; Crepaldi, M.; Bocchini, S.; Bejtka, K.; Verna, A.; Pirri, C.F. Ultraviolet mem-sensors: Flexible anisotropic composites featuring giant photocurrent enhancement. *Nano Res.* **2015**, *8*, 1956–1963.

41. Laurenti, M.; Verna, A.; Fontana, M.; Quaglio, M.; Porro, S. Selective growth of ZnO nanowires on substrates patterned

by photolithography and inkjet printing. *Appl. Phys. A Mater. Sci. Proces.* **2014**, *117*, 901–907.

42. Yi, G.C.; Wang, C.; Park, W.I. ZnO nanorods: Synthesis, characterization and applications. *Semicond. Sci. Technol.* **2005**, *20*, S22–S34.

43. Xi, Y.; Hu, C.G.; Han, X.Y.; Xiong, Y.F.; Gao, P.X.; Liu, G.B. Hydrothermal synthesis of ZnO nanobelts and gas sensitivity property. *Solid State Commun.* **2007**, *141*, 506–509.

44. Gao, W.; Li, Z. ZnO thin films produced by magnetron sputtering. *Ceram. Int.* **2004**, *30*, 1155–1159.

45. Tsoutsouva, M.G.; Panagopoulos, C.N.; Papadimitriou, D.; Fasaki, I.; Kompitsas, M. ZnO thin films prepared by pulsed laser deposition. *Mater. Sci. Eng. B* **2011**, *176*, 480–483.

46. Znaidi, L. Sol–gel-deposited ZnO thin films: A review. *Mater. Sci. Eng. B* **2010**, *174*, 18–30.

47. Shishodia, P.K.; Kim, H.J.; Wakahara, A.; Yoshida, A.; Shishodia, G.; Mehra, R.M. Plasma enhanced chemical vapor deposition of ZnO thin films. *J. Non-Cryst. Solids* **2006**, *352*, 2343–2346.

48. Malm, J.; Sahramo, E.; Perälä, J.; Sajavaara, T.; Karppinen, M. Low-temperature atomic layer deposition of ZnO thin films: Control of crystallinity and orientation. *Thin Solid Films* **2011**, *519*, 5319–5322.

49. Laurenti, M.; Canavese, G.; Sacco, A.; Fontana, M.; Bejtka, K.; Castellino, M.; Pirri, C.F.; Cauda, V. Nanobranched ZnO structure: *p*-type doping induces piezoelectric voltage generation and ferroelectric-photovoltaic effect. *Adv. Mater.* **2015**, *27*, 4218–4223.

50. Laurenti, M.; Garino, N.; Porro, S.; Fontana, M.; Gerbaldi, C. Zinc oxide nanostructures by chemical vapour deposition as anodes for Li-ion batteries. *J. Alloys Compd.* **2015**, *640*, 321–326.

51. Lamberti, A.; Sacco, A.; Laurenti, M.; Fontana, M.; Pirri, C.F.; Bianco, S. Sponge-like ZnO nanostructures by low temperature water vapor-oxidation method as dye-sensitized

solar cell photoanodes. *J. Alloys Compd.* **2014**, *615*, S487–S490.

52. Laurenti, M.; Cauda, V.; Gazia, R.; Fontana, M.; Rivera, V.F.; Bianco, S.; Canavese, G. Wettability control on ZnO nanowires driven by seed layer properties. *Eur. J. Inorg. Chem.* **2013**, *2013*, 2520–2527.

53. Dodds, J.S.; Meyers, F.N.; Loh, K.J. Piezoelectric characterization of PVDF-TrFE thin films enhanced with ZnO nanoparticles. *IEEE Sens. J.* **2012**, *12*, 1889–1890.

54. Dodds, J.S.; Meyers, F.N.; Loh, K.J. Enhancing the piezoelectric performance of PVDF-TrFE thin films using zinc oxide nanoparticles. In Proceedings of the Sensors and Smart Structures Technologies for Civil, Mechanical, and Aerospace Systems, San Diego, CA, USA, 11 March 2012.

55. Gullapalli, H.; Vemuru, V.S.; Kumar, A.; Botello-Mendez, A.; Vajtai, R.; Nagarajaiah, S.; Terrones, M.; Ajayan, P.M. Flexible piezoelectric ZnO-paper nanocomposite strain sensor. *Small* **2010**, *6*, 1641–1646.

56. Loh, K.; Chang, D. Zinc oxide nanoparticle-polymeric thin films for dynamic strain sensing. *J. Mater. Sci.* **2010**, *46*, 228–237.

57. Laurenti, M.; Verna, A.; Chiolerio, A. Evidence of negative capacitance in piezoelectric ZnO thin films sputtered on interdigital electrodes. *ACS Appl. Mater. Interfaces* **2015**. in press.

58. Prepelita, P.; Craciun, V.; Filipescu, M.; Garoi, F. Sputtered zinc oxide thin films deposited on polyimide substrate and annealing effect on the physical properties. *Thin Solid Films* **2013**, *545*, 564–570.

59. Changji, H.; Zhenhui, H.; Weichun, F.; Qi, Z. Influence of deposition pressure on the adhesion of ZnO thin films deposited by cathodic vacuum arc deposition on polyimide foil substrates. *J. Phys. D Appl. Phys.* **2009**, *42*, 185303.

60. Mirica, E.; Kowach, G.; Evans, P.; Du, H. Morphological evolution of ZnO thin films deposited by reactive sputtering. *Cryst. Growth Des.* **2004**, *4*, 147–156.

61. Gardeniers, J.G.E.; Rittersma, Z.M.; Burger, G.J. Preferred orientation and piezoelectricity in sputtered ZnO films. *J. Appl. Phys.* **1998**, *83*, 7844–7854.

62. Bang, S.; Lee, S.; Park, J.; Park, S.; Jeong, W.; Jeon, H. Investigation of the effects of interface carrier concentration on ZnO thin film transistors fabricated by atomic layer deposition. *J. Phys. D Appl. Phys.* **2009**, *42*, 235102.

63. Trolier-McKinstry, S.; Muralt, P. Thin film piezoelectrics for MEMS. *J. Electroceramics***2004**, *12*, 7–17.

64. Karanth, D.; Fu, H. Large electromechanical response in ZnO and its microscopic origin.*Phys. Rev. B* **2005**, *72*, 064116.

Chapter 10

Electrical Properties of CuO-Doped PZT-PZN-PMnN Piezoelectric Ceramics Sintered at Low Temperature

Phan Dinh Gio[1*], Le Dai Vuong[1,2], Ho Thi Thanh Hoa[1]

[1]*Department of Physics, College of Sciences, Huế University, Huế, Vietnam*

[2]*Faculty of Chemical and Environmental Engineering, Huế Industry College, Huế, Vietnam*

ABSTRACT

The $0.8Pb(Zr_{0.48}Ti_{0.52})O_3$-$0.125Pb(Zn_{1/3}Nb_{2/3})O_3$-$0.075Pb(Mn_{1/3}Nb_{2/3})O_3$ (PZT-PZN-PMnN) + x wt% CuO piezoelectric ceramics, where x = 0.0, 0.05, 0.075, 0.10, 0.125, 0.150, and 0.175, have been fabricated by the conventional solid-state reaction method and the B-site Oxide mixing technique (BO). The effect of CuO on the sinterability, structure, and electrical properties of PZT-PZN-PMnN ceramics was systematically studied. The CuO addition significantly reduced the sintering temperature of the ceramics from 1150°C to 850°C. Experimental results showed that with the doping of CuO, all the ceramics could be well sintered and exhibit a dense, pure perovskite structure. The specimen containing 0.125 wt% CuO sintered at 850°C showed the good electrical properties: the density of 7.91 g/cm^3; the electromechanical coupling factor, k_p = 0.55 and k_t = 0.46; the dielectric constant, ε = 1179; the dielectric loss (tand) of 0.006; the mechanical quality factor (Q_m) of 1174; the piezoelectric constant (d_{31}) of 112 pC/N.

Keywords: Crystal Structure, Dielectric, Piezoelectric Constant, Mechanical Quality Factor

1. INTRODUCTION

In recent years, many material scientists are interested in research and application of multi-component piezoelectric ceramics combine PZT with relaxor ferroelectrics, such as $Pb(Zr_{0.48}Ti_{0.52})O_3$-$Pb(Zn_{1/3}Nb_{2/3})O_3$ (PZT- PZN), $Pb(Zr_{0.48}Ti_{0.52})O_3$-$Pb(Mg_{1/3}Nb_{2/3})O_3$ (PZT-PMN), $Pb(Zr_{0.48}Ti_{0.52})O_3$-$Pb(Zn_{1/3}Nb_{2/3})O_3$-$Pb(Mn_{1/3}Nb_{2/3})O_3$ (PZT-PZN-PMnN), etc., due to their excellent piezoelectric properties and many applications in piezoelectric actuators and transformers [1] -[6] . These ceramics often have large dielectric constant e, high mechanical quality factor (Q_m), high electromechanical coupling factor (k_p), high Curie temperature and low dielectric loss factor (tand). However, the sintering temperature of the ceramics is quite high (>1150°C), which leads evaporation of PbO during sintering process, resulting in reducing properties of ceramic compositions and environmental pollution. Therefore, lowering sintering temperature of PZT based ceramics is very necessary.

There are many methods to lower the sintering temperature, however, the most common and effective method to reduce the sintering temperature of PZT based ceramics is to add the low-temperature melting oxides or compounds for liquid phase sintering at a low temperature. Many researchers have successfully decreased the sintering temperature of PZT based ceramics by using various additives such as B_2O_3, Bi_2O_3, Li_2CO_3, $BiFeO_3$, CuO, CuO + Bi_2O_3, etc. [7] - [14] .

In some cases, these additives can facilitate lower the sintering temperature, but decrease simultaneously the piezoelectric properties of ceramics due to the formation of piezoelectrically inactive phases in the grain boundary regions. Therefore, the research and fabrication ceramics sintered at low temperature, while improving or not reducing the piezoelectric properties of ceramics system are very important.

Recently, we studied the effect of Li_2CO_3 addition on the sintering behaviour and physical properties of PZT- PZN-PMnN ceramics. We decreased the sintering temperature of ceramics from 1150°C to 930°C and maintained good electrical properties: the electromechanical coupling factor k_p = 0.64, the dielectric constant ε = 1320, the dielectric loss (tanδ) of 0.005, the mechanical quality factor (Q_m) of 1150, the piezoelectric constant (d_{31}) of 145 pC/N, and the remanent polarization (P_r) of 30.5 μC/cm^2 [7].

In this work, we present some research results on the effect of CuO on the sinterability, structure, and electrical properties of PZT-PZN-PMnN ceramics.

2. EXPERIMENTAL PROCEDURE

The general formula of the studied material was $0.8Pb(Zr_{0.48}Ti_{0.52})O_3$-$0.125Pb(Zn_{1/3}Nb_{2/3})O_3$-$0.075Pb(Mn_{1/3}Nb_{2/3})O_3$ (PZT-PZN-PMnN) + x wt% CuO, where x = 0.0, 0.05, 0.075, 0.10, 0.125, 0.150, and 0.175. Reagent grade oxide powders of PbO, ZnO, MnO_2, Nb_2O_5, ZrO_2, $TiO_2$and CuO (purity ≥ 99%) were used as starting materials.

The PZN based ceramic materials with the a pure perovskite structure are very difficult to prepare by conventional ceramic processing, because of the high polarizability of Pb^{2+} and its interaction with Zn^{2+} cations, resulting the formation of the pyrochore phase and they will be detrimental to the physical properties of lead based ferroelectric ceramics [15] . Therefore, in this work, we combined the conventional solid-state reaction method and the B-site Oxide mixing technique (BO) for fabricate ceramic samples. Firstly, the mixture of $(Zn_{0.125}Mn_{0.075})Nb_{0.4}$ $[(Zr_{0.48}Ti_{0.52})_{2.4}]O_6$ (BO) were prepared by reactions of ZnO, MnO_2, Nb_2O_5, ZrO_2 and TiO_2 at temperature 1100°C for 2 h. Then, the powders of BO and PbO were weighed and milled for 8 h. The powders were calcined at temperature 850°C for 2 h, producing the PZT-PZN-PMnN compound. Thereafter CuO were mixed with the calcined PZT-PZN-PMnN powder, and powders milled for 16 h. The ground materials were pressed into disk 12 mm in diameter and 1.5 mm in thick under 100 MPa. These pellets were coated with $PbZrO_3$ powder then were sintered in a sealed alumina crucible at the temperature of 800°C, 830°C, 850°C, and 870°C for 4 h.

The crystal structure of the sintered samples was examined by X-ray diffraction (XRD, D8 ADVANCE). The microstructure of the samples was examined by using a scanning electron microscope (SEM) (Hitachi S-4800). The densities of samples were measured by Archimedes method. The ceramic samples were polled in a silicone oil bath at 120°C by applying dc field of 30 kV/cm for 20 min then cooling down to room temperature under the same electric field. They were aged for 24 h prior to testing. The piezoelectric properties were determined by the resonance and antiresonance frequencies using an impedance analyzer (HP 4193A and RLC HIOKI 3532). Temperature dependence of dielectric constant was determined using RLC HIOKI 3532 with automatic programming; temperature of the samples was measured using Digital Multimeter 7562. The ferroelectric property was measured by Sawyer-Tower method.

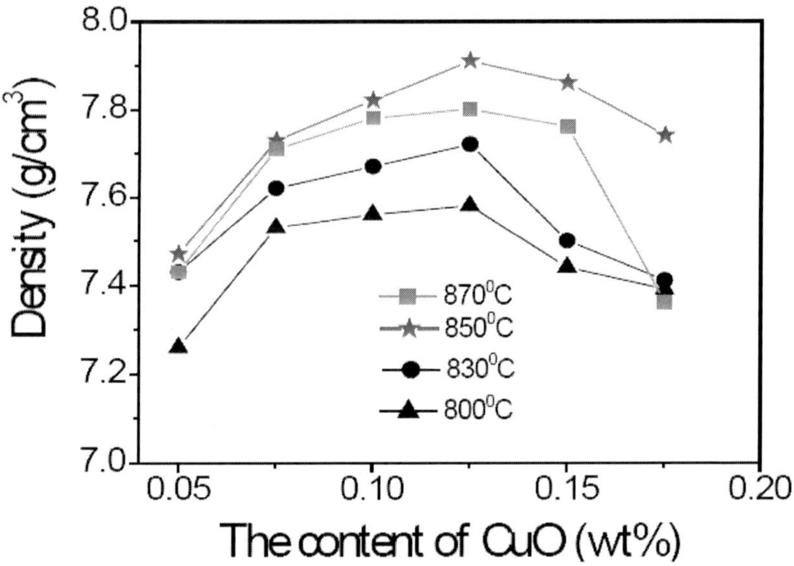

Figure 1. The density of PZT-PZN-PMnN ceramics with different amounts of CuO additive sintered at different temperatures: 800°C, 830°C, 850°C, and 870°C.

3. RESULTS AND DISCUSSION

Figure 1 shows the densities as a function of sintering temperature for PZT-PZN-PMnN ceramics with various CuO additions. With increasing of sintering temperature and CuO content, the density increases and reaches the maximum value (7.91 g/cm^3) at 850°C sintering temperature and at 0.125 wt% CuO content, then decreases. In our previous work [7] , the sintering temperature of undoped PZT-PZN-PMnN ceramics was as high as 1150°C (the density of 7.82 g/cm^3). Thus, the addition of CuO improved the sinterability, reduced the sintering temperature of 300°C compared with pure samples and increasing density of the ceramic samples. The above results consistent with the work of the authors Kim and co-worker studied the effect of CuO on structure and electrical properties of 0.4 $Pb(Mg_{1/3}Nb_{2/3})O_3$-0.25$PbZrO_3$-0.35$PbTiO_3$ ceramic system [16] .

Figure 2 shows the SEM images of the PZT-PZN-PMnN + x wt% CuO ceramics sintered at 850°C: (a) x = 0.0, (b) x = 0.05, (c) x = 0.075, (d) x = 0.10, (e) x = 0.125 and (g) x = 0.150. Figure 2(a) shows the microstructure of discrete grains, porous, not sintered materials. However, the microstructure of samples becomes denser and grain size increases as the CuO sintering aid is increased (Figures 2(b)-(e)). A homogeneous microstructure developed for the sample with 0.125 wt% CuO added (Figure 2(e)). However, Figure 2(g) also shows that further increasing the CuO content to 0.125 wt% gives rise to an abnormal grain boundary, porous appeared, the average grain size reduces. Such with the 0.125 wt% CuO-added sample, the highly dense and homogeneous microstructure was obtained, which may expect improved properties of ceramics.

The lowering of the sintering temperature and grain growth with CuO addition can be explained by liquid phase sintering. The phase diagram of Hitoshi Kitaguchi [17] has shown that CuO and PbO form the liquid phase at point eutectic 789°C. So when CuO doped in PZT-PZN-PMnN ceramics, CuO reacted with PbO and formed a liquid phase during the sintering, which assisted the densification of the specimens and increasing grain size. The second phase occurs at grain boundaries related to the limited solubility of CuO.

Figure 3 shows the XRD patterns of the PZT-PZN-PMnN + x wt% CuO ceramic samples sintered at 850°C for 4 h. It can be seen that all samples exhibit a perovskite structure, and not detect a second phase until x = 0.125. However, when the CuO content increases over 0.125 wt%, second phase peak was observed. Crystal structure of the samples is modified significantly by CuO additions. For the pure sample (x = 0.0), the rhombohedral structure developed. All the samples with the addition of CuO had a tetragonal structure as indicated by the splitting of (002) and (200) peaks at 2q » 44°. However, phase transition did not appear. This result suggests that Cu^{2+}ions are substituted for B-site of perovskite structure ABO_3 which lead to the distortion of crystal lattice. The results are consistent with several published works [18] - [21] .

Figure 4 shows temperature dependence of dielectric constant ε and dielectric loss tand as a function of CuO content. With increasing CuO doping, Curie temperature (T_c) of PZT-PZN-PMnN ceramics slight decreases. The composition with 0.125 wt% CuO content shows highest peak dielectric constants (12,000), which appears at about 266°C.

Figure 5 shows temperature dependence of dielectric constant ε of 0.125 wt% CuO-doped ceramic sample measured at frequency of 1 kHz, 10 kHz, 100 kHz and 1 MHz. It can be seen that the shape of the ε (T) peaks are broad, which is typical of a case diffuse transition with frequency dispersion. When the measured frequency increases, the maximum of ε_{max} decreases and shifts toward higher temperatures while dielectric loss increases near the Curie point, which is typical of a relaxor material [1] [7] .

To determine piezoelectric properties of ceramics, resonant vibration spectra of samples were measured at room temperature (Figure 6). From these resonant spectra, piezoelectric parameters of samples were determined.

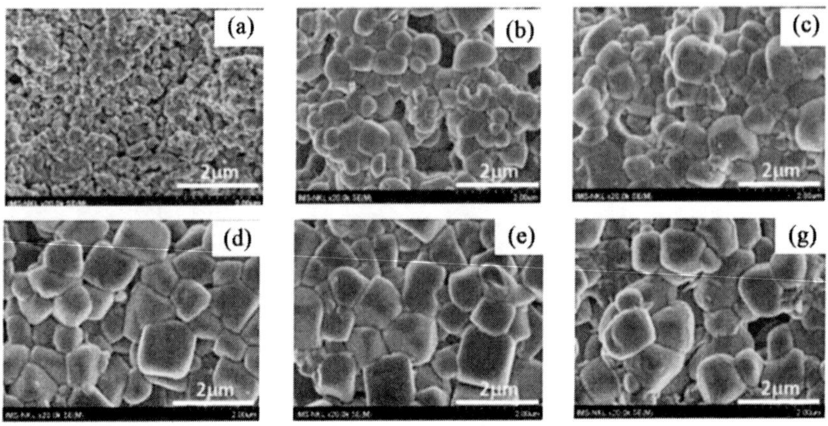

Figure 2. SEM images of PZT-PZN-PMnN with different amounts of CuO additive sintered at 850°C: (a) x = 0.0; (b) x = 0.05; (c) x= 0.075; (d) x = 0.10; (e) x = 0.125; (g) x = 0.150.

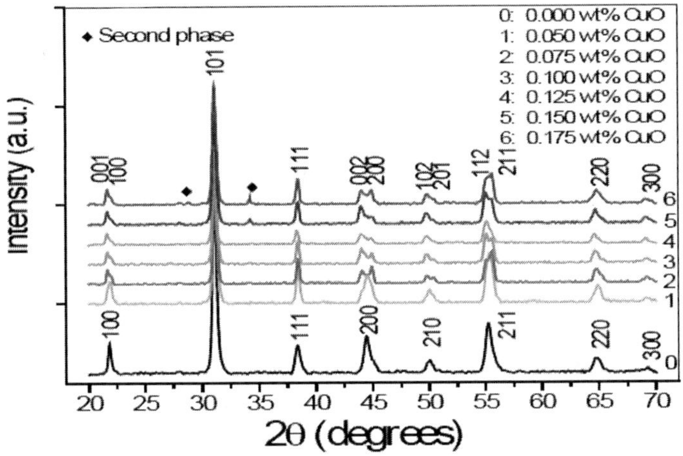

Figure 3. X-ray diffraction patterns of the PZT-PZN-PMnN with different amounts of CuO additive sintered at 850˚C: (0) x = 0.0; (1) x = 0.05; (2) x= 0.075; (3) x = 0.10; (4) x = 0.125; (5) x = 0.150; (6) x = 0.175.

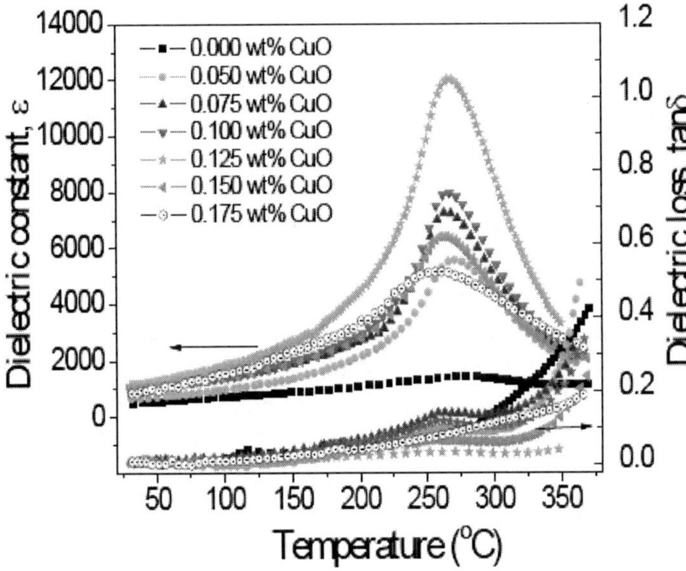

Figure 4. Temperature dependence of dielectric constant ε and dielectric loss tand of the PZT-PZN-PMnN ceramics with different amounts of CuO additive sintered at 850˚C.

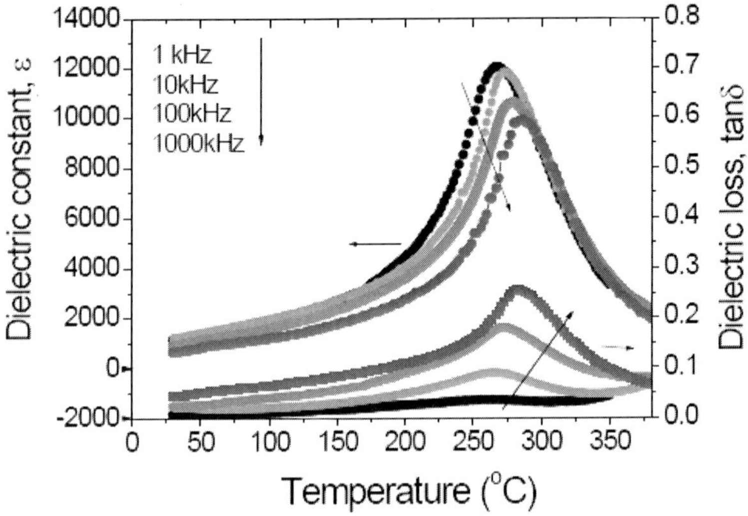

Figure 5. Temperature dependence of dielectric constant ε of 0.125 wt% CuO-doped sample measured at different frequencies: 1 kHz, 10 kHz, 100 kHz and 1 MHz.

Figure 7 and Figure 8 show the electromechanical coupling factor (k_p, k_t), the piezoelectric constant (d_{31}), the mechanical quality factor (Q_m), the dielectric constant (e) and dielectric loss (tanδ) change as a function of the CuO content. When the CuO content is lower than 0.125 wt%, the values of k_p, k_t, d_{31}, e and Q_m are rapidly increase with increasing content of CuO, while the dielectric loss tanδ are strong decrease. The largest values for k_p of 0.55, k_t of 0.46, d_{31} of 112 pC/N, e of 1179, Q_m of 1174 and minimum value of the dielectric loss tanδ is 0.006 were obtained at x = 0.125. These are probably related to characteristics of the density and the increasing grain size. During sintering, the presence of liquid phase enhances the density and grain size, which leads to the decrease of the energy loss and improvement of the electrical properties. Chao et al. [21] investigated CuO- doped PZT-PMN-PZN ceramics with compositions close to the morphotropic phase boundary (MPB) sintered at 920°C. The optimized results of k_p (0.53), e (982) and Q_m(1645) were obtained at 0.2 wt% CuO.

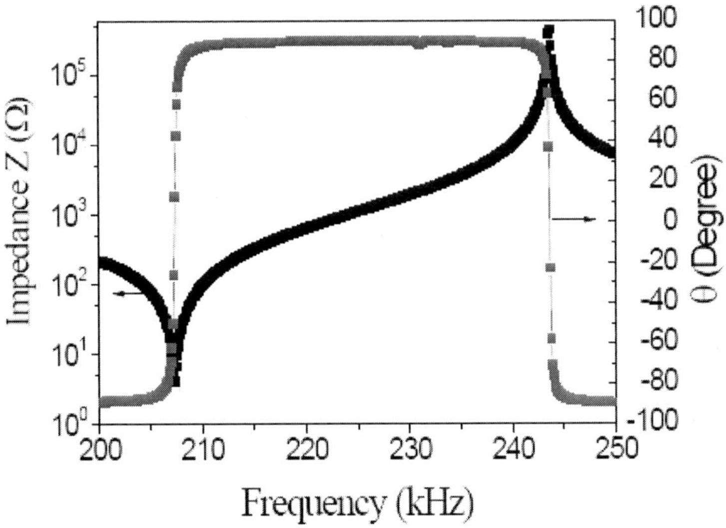

Figure 6. Spectrum of radial resonance of PZT-PZN-PMnN + 0.125 wt% CuO ceramic sample sintered at 850˚C.

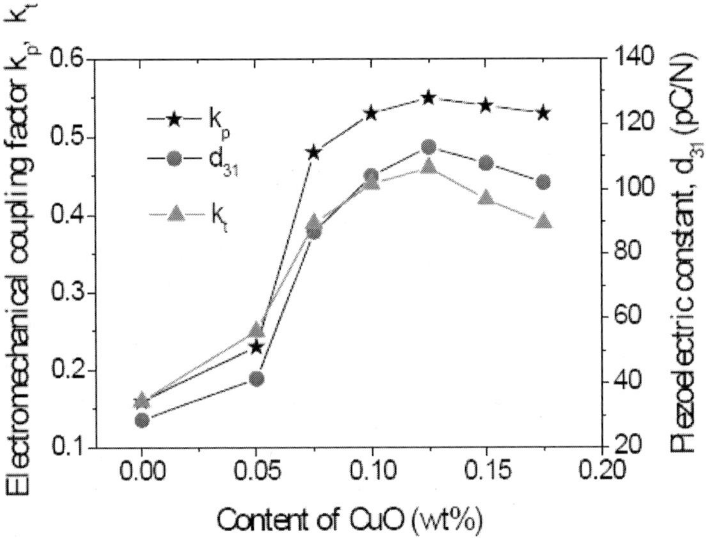

Figure 7. Electromechanical coupling factor k_p, k_t and piezoelectric constant d_{31} of PZT-PZN-PMnN ceramics sintered at 850˚C as a function of CuO content.

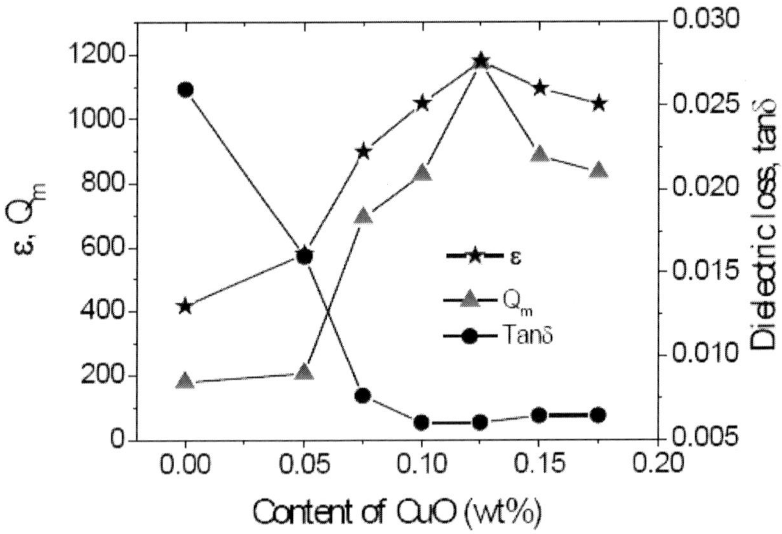

Figure 8. Dielectric constant e, dielectric loss tand and mechanical quality factor Q_m of PZT-PZN-PMnN ceramics sintered at 850˚C as a function of CuO content.

Nam et al. [14] also showed that CuO could increase the piezoelectric properties and reducing the sintering temperature of MnO_2-doped $0.75Pb(Zr_{0.47}Ti_{0.53})O_3$-$0.25Pb(Zn_{1/3}Nb_{2/3})O_3$ ceramics from 930˚C down to 850˚C The optimized values for k_p of 0.50, Q_m of 1000 were obtained at at 0.5 wt% CuO. The ceramic samples have a lower k_p and Q_m values than that of our ceramic samples. Accordingly, compared with the research results of F. Gao and co-workers on the PZT-PZN-PMnN ceramics [3] , in our work with a small amount of CuO was added to PZT-PZN-PMnN ceramics reduced the sintering temperature of 300˚C, with the retention of good piezoelectric properties.

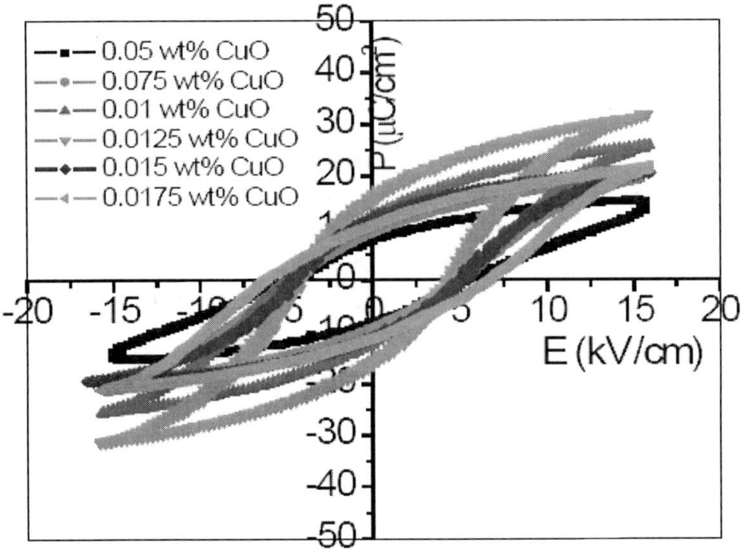

Figure 9. Hysteresis loops of PZT-PZN-PMnN + x wt% CuO ceramic samples.

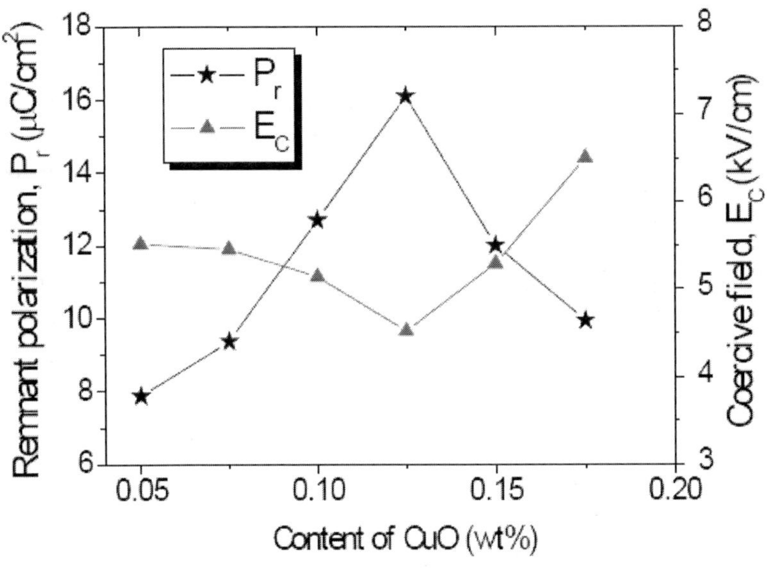

Figure 10. The remnant polarization (P_r) and coercive field (Ec) as a function of the CuO content.

Figure 9 shows the shape of feroelectric hysteresis loops of the samples PZT-PZN-PMnN + x wt% CuO measured at room temperature. From the shape of these loops, the remanent polarization (P_r) and the coercive field (E_C) were determined, as shown in Figure 10. With increasing of CuO content, a sharp increases in P_r was observed for samples until x = 0.125, reaches the highest value (16 μC/cm^2) at x = 0.125, and then decreases, while the coercive field E_C strong decreases and reaches smallest value (4.5kV/cm) at x = 0.125. These results are in good agreement with the studied dielectric and piezoelectric properties of the ceramic samples.

4. CONCLUSION

The effect of CuO addition on the sintering behavior and physical properties of $0.8Pb(Zr_{0.48}Ti_{0.52})O_3$- $0.125Pb(Zn_{1/3}Nb_{2/3})O_3$- $0.075Pb(Mn_{1/3}Nb_{2/3})O_3$ + x wt% CuO (x = 0.0 ÷ 0.175) ceramics was investigated. The addition of CuO improved the sinterability of the samples and caused an increase in the density and grain size at low sintering temperature (850°C). All samples have pure perovskite phase. Crystal structure of the ceramics is modified significantly from rhombohedral structure to tetragonal structure by CuO additions. At the CuO content of 0.125 wt%, physical properties of ceramics are best: the density of 7.91 g/cm^3; the electromechanical coupling factor, k_p = 0.55 and k_t = 0.46; the dielectric constant, ε = 1179; the dielectric loss (tanδ) of 0.006; the mechanical quality factor (Q_m) of 1174; the piezoelectric constant (d_{31}) of 112 pC/N; the remanent polarization (P_r) of 16 μC/cm^2. The improvement of the electrical properties of the ceramics after adding CuO is mainly due to the presence of liquid phase in during sintering enhances the density and grain size, which leads to the decrease of the energy loss and improvement of the electrical properties.

REFERENCES

1. Xu, Y. (1991) Ferroelctric Materials and Their Applications. North-Holland, Amsterdam-London-New York-Tokyo.

2. Hou, Y.D., Zhu, M.K., Tian, C.S. and Yan, H. (2004) Structure and Electrical Properties of PMZN-PZT Quaternary Ceramics for Piezoelectric Transformers. Sensors and Actuators A: Physical, 116, 455-460.

3. Gao, F., Cheng, L., Hong, R., Liu, J., Wang, C. and Tian, C. (2009) Crystal Structure and Piezoelectric Properties of $xPb(Mn_{1/3}Nb_{2/3})O_3$-(0.2 − x)$Pb(Zn_{1/3}Nb_{2/3})O_3$-0.8$Pb(Zr_{0.52}Ti_{0.48})O_3$ Ceramics. Ceramics International, 35, 1719-1723.

4. Hou, Y.D, Chang, L.M., Zhu, M.K., Song, X.M. and Yan, H. (2007) Effect of Li_2CO_3Addition on the Dielectric and Piezoelectric Responses in the Low-Temperature Sintered 0.5PZN-0.5PZT Systems. Journal of Applied Physics, 102, Article ID: 084507.

5. Jin, B.M., Lee, D.S., Kimb, I.W., Kwon, J.H., Lee, J.S., Song, J.S. and Jeong, S.J. (2004) The Additives for Improving Piezoelectric and Ferroelectric Properties of 0.2$Pb(Mg_{1/3}Nb_{2/3})O_3$-0.8[$PbZrO_3$-$PbTiO_3$] Ceramics. Ceramics International, 30, 1449-1451.

6. Lee, J.S., Choi, M.S., Hung, N.V., Kim, Y.S., Kim, I.W., Park, E.C., Jeong, S.J. and Song, J.S. (2007) Effects of High Energy Ball-Milling on the Sintering Behavior and Piezoelectric Properties of PZT-Based Ceramics. Ceramics International, 33, 1283-1286.

7. Vuong, L.D. and Gio, P.D. (2013) Effect of Li_2CO_3 Addition on the Sintering Behavior and Physical Properties of PZT-PZN-Pmnn Ceramics. International Journal of Materials Science and Applications, 2, 89-93.

8. Kim, J.M., Kim, J.S. and Cheon, C.I. (2011) Low-Temperature Sintering and Electrical Properties of PGO-Doped PNN-PZT Ceramics. Journal of Ceramic Processing Research, 12, 12-15.

9. Yoo, J. and Lee, S. (2009) Piezoelectric and Dielectric Properties of Low Temperature Sintered $Pb(Mn_{1/3}Nb_{2/3})$ 0.02$(Ni_{1/3}Nb_{2/3})$0.12(ZrxTi1-x)0.86O_3 System Ceramics. Transactions on Electrical and Electronic Materials, 10, 121-125.

10. Chao, X., Yang, Z., Kang, C. and Chang, Y. (2008) Effects of $BiFeO_3$ Addition on Electrical Properties and Temperature Stability of Low Temperature Sintered PZT-PFW-PMN Ceramics. Sensors and Actuators A: Physical, 141, 482- 488.

11. Han, H.S., Park, E.C., Lee, J.S., Yoon, J.I. and Ahn, K.K. (2011) Low-Firing $Pb(Zr,Ti)O_3$-Based Multilayer Ceramic Actuators Using Ag Inner Electrode. Transactions on Electrical and Electronic Materials, 12, 249-252.

12. Yoo, J., Lee, I., Paik, D.S. and Park, Y.W. (2009) Piezoelectric and Dielectric Properties of Low Temperature Sintering $Pb(Mn_{1/3}Nb_{2/3})O_3$-$Pb(Zn_{1/3}Nb_{2/3})O_3$-$Pb(Zr_{0.48}Ti_{0.52})O_3$Ceramics with Variation of Sintering Time. Journal of Electroceramics, 23, 519-523.

13. Yoon, S.J., Choi, J.W., Choi, J.Y., Wan, D., Li, Q. and Yang, Y. (2010) Influences of Donor Dopants on the Properties of PZT-PMS-PZN Piezoelectric Ceramics Sintered at Low Temperatures. Journal of the Korean Physical Society, 57, 863-867.

14. Nam, C.H., Park, H.Y., Seo, I.T., Choi, J.H., Nahm, S. and Lee, H.G. (2011) Effect of CuO on the Sintering Temperature and Piezoelectric Properties of MnO_2-Doped $0.75Pb(Zr_{0.47}Ti_{0.53})O_3$-$0.25Pb(Zn_{1/3}Nb_{2/3})O_3$ Ceramics. Journal of Alloys and Compounds, 509, 3686-3689.

15. Fan, H. and Kim, H.E. (2002) Perovskite Stabilization and Electromechanical Properties of Polycrystalline Lead Zinc Niobate-Lead Zirconate Titanate. Journal of Applied Physics, 91, 317-322.

16. Kim, M.S., Jeon, S., Jeong, S.J., Kim, I.S. and Song, J.S. (2008) Effect of CuO Additions on Microstructures and Electromechanical Properties of $0.4Pb(Mg_{1/3}Nb_{2/3})O_3$-$0.25PbZrO_3$-$0.35PbTiO_3$ Ceramics. Electronic Materials Letters, 4, 189-192.

17. Kitaguchi, H., Takada, J., Oda, K. and Miura, Y. (1990) Equilibrium Phase Diagram for the System PbO-CaO-CuO. Journal of Materials Research, 5, 829-931.

18. Kim, Y.H., Ryu, H., Cho, Y.K., Lee, H.J. and Nahm, S. (2013) TEM Observations on $0.65Pb(Zr_{0.42}Ti_{0.58})O_3$-$0.35Pb(Ni_{0.33}Nb_{0.67})O_3$ Ceramics with CuO Additive. Journal of the American Ceramic Society, 96, 312-317.

19. Jian, H.L. (2013) Effect of CuO Addition on Structure and Electrical Properties of Low Temperature Sintered Quaternary Piezoelectric Ceramics. Bulletin of Materials Science, 36, 877-881.

20. Nam, C.H., Park, H.Y., Seo, I.T., Choi, J.H., Joung, M.R., Nahm, S., et al. (2011) Low-Temperature Sintering and Pie- zoelectric Properties of $0.65Pb(Zr_{1-x}Ti_x)O_3$-$0.35Pb(Ni_{0.33}Nb_{0.67})O_3$ Ceramics. Journal of the American Ceramic Society, 94, 3442-3448.

21. Chao, X., Ma, D., Gu, R. and Yang, Z. (2010) Effects of CuO Addition on the Electrical Responses of the Low-Tem-perature Sintered $Pb(Zr_{0.52}Ti_{0.48})O_3$-$Pb(Mg_{1/3}Nb_{2/3})O_3$-$Pb(Zn_{1/3}Nb_{2/3})O_3$ Ceramics. Journal of Alloys and Compounds, 491, 698-702.

Chapter 11

A Study on the Mode Shape Piezoelectric Motor

Jwo Ming Jou

Department of Mechanical Engineering, Cheng Shiu University, Kaohsiung City, Taiwan

ABSTRACT

In this paper, we try to use the coating of effective electrode surface and change the direction of polarization to design the mode shape piezoelectric motors of the first three modes. We also compare the gain of the mode shape piezoelectric motors with respect to the normal shape piezoelectric motor, including rotational speed, loading ability, torque, phase angle conversion and efficiency. According to the results of theoretical and simulation analysis, we have found that the gain of the mode shape piezoelectric stators are larger than the normal shape piezoelectric stator on average. According to the results of experiments, we found that the gain of the rotational speed, loading ability, torque, driving phase angle conversion and efficiency of the mode shape (MS1 - 3) piezoelectric motors are higher than the normal shape piezoelectric motor (NS) under driving condition of the second vibration mode. Also, the gain of the rotational speed and loading ability of the mode shape 2 (MS2) piezoelectric motor are higher than other shapes piezoelectric motors (NS, MS1 and MS3) under driving condition of the second vibration mode. The used maximum rotational speed of the mode shape 2 (MS2) piezoelectric motor is up to 946 rpm under conditions of 180 $V_{p\text{-}p}$ driving voltage, 10.7 kHz driving frequency, 0° driving phase angle and

13.0 gw net weight. The maximum loading ability and torque of the mode shape 2 (MS2) piezoelectric motor is respectively 451 gw and 0.91 mkgw-m under conditions of 180 V_{p-p} driving voltage, 10.7 kHz driving frequency, 0° driving phase angle and 173 rpm rotational speed. And the gain of efficiency (output power) and maximum loading ability (torque) of the mode shape 2 (MS2) piezoelectric motor are respectively 2.28 and 1.54 with respect to the normal shape piezoelectric motor under conditions of 180 V_{p-p} driving voltage, 10.7 kHz driving frequency and 0° driving phase angle. According to the results of the experiments, we have finally found that the piezoelectric motors (NS and MS1 - 3) can be driven only by the second vibration mode because the stator can produce elliptical motion and allows the rotor to generate orientation rotation. However, the first vibration mode can allow the rotor to be rotated very fast but it can't make the rotation of the rotor orientation. Furthermore, we also found that the rotor can't rotate by the third vibration mode because its vibration energy is absorbed by the structure itself, so causing the rotor stagnation.

Keywords: Mode Shape (MS), Normal Shape (NS), Piezoelectric Motor, Piezoelectric Stator, Gain

1. INTRODUCTION

The main object for study is a mode shape piezoelectric motor and its comparison object is the normal shape piezoelectric motor in this paper. Its appearance is similar to the tubular piezoelectric motors or ultrasonic motors. For comparison, the found experimental data are listed in Table 1 [1] - [10]. The main difference between the normal shape and mode shape piezoelectric motor lies in the different coating method of effective electrode surface. Due to the normal shape, the tubular piezoelectric motors have many advantages, including small size, fast rotational or movement speed and response, high torque or high loading ability. So they are very useful for optical and electro-mechanical systems. For example, they may be applied to autofocus system of the optical lens, and let them to be embedded within the camera or mobile phones. Also, they are the main transmission mode generated by the actuating force or frictional force of the piezoelectric stator to drive the rotor in order to rotate or slide movement. The size of the actuating force or frictional force depends on the composition of the stator and motor, the coating of effective electrode surface, and driving conditions. The tubular piezoelectric motor was first described in 2002 [1]. It is primarily a metal tube on both sides of two piezoelectric ceramic affixed. The stator size is D2.4 mm × L10 mm. The maximum angular or rotational speed of tubular piezoelectric motor is 60 rad/s or 573 rpm under conditions of 120 $V_{p\text{-}p}$ driving voltage and 69.5 kHz driving frequency. And the maximum torque, output power and efficiency are 1.8 mNm, 60 mW, and 25% respectively. The following year, a smaller tubular piezoelectric motor has been proposed for publication [2], which size of the stator is D1.6 mm × L6 mm. But the maximum angular or rotational speed is only 45 rad/s or 430 rpm under conditions of 100 $V_{p\text{-}p}$ driving voltage and 131.8 kHz driving frequency. And the maximum torque, output power and efficiency are 0.5 mNm, 45 mW and 16% respectively. Then again the next year, another smaller tubular piezoelectric motor is proposed [3], which size of the stator is D0.8 mm × L2.2 mm. The maximum rotational speed is up to 3850 rpm under conditions of 40 $V_{p\text{-}p}$ driving voltage and 70.0 kHz driving frequency. But the maximum torque and loading are only 25 nNm and 0.5 mN respectively. In 2005, a miniature tubular piezoelectric motor is published [4], which size of the stator is D4.0 mm × L5.0 mm. The maximum rotational speed is up to 4000 rpm under conditions of 25 $V_{p\text{-}p}$ driving voltage and 67.0 kHz driving frequency. But the maximum torque is only 1.6 μNm. In 2006, another miniature tubular piezoelectric motor is proposed [5], which size ratio of L/D of the stator is 3.50. The maximum rotational speed is only 385 rpm under conditions of 200 $V_{p\text{-}p}$ driving voltage and 49.4 kHz driving frequency. And the maximum torque is 3.1 Nm. In 2007, another one miniature tubular piezoelectric motor is published [6], which size of the stator is D4.0 mm × L5.0 mm. The maximum rotational speed is up to 9600 rpm under conditions of 30 $V_{p\text{-}p}$ driving voltage and 314.5 kHz driving frequency. But the maximum torque is 5.5 μNm. In 2009, a metal tube type piezoelectric motor is applied to push the optical lens [7] , which size of the stator is D3.0 mm × L12.0 mm. The maximum rotational speed is 2075 rpm under conditions of 140 $V_{p\text{-}p}$ driving voltage and 62.0 kHz driving frequency. And the maximum loading is 420 mN. In 2011, a single vibration mode tubular piezoelectric motor is published [8] , which size of the stator is D6.6 mm × L25.4 mm. The maximum rotational speed is 400 rpm under conditions of 100 $V_{p\text{-}p}$ driving voltage and 23.5 kHz driving frequency. But the maximum torque is only 0.3 mNm. Next year, a rotary-linear piezoelectric motor is proposed [9] , which size of the stator is D12.0 mm × L55.0 mm. The maximum movement speed is only 11 mm/s under conditions of 150 V_{p-p} driving voltage and 19.7 kHz driving frequency. But the maximum torque is up to 500 mN. Recently, a tubular piezoelectric motor is proposed [10], which size of the stator is D5.0 mm × L15.0 mm. The maximum rotational speed is only 208 rpm under conditions of 100 V_{p-p} driving voltage and 97.0 kHz driving frequency. But the maximum torque and loading are only 3.6 μNm and 5.0 mN. At this point, we can find that the tubular piezoelectric motors have characteristics of high speed and low torque or low speed and high torque. However, if the torque or loading of the tubular piezoelectric motor is too small, then a big speed is still not being applied to for optical or electro-mechanical systems. Therefore, the appropriate composition, size, torque, loading ability and rotational or movement speed of the tubular piezoelectric motor must be carefully considered.

Table 1. The performance comparison of the tubular piezoelectric motors [1] -[10].

References	Speed (rpm or mm/s)	Torque (mNm) or Loading (mN)	Voltage (V_{p-p})	Frequency (kHz)	Size (mm)
[1]	573	1.8	120	69.5	D2.4 × L10.0
[2]	430	0.5	100	131.8	D1.6 × L6.0
[3]	3850	2.5E−5 or 0.5	40	70.0	D0.8 × L2.2
[4]	4000	1.6E−3	25	67.0	D4.0 × L5.0
[5]	380	3.1	200	49.4	L/D = 3.5
[6]	9600	5.5E-3	30	314.5	D4.0 × L5.0
[7]	2075	420	140	62.0	D3.0 × L12
[8]	400	0.3	100	23.5	D6.6 × L25.4
[9]	11	500	150	19.7	D12.0 × L55.0
[10]	208	3.6E−3 or 5.0	100	97.0	D5.0 × L15.0

2. THE COMPOSITION AND OPERATION PRINCIPLE

In this paper, the composition and operation principle of the normal shape and mode shape piezoelectric motors as shown Figures 1-3. Wherein the composition of the normal shape or mode shape piezoelectric motors which includes a piezoelectric stator and a rotor. The piezoelectric stator is composed of a metal tube and four piezoelectric ceramics which have the normal shape and mode shape effective electrode surface, as shown Figure 1 and Figure 3. And the full size of the stator piezoelectric is 50 mm × 7.4 mm × 7.4 mm. In operation, as long as to drive the piezoelectric stator by appropriate frequency and voltage, to make the free end of stator generate elliptical trajectory, thereby to push the rotor rotates, as shown Figure 2.

As for the coating method of effective electrode surface of the mode shape piezoelectric ceramics are based on the vibration modes of the cantilever beam or piezoelectric stator, shown as Figure 4. In which the equation of dimensionless vibration modes $\bar{\phi}_n(\bar{x})$ as follows:

$$\bar{\phi}_n(\bar{x}) = \cosh\bar{\lambda}_n\bar{x} - \cos\bar{\lambda}_n\bar{x} - \bar{\sigma}_n\left(\sinh\bar{\lambda}_n\bar{x} - \sin\bar{\lambda}_n\bar{x}\right),\ 0 \le \bar{x} \le 1;\ n = 1,\ 2,\ 3, \cdots, \infty \quad (1)$$

where the dimensionless eigenvalues $\bar{\lambda}_n$ and parameters $\bar{\sigma}_n$ are defined as:

$$\cosh\bar{\lambda}_n \cos\bar{\lambda}_n + 1 = 0 \quad (2)$$

and

$$\bar{\sigma}_n = \frac{\sinh\bar{\lambda}_n - \sin\bar{\lambda}_n}{\cosh\bar{\lambda}_n + \cos\bar{\lambda}_n} \quad (3)$$

As for the effective electrode surfaces $FP_n(\bar{x})$ of the normal shape and mode shape piezoelectric ceramics can be defined as:

$$FP_n(\bar{x}) = \begin{cases} \left[FP_n(\bar{x})\right]_{NS} = 0 \\ \left[FP_n(\bar{x})\right]_{MS} = \dfrac{1}{\bar{\lambda}_n^2}\dfrac{\partial^2\phi_n(\bar{x})}{\partial\bar{x}^2} \end{cases} \quad (4)$$

Because of the high frequency vibrations are easily absorbed by the structure itself, so we only focus on the first three vibration modes and effective electrode surfaces of the mode shape piezoelectric ceramics, as shown Table 2 and Figure 5.

3. Equation of Motion

The equation of motion of the normal shape and mode shape piezoelectric stator can be obtained from the previous paper as follows [11]:

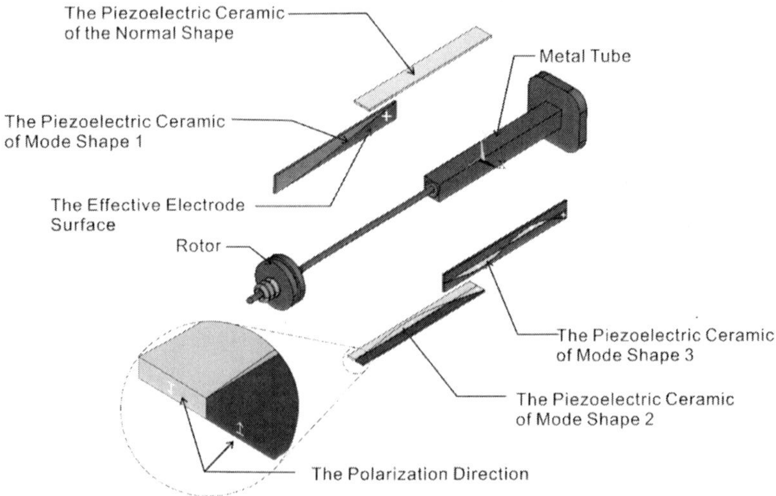

Figure 1. The composition of the normal shape or mode shape piezoelectric motor.

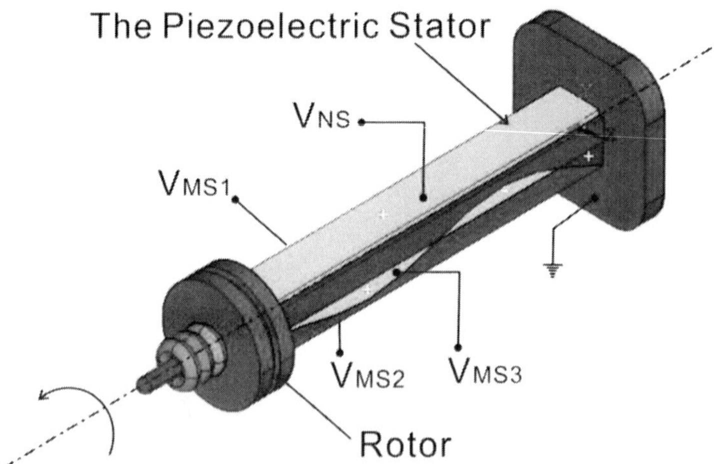

Figure 2. The operation principle of the normal shape or mode shape piezoelectric motor.

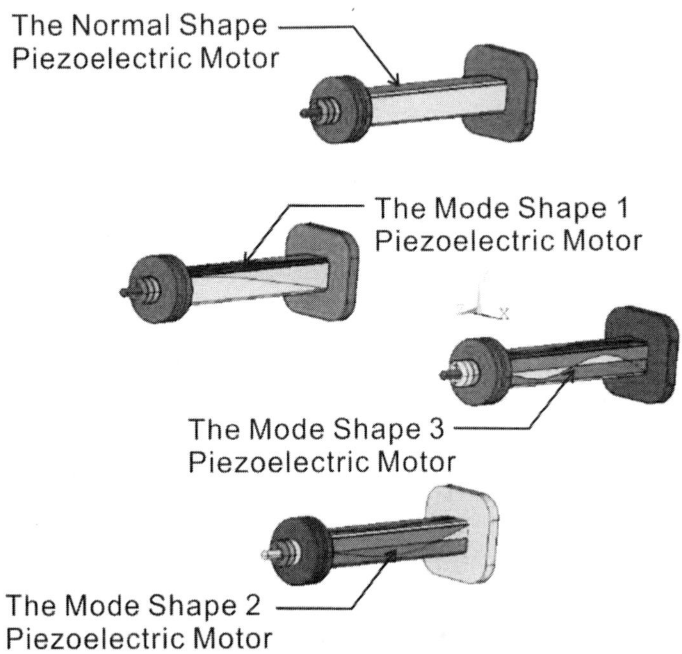

Figure 3. The oblique view of the normal shape and mode shape piezoelectric motors.

Table 2. The first three vibration modes and effective electrode surfaces of the mode shape piezoelectric ceramics.

x/L	$\bar{\phi}_1$	$\bar{\phi}_2$	$\bar{\phi}_3$	FP_1	FP_2	FP_3
0.0	0.00	0.00	0.00	2.00	2.00	2.00
0.1	0.03	0.19	0.46	1.70	1.00	0.50
0.2	0.13	0.60	1.21	1.50	0.10	−0.80
0.3	0.27	1.05	1.51	1.20	−0.60	−1.30
0.4	0.46	1.37	1.05	0.90	−1.20	−0.90
0.5	0.68	1.43	0.04	0.70	−1.40	0.00
0.6	0.92	1.18	−0.95	0.50	−1.40	1.10
0.7	1.18	0.63	−1.31	0.30	−1.10	1.50
0.8	1.45	−0.14	−0.79	0.10	−0.60	1.20
0.9	1.72	−1.05	0.46	0.00	−0.20	0.50
1.0	2.00	−2.00	2.00	0.00	0.00	0.00

where $\bar{\lambda}_1 = 1.8751$, $\bar{\lambda}_2 = 4.6941$, $\bar{\lambda}_3 = 7.8548$; $\bar{\sigma}_1 = 0.7341$, $\bar{\sigma}_2 = 1.0185$, $\bar{\sigma}_3 = 0.9992$.

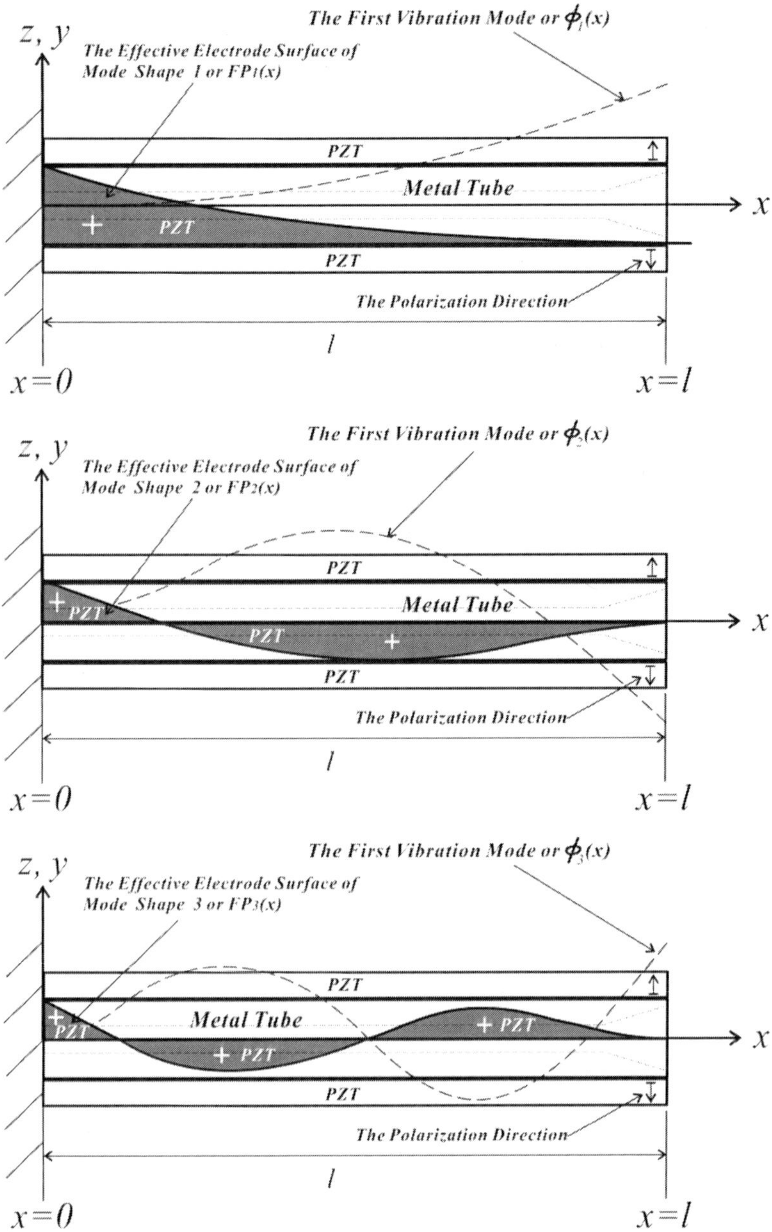

Figure 4. The first three vibration modes and effective electrode surface of mode shape piezoelectric stator.

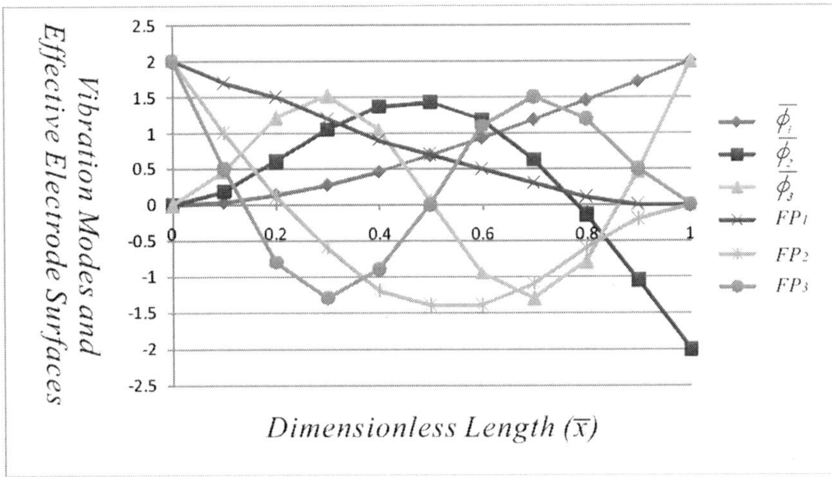

Figure 5. The first three vibration modes and effective electrode surfaces of the mode shape piezoelectric ceramics.

where $\overline{\lambda}_1 = 1.8751,\ \overline{\lambda}_2 = 4.6941,\ \overline{\lambda}_3 = 7.8548;\ \overline{\sigma}_1 = 0.7341,\ \overline{\sigma}_2 = 1.0185,\ \overline{\sigma}_2 = 0.9992$.

$$\left[\frac{\partial^4 \overline{w}(\overline{x},t)}{\partial \overline{x}^4} + \overline{\lambda}_i^4 \frac{\partial^2 \overline{w}(\overline{x},t)}{\partial t^2} = 0\right]_{NS}. \tag{5}$$

and

$$\left[\frac{\partial^4 \overline{w}(\overline{x},t)}{\partial \overline{x}^4} + \overline{\lambda}_i^4 \frac{\partial^2 \overline{w}(\overline{x},t)}{\partial t^2} = \frac{\Lambda_3(t)}{\overline{\lambda}_n^2 D_{11}} \frac{\partial^4 \overline{\phi}_n(\overline{x})}{\partial \overline{x}^4}\right]_{MS}. \tag{6}$$

where the definition of constants in the above Equation (5)-(6) as follow:

$$D_{11} = \frac{1}{3}\sum_{k=1}^{n}\left[\left(c_{11}\right)_k \left(z_k^3 - z_{k-1}^3\right)\right]. \tag{7}$$

$$\rho h = \sum_{k=1}^{n} (\rho_k h_k) \quad (8)$$

$$h_k = z_k - z_{k-1} \quad (9)$$

$$\bar{\lambda}_i^4 = \frac{\rho h l^4 \omega_i^2}{D_{11}} = \frac{4\rho h l^4 \pi^2 f_i^2}{D_{11}} \quad (10)$$

$$\bar{\lambda}_n^4 = \frac{\rho h l^4 \omega_n^2}{D_{11}} = \frac{4\rho h l^4 \pi^2 f_n^2}{D_{11}} \quad (11)$$

where in above the symbols of

c_{11}, D_{11}, e_{31}, f_i, f_n, h, V_3, W, t, x, z, ω_i, ω_n, $\bar{\lambda}_i$ and $\bar{\lambda}_n$ are defined as the Young's modulus, bending stiffness constant per unit width, an arbitrary frequency, the resonance frequency, thickness, piezoelectric charge constant, transverse driving voltage, transverse displacement, time, longitudinal coordinate, transverse coordinate, density, the angle velocity of arbitrary frequency, the angle velocity of resonance frequency, the eigenvalues of arbitrary frequency and the eigenvalues of resonance frequency.

Furthermore we can find the general solution of mode shape and normal shape piezoelectric stator from the previous paper as follow [11]:

$$\bar{w}_{NS}(\bar{x},t) = \left[C_1 \sinh \bar{\lambda}_i \bar{x} + C_2 \cosh \bar{\lambda}_i \bar{x} + C_3 \sin \bar{\lambda}_i \bar{x} + C_4 \cos \bar{\lambda}_i \bar{x}\right] e^{j\omega_i t} \quad (12)$$

and

$$\bar{w}_{MS}(\bar{x},t) = \left[C_1 \sinh \bar{\lambda}_i \bar{x} + C_2 \cosh \bar{\lambda}_i \bar{x} + C_3 \sin \bar{\lambda}_i \bar{x} + C_4 \cos \bar{\lambda}_i \bar{x} + A_n \bar{\phi}_n(\bar{x})\right] e^{j\omega_i t} \quad (13)$$

where above constants can be determined by the following electro-mechanical boundary conditions:

$$\left[\bar{w}(0,t)=\bar{w}'(0,t)=0;\ \bar{w}''(1,t)=\frac{\Pi_3 e^{j\omega_i t}}{\bar{\lambda}_n^2 D_{11}};\ \bar{w}'''(1,t)=0\right]_{NS}. \quad (14)$$

and

$$\left[\bar{w}(0,t)=\bar{w}'(0,t)=0;\ \bar{w}''(1,t)=\frac{\Pi_3 e^{j\omega_i t}}{\bar{\lambda}_n^2 D_{11}}\frac{\partial^2\bar{\phi}_n(\bar{x})}{\partial\bar{x}^2}\bigg|_{\bar{x}=1};\ \bar{w}'''(1,t)=\frac{\Pi_3 e^{j\omega_i t}}{\bar{\lambda}_n^2 D_{11}}\frac{\partial^3\bar{\phi}_n(\bar{x})}{\partial\bar{x}^3}\bigg|_{\bar{x}=1}\right]_{MS}. \quad (15)$$

Let Equation (12) and Equation (13) are substituted into Equation (14) and Equation (15), we can get the transverse displacement solution of the first three vibration modes of free end of the normal shape and mode shape piezoelectric stators on xz-plane as follow:

$$\left[\bar{w}(\bar{x},t)\right]_{NS}=\frac{\Pi_3 e^{j\omega_i t}}{D_{11}}\frac{\cosh\bar{\lambda}_i+\cos\bar{\lambda}_i}{\bar{\lambda}_i^2\left(1+\cosh\bar{\lambda}_i\cos\bar{\lambda}_i\right)}\bar{\phi}_i(1). \quad (16)$$

and

$$\left[\bar{w}(\bar{x},t)\right]_{MS}=\frac{\Pi_3 e^{j\omega_i t}}{D_{11}}\frac{\bar{\lambda}_n^4}{\bar{\lambda}_n^2\left(\bar{\lambda}_n^4-\bar{\lambda}_i^4\right)}\bar{\phi}_n(1). \quad (17)$$

where the eigenvalues $\bar{\lambda}_i \neq \bar{\lambda}_n$, $\bar{\phi}_i(1)\neq\bar{\phi}_n(1)$ and $i \in R$ and $n \in I$.

Similarly, we can get the solution of vertical displacement of the first three vibration modes of the normal shape and mode shape piezoelectric stators at of free end on xy-plane as follow:

$$\left[\bar{v}(\bar{x},t)\right]_{NS}=\frac{\Pi_3 e^{j\omega_i t}}{D_{11}}\frac{\cosh\bar{\lambda}_i+\cos\bar{\lambda}_i}{\bar{\lambda}_i^2\left(1+\cosh\bar{\lambda}_i\cos\bar{\lambda}_i\right)}\bar{\phi}_i(1). \quad (18)$$

and

$$\left[\bar{v}(\bar{x},t)\right]_{MS}=\frac{\Pi_3 e^{j\omega_i t}}{D_{11}}\frac{\bar{\lambda}_n^4}{\bar{\lambda}_n^2\left(\bar{\lambda}_n^4-\bar{\lambda}_i^4\right)}\bar{\phi}_n(1). \quad (19)$$

Under steady-state operating mode, above Equation (16-19) can be rewritten as follow:

$$\left[\overline{w}(\overline{x},t)\right]_{NS}=\frac{\Pi_3\overline{\phi}_i(1)}{D_{11}}\frac{\cosh\overline{\lambda}_i+\cos\overline{\lambda}_i}{\overline{\lambda}_i^2\left(1+\cosh\overline{\lambda}_i\cos\overline{\lambda}_i\right)}\cos\left(\omega_i t+\Phi\right)=\overline{W}_{NS}\cos\left(\omega_i t+\Phi\right). \quad (20)$$

$$\left[\overline{w}(\overline{x},t)\right]_{MS}=\frac{\Pi_3\overline{\phi}_n(1)}{D_{11}}\frac{\overline{\lambda}_n^4}{\overline{\lambda}_n^2\left(\overline{\lambda}_n^4-\overline{\lambda}_i^4\right)}\cos\left(\omega_i t+\Phi\right)=\overline{W}_{MS}\cos\left(\omega_i t+\Phi\right). \quad (21)$$

$$\left[\overline{v}(\overline{x},t)\right]_{NS}=\frac{\Pi_3\overline{\phi}_i(1)}{D_{11}}\frac{\cosh\overline{\lambda}_i+\cos\overline{\lambda}_i}{\overline{\lambda}_i^2\left(1+\cosh\overline{\lambda}_i\cos\overline{\lambda}_i\right)}\cos\left(\omega_i t+\Phi\right)=\overline{V}_{NS}\cos\left(\omega_i t+\Phi\right). \quad (22)$$

and

$$\left[\overline{v}(\overline{x},t)\right]_{MS}=\frac{\Pi_3\overline{\phi}_n(1)}{D_{11}}\frac{\overline{\lambda}_n^4}{\overline{\lambda}_n^2\left(\overline{\lambda}_n^4-\overline{\lambda}_i^4\right)}\cos\left(\omega_i t+\Phi\right)=\overline{V}_{MS}\cos\left(\omega_i t+\Phi\right). \quad (23)$$

where Φ is defined as driving phase angle. And $\overline{W}_{NS}$, $\overline{W}_{MS}$, $\overline{V}_{NS}$ and $\overline{V}_{MS}$ representative amplitude of the normal shape and mode shape piezoelectric stators at the transverse and vertical direction respectively.

In order to get an elliptical trajectory of the normal shape and mode shape piezoelectric stators at the free end, we can make driving phase angle difference 90° or use a different driving waveforms under different sides of piezoelectric stators. Therefore, above Equations (23)-(24) can be rewritten again as follow:

$$\left[\overline{v}(\overline{x},t)\right]_{NS}=\frac{\Pi_3\overline{\phi}_i(1)}{D_{11}}\frac{\cosh\overline{\lambda}_i+\cos\overline{\lambda}_i}{\overline{\lambda}_i^2\left(1+\cosh\overline{\lambda}_i\cos\overline{\lambda}_i\right)}\sin\left(\omega_i t+\Phi\right)=\overline{V}_{NS}\sin\left(\omega_i t+\Phi\right). \quad (24)$$

and

$$\left[\overline{v}(\overline{x},t)\right]_{MS}=\frac{\Pi_3\overline{\phi}_n(1)}{D_{11}}\frac{\overline{\lambda}_n^4}{\overline{\lambda}_n^2\left(\overline{\lambda}_n^4-\overline{\lambda}_i^4\right)}\sin\left(\omega_i t+\Phi\right)=\overline{V}_{MS}\sin\left(\omega_i t+\Phi\right). \quad (25)$$

Follow by, we can get an elliptical trajectory of the normal shape and mode shape piezoelectric stators at the free end on the zy-plane from Equations (20)-(21) and Equations (24)-(25), shown as Figure 6.

If we further consider the effect of axial or horizontal vibration on the normal shape and mode shape piezoelectric stators, then above equations can be rewritten as again:

$$\left[\bar{w}(\bar{x},t)\right]_{NS}=\frac{\Pi_3\bar{\phi}_i(1)}{D_{11}}\frac{\cosh\bar{\lambda}_i+\cos\bar{\lambda}_i}{\bar{\lambda}_i^2\left(1+\cosh\bar{\lambda}_i\cos\bar{\lambda}_i\right)}\cos\left(\omega_i t+\Phi\right)=\bar{W}_{NS}\cos\left(\omega_i t+\Phi\right)\sin\omega_i t \quad (26)$$

$$\left[\bar{w}(\bar{x},t)\right]_{MS}=\frac{\Pi_3\bar{\phi}_n(1)}{D_{11}}\frac{\bar{\lambda}_n^4}{\bar{\lambda}_n^2\left(\bar{\lambda}_n^4-\bar{\lambda}_i^4\right)}\cos\left(\omega_i t+\Phi\right)=\bar{W}_{MS}\cos\left(\omega_i t+\Phi\right)\sin\omega_i t \quad (27)$$

$$\left[\bar{v}(\bar{x},t)\right]_{NS}=\frac{\Pi_3\bar{\phi}_i(1)}{D_{11}}\frac{\cosh\bar{\lambda}_i+\cos\bar{\lambda}_i}{\bar{\lambda}_i^2\left(1+\cosh\bar{\lambda}_i\cos\bar{\lambda}_i\right)}\sin\left(\omega_i t+\Phi\right)=\bar{V}_{NS}\sin\left(\omega_i t+\Phi\right)\sin\omega_i t \quad (28)$$

and

$$\left[\bar{v}(\bar{x},t)\right]_{MS}=\frac{\Pi_3\bar{\phi}_n(1)}{D_{11}}\frac{\bar{\lambda}_n^4}{\bar{\lambda}_n^2\left(\bar{\lambda}_n^4-\bar{\lambda}_i^4\right)}\sin\left(\omega_i t+\Phi\right)=\bar{V}_{MS}\sin\left(\omega_i t+\Phi\right)\sin\omega_i t \quad (29)$$

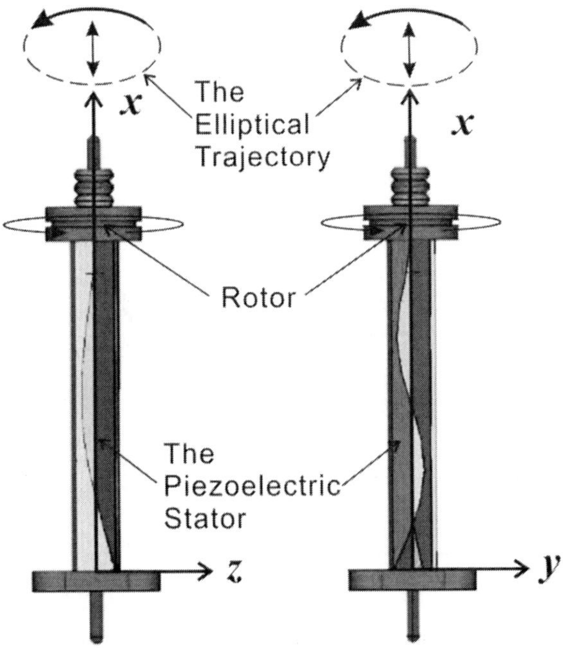

Figure 6. The elliptical trajectory of the normal shape or mode shape piezoelectric stator on xz-plane and xy-plane.

where the axial or horizontal vibration displacement of the normal shape and mode shape piezoelectric stators as follow:

$$\left[\bar{u}\left(\bar{x},t\right)\right]_{NS}=\bar{U}_{NS}\cos\omega_i t \quad (30)$$

and

$$\left[\bar{u}\left(\bar{x},t\right)\right]_{MS}=\bar{U}_{MS}\cos\omega_i t \quad (31)$$

wherein $\bar{U}_{NS}$ and $\bar{U}_{MS}$ representative amplitude of the normal shape and mode shape piezoelectric stators at axial or horizontal direction respectively.

So far, we can get an oval ball of the normal shape and mode shape piezoelectric stators at the free end, as follow:

$$\left(\frac{\bar{u}}{\bar{\bar{U}}}\right)_{NS}^{2}+\left(\frac{\bar{v}}{\bar{\bar{V}}}\right)_{NS}^{2}+\left(\frac{\bar{w}}{\bar{\bar{W}}}\right)_{NS}^{2}=1 \quad (32)$$

and

$$\left(\frac{\bar{u}}{\bar{\bar{U}}}\right)_{MS}^{2}+\left(\frac{\bar{v}}{\bar{\bar{V}}}\right)_{MS}^{2}+\left(\frac{\bar{w}}{\bar{\bar{W}}}\right)_{MS}^{2}=1 \quad (33)$$

In fact, $\bar{U}_{\mathrm{NS}} \ll \bar{V}_{\mathrm{NS}} \approx \bar{W}_{\mathrm{NS}}$ and $\bar{U}_{\mathrm{MS}} \ll \bar{V}_{\mathrm{MS}} \approx \bar{W}_{\mathrm{MS}}$ for 3D piezoelectric stators.

4. GAIN AND EFFICIENCY

In this paper, we can get the gain of vertical and transverse displacement or velocity ($\dot{\bar{V}}_{NS \text{ or } MS}$ or $\dot{\bar{W}}_{NS \text{ or } MS}$) of the normal shape and mode shape piezoelectric stators or motors from Equations (24)-(27) under the same driving conditions, as follow:

$$G_{\bar{V}1}=\left[\bar{V}\left(\bar{\lambda}_i\right)\right]_{MS}:\left[\bar{V}\left(\bar{\lambda}_i\right)\right]_{NS}=\left[\frac{1}{\bar{\lambda}_n^2}\frac{\bar{\lambda}_n^4}{\bar{\lambda}_n^4-\bar{\lambda}_i^4}\bar{\phi}_n(1)\right]_{MS}:\left[\frac{\left(\cosh\bar{\lambda}_i+\cos\bar{\lambda}_i\right)}{\bar{\lambda}_i^2\left(1+\cosh\bar{\lambda}_i\cos\bar{\lambda}_i\right)}\bar{\phi}_i(1)\right]_{NS}. \quad (34)$$

or

$$G_{\dot{V}1} = \left[\dot{\bar{V}}(\bar{\lambda}_i)\right]_{MS} : \left[\dot{\bar{V}}(\bar{\lambda}_i)\right]_{NS} = \left[\frac{1}{\bar{\lambda}_n^2}\frac{\bar{\lambda}_n^4}{\bar{\lambda}_n^4 - \bar{\lambda}_i^4}\bar{\phi}_n(1)\right]_{MS} : \left[\frac{\left(\cosh\bar{\lambda}_i + \cos\bar{\lambda}_i\right)}{\bar{\lambda}_i^2\left(1 + \cosh\bar{\lambda}_i\cos\bar{\lambda}_i\right)}\bar{\phi}_i(1)\right]_{NS}. \quad (35)$$

and

$$G_{\bar{W}1} = \left[\bar{W}(\bar{\lambda}_i)\right]_{MS} : \left[\bar{W}(\bar{\lambda}_i)\right]_{NS} = \left[\frac{1}{\bar{\lambda}_n^2}\frac{\bar{\lambda}_n^4}{\bar{\lambda}_n^4 - \bar{\lambda}_i^4}\bar{\phi}_n(1)\right]_{MS} : \left[\frac{\left(\cosh\bar{\lambda}_i + \cos\bar{\lambda}_i\right)}{\bar{\lambda}_i^2\left(1 + \cosh\bar{\lambda}_i\cos\bar{\lambda}_i\right)}\bar{\phi}_i(1)\right]_{NS}. \quad (36)$$

or

$$G_{\dot{\bar{W}}1} = \left[\bar{W}(\bar{\lambda}_i)\right]_{MS} : \left[\bar{W}(\bar{\lambda}_i)\right]_{NS} = \left[\frac{1}{\bar{\lambda}_n^2}\frac{\bar{\lambda}_n^4}{\bar{\lambda}_n^4 - \bar{\lambda}_i^4}\bar{\phi}_n(1)\right]_{MS} : \left[\frac{\left(\cosh\bar{\lambda}_i + \cos\bar{\lambda}_i\right)}{\bar{\lambda}_i^2\left(1 + \cosh\bar{\lambda}_i\cos\bar{\lambda}_i\right)}\bar{\phi}_i(1)\right]_{NS}. \quad (37)$$

So for, we can predict the gain of the normal shape and mode shape piezoelectric stators from Equations (34)-(37) under the same driving conditions.

In addition, the conversion efficiency of the normal shape and mode shape piezoelectric stators or motors can be defined as:

$$\eta = \frac{P_{Out}}{P_{in}} \le 1.0 \quad . \quad (38)$$

where the definition of the input power as:

$$P_{in} = (IV)_{in} = (IV)_{p-p}\big/\sqrt{2} = (IV)_{rms} \quad . \quad (39)$$

and the definition of the output power as:

$$\begin{aligned} P_{Out} &= F_f \dot{\bar{V}}_{NS\ \mathrm{or}\ MS} \approx F_f \dot{\bar{W}}_{NS\ \mathrm{or}\ MS} = \mu_d W_R r(\omega_n)_{NS\ or\ MS} = T_q(\omega_n)_{NS\ or\ MS} \\ &= \mu_d W_R r(2\pi f_n)_{NS\ or\ MS} = \mu_d W_R r\left(2\pi\frac{N_n}{60}\right)_{NS\ or\ MS} = \frac{2\pi}{60}\mu_d r W_R (N_n)_{NS\ or\ MS} \quad . \end{aligned} \quad (40)$$

where f_n, F_f, I_{in}, r, N_n, T_q, V_{in}, W_R, ω_i, μ_d and η representative the modal resonance frequency, frictional force, input driving current, radius of rotor, rotational speed, torque, input driving voltage, weight or loading of rotor, the modal angle velocity, dynamic friction coefficient and efficiency of the normal shape and mode shape piezoelectric stators or motors.

5. CASE STUDY—THEORETICAL ANALYSIS, SIMULATION ANALYSIS AND EXPERIMENTS

In this paper, we have to understand the differences and gain between the normal shape and mode shape piezoelectric motors or stators, in particular, through theoretical analysis, simulation analysis and experiments to validate. And the material properties are used in the analysis and the experimental procedure, as shown in Table 3.

wherein the step of the theoretical analysis is as follows:

(5-1-1) Using the simple approximate solution of one-dimensional spatial structure to do analysis, and only do a comparative analysis of dimensionless vibration displacement or velocity by Equations (38-39).

(5-1-2) Using the minimum frequency spacing ($\Delta f_i = 1\,\text{Hz}$) to analyze the gain of the first three modes of the mode shape and normal shape piezoelectric stators from 1 kHz to 40 kHz.

As for the simulation analysis procedure is as follows:

(5-2-1) Modeling of the mode shape and normal shape piezoelectric stators respectively, including select element type, enter the physical properties, as well as coordinate system conversion, as shown Figure 7.

(5-2-2) Meshing of the mode shape and normal shape piezoelectric stators respectively, including select the most sophisticated cutting of mesh or select the smart size 1, as shown Figure 8.

(5-2-3) Simulation Analysis: Solving of the mode shape and normal shape piezoelectric stators respectively, including setting boundary conditions of electro-mechanical, as shown Figure 9. All of which driving voltage on the effective electrode surface is 1.0 V.

(5-2-4) Post-processing of the mode shape and normal shape piezoelectric stators respectively, includes processing the first three modes, the maximum deformation or electric potential, as shown Figures 10-13.

As regards the experimental step is as follows:

(5-3-1) Cutting the piezoelectric ceramics with the effective electrode surface, and making the piezoelectric stators of normal shape and mode shape, shown as Figure 14.

(5-3-2) Using the dual channel arbitrary function generator (Model: A-303, AA Lab. Systems Ltd. Co.) and the power amplifier (Model: AFG-3022, Tektronix Co.) to drive the normal shape and mode shape piezoelectric motors, shown as Figure 15.

Table 3. The material properties and size of piezoelectric stator.

Dimension _ Material Name	1D_PZT	2D_PZT	3D_PZT	1D_Al	2D_Al	3D_Al
Relative Permittivity (N.A.)	ε11 = 1730	ε11 = ε22 = 1730, ε33 = 1700	ε11 = ε22 = 1730, ε33 = 1700	0	0	0
Piezoelectric Stress Constant (V/Nm)	e31 = −5.3	e31 = e32 = −5.3	e31 = e32 = −5.3, e33 = 15.8, e24 = e15 = 12.3.	0	0	0
Young's Modulus (Pa)	c11 = 1.2e11	c11 = c22 = 1.2e11	c11 = c22 = 1.2e11, c12 = c21 = 7.52e10, c13 = c31 = c23 = c32 = 7.51e10, c33 = 1.11e11, c44 = 3.0e10, c55 = c66 = 2.6e10	c11 = 7E10	c11 = c22 = 7E10	c11 = c22 = c33 = 7E10
Density (kg/m^3)	7600	7600	7600	2700	2700	2700
Poisson Ratio (N.A.)	0	0.33	0.33	0	0.35	0.35
Size (mm^3)	50 × 6 × 0.7	50 × 6 × 0.7	50 × 6 × 0.7	50 × 6 × 6	50 × 6 × 6	50 × 6 × 6

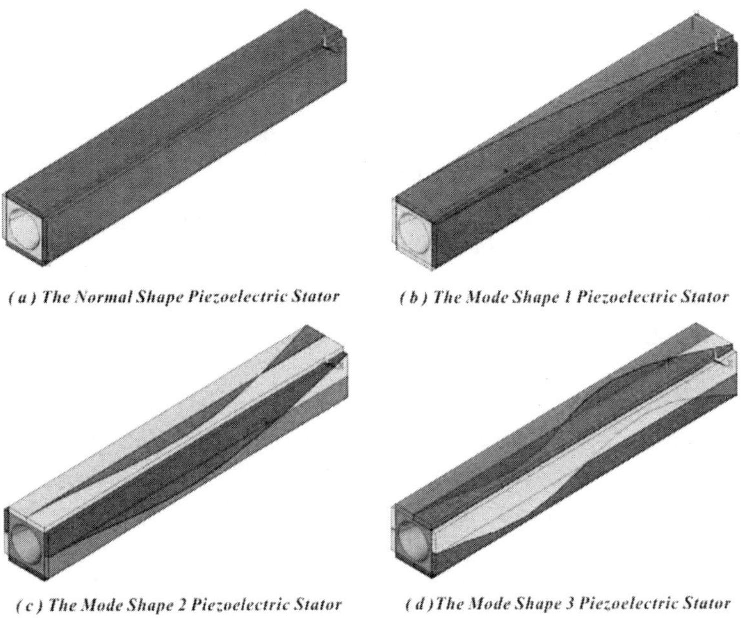

Figure 7. Modeling and setting the material properties of the normal shape and mode shape piezoelectric stators.

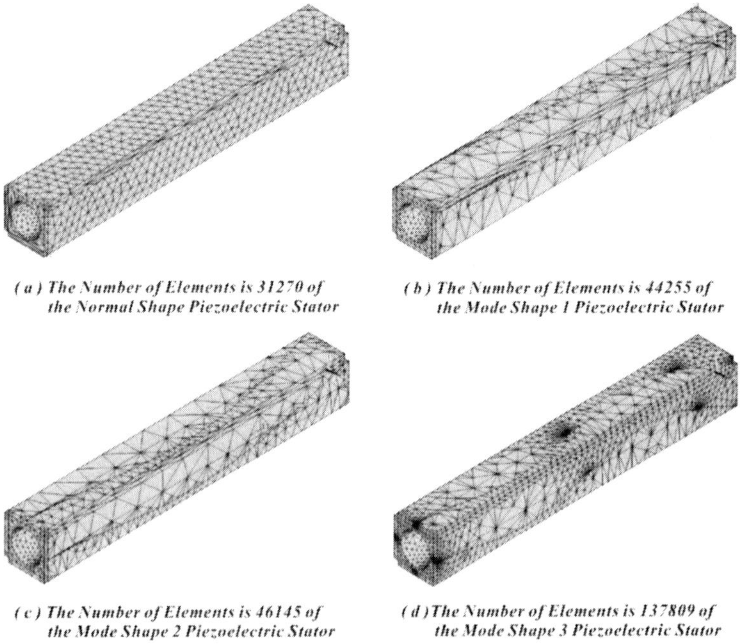

Figure 8. Meshing the material properties of the normal shape and mode shape piezoelectric stators using the most fine mesh.

(5-3-3) Using the digital tachometer (Model: RM-1501, TES Electrical Electronic Co.) and the sound level meter (Model: TES 1350A , TES Electrical Electronic Co.) to measure the rotational speed and noise of the normal shape and mode shape piezoelectric motors, shown as Figure 15.

(5-3-4) Using the loading test weight and rotors to test the loading ability of the normal shape and mode shape piezoelectric motors, shown as Figure 16. Which the material of the loading test weight or mass is aluminum and copper. Their size and weight are D50 mm ´ H10 mm and 53 gw or 165 gw. And the material of the rotor is copper. Its size and weight are D15 mm ´ H10 mm and 13 gw. Due to the material of metal tube of the piezoelectric stator is aluminum, and the material of rotor is copper. So the dynamic friction coefficient of metal tube and rotor is about 0.22 - 0.27 or $\mu_d|_{Al-Cu} = 0.22 - 0.27$. As for the transform coefficient of Equation (42) must be determined after the experiment, it is usually much less than 1.0 or $\mu_T \ll 1.0$.

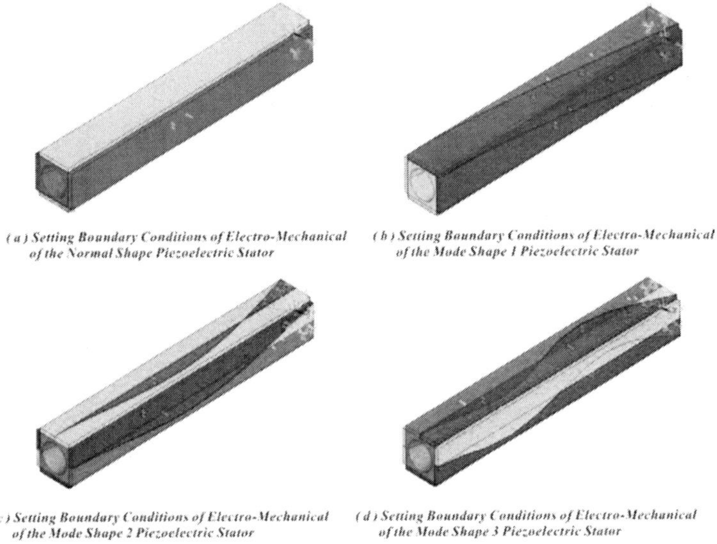

Figure 9. Defined loads or setting boundary conditions of electrode-mechani- cal of the normal shape and mode shape piezoelectric stators under clamped- free mechanical boundary conditions and 1.0 V_{p-p} driving voltage electrical boundary conditions.

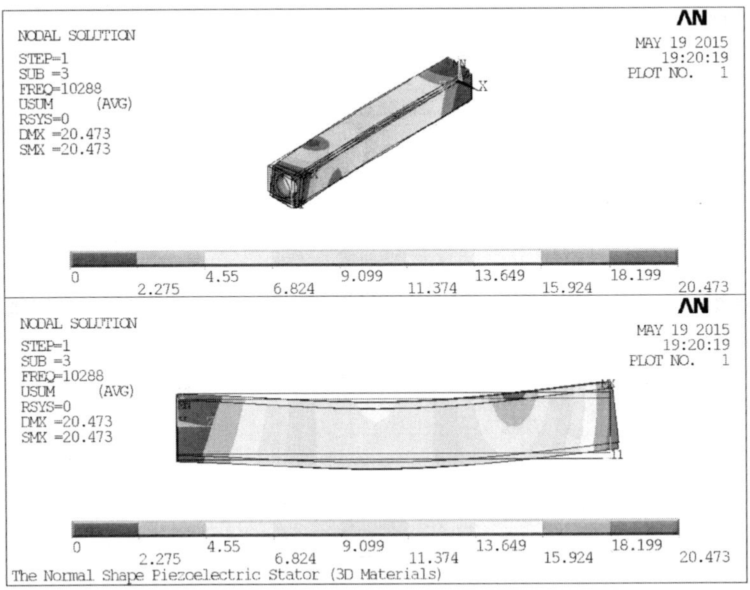

Figure 10. The first three vibration modes of the normal shape piezoelectric stators on the yz plane using 3D Materials.

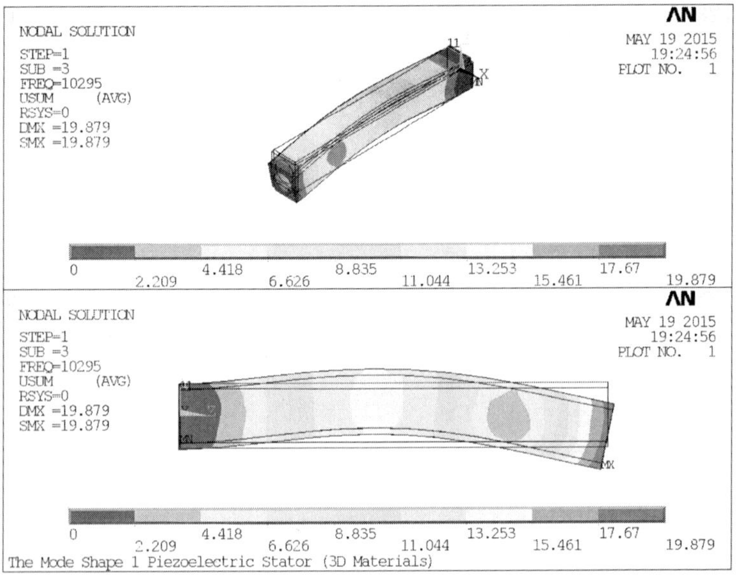

Figure 11. The first three vibration modes of the mode shape 1 piezoelectric stators on the yz plane using 3D Materials.

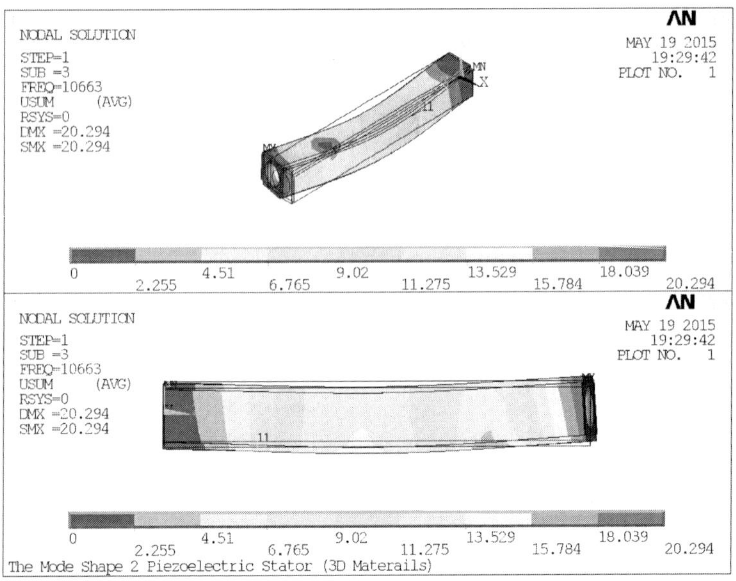

Figure 12. The first three vibration modes of the mode shape 2 piezoelectric stators on the yz plane using 3D Materials.

6. RESULTS AND DISCUSSION

According to the results of theoretical analysis, simulation analysis and experiments, we found:

1) Under conditions of theoretical analysis of one-dimensional (1D) approximation solution and using the minimum frequency spacing of 1Hz, the gain of the maximum dimensionless vibration displacement or velocity of the all mode shape 1 - 3 (MS1 - MS3) piezoelectric stators are far than the normal shape (NS) piezoelectric stator, as shown Figure 17 and Table 4. Where the mode shape 3 (MS3) piezoelectric stator has the largest gain which is about 20.06. Followed by, the gain of the mode shape 1 (MS1) piezoelectric stator is about 19.70. Furthermore is the gain of the mode shape 2 (MS2) piezoelectric stator is about 3.33.

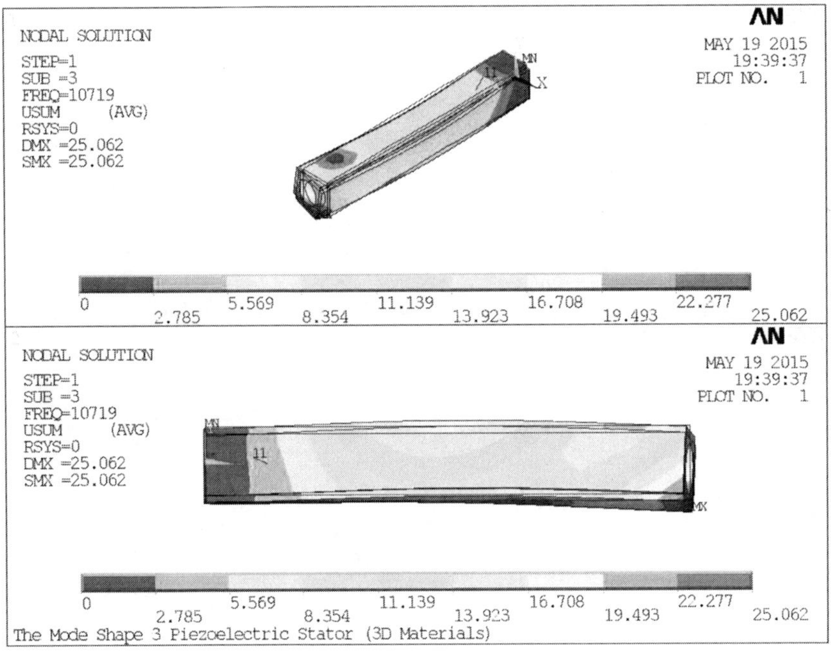

Figure 13. The first three vibration modes of the mode shape 3 piezoelectric stators on the yz plane using 3D Materials.

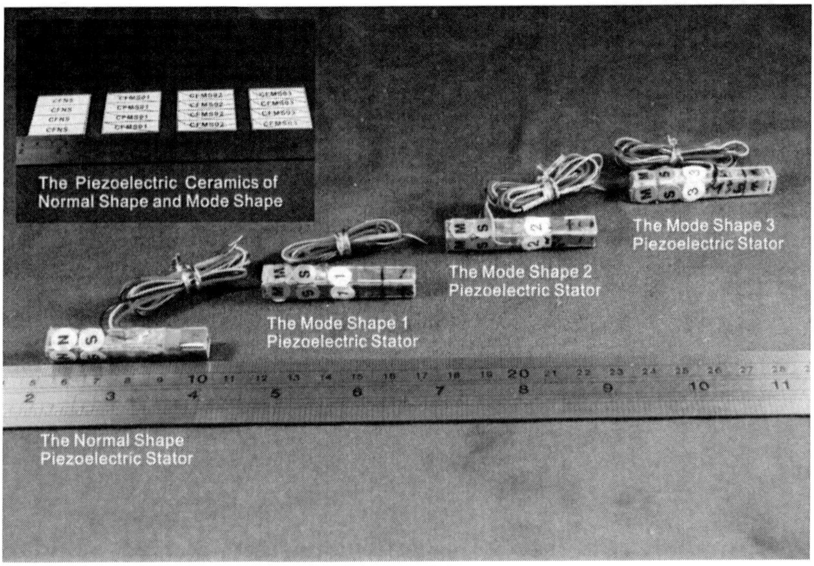

Figure 14. The piezoelectric ceramics and stators of the normal shape and mode shape.

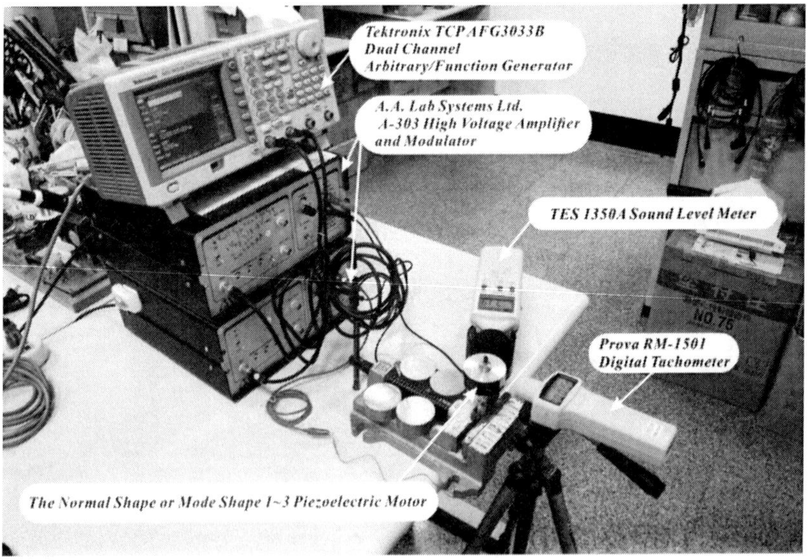

Figure 15. The experimental structure of the normal shape or mode shape piezoelectric motor.

Figure 16. The normal shape and mode shape piezoelectric motors.

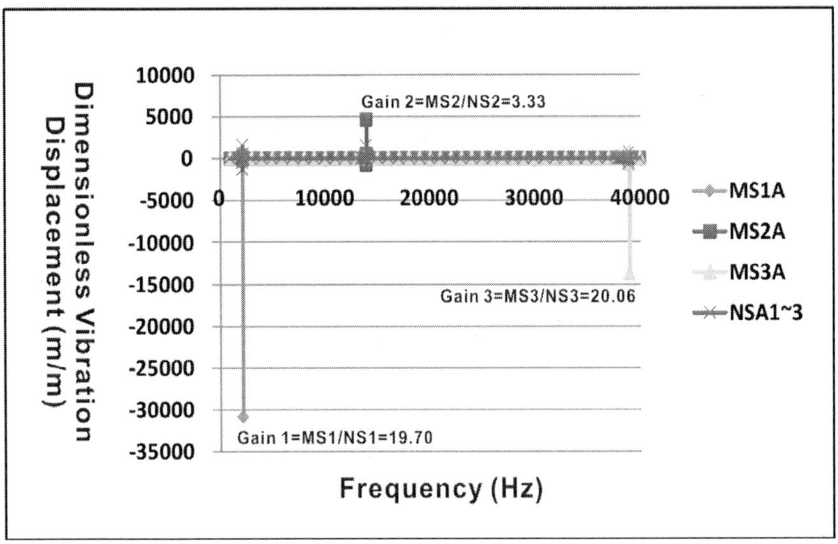

Figure 17. The gain and dimensionless vibration displacement or velocity of the first three modes of the normal shape and mode shape piezoelectric stators by 1D approximate solution.

Table 4. The gain of the maximum dimensionless vibration displacement or rotational speed of the normal shape and mode shape piezoelectric stators under conditions of theory analysis, simulation analysis and experiments.

Mode	G1_Th.	G2_Th.	G3_Th.	G1_1D_Si.	G2_1D_Si.	G3_1D_Si.	G1_2D_Si.	G2_2D_Si.	G3_2D_Si.	G1_3D_Si.	G2_3D_Si.	G3_3D_Si.	G1_Ex.	G2_Ex.	G3_Ex.
1	19.70	0.00	0.00	1.10	1.00	0.80	1.30	1.30	1.30	1.20	1.30	1.00	N.A.	N.A.	N.A.
2	0.00	3.30	0.00	1.20	0.90	0.80	1.40	1.40	1.40	1.20	1.30	1.30	1.19	1.31	1.29
3	0.00	0.00	20.10	1.40	1.10	0.80	1.50	1.50	1.50	1.10	1.50	1.50	0.00	0.00	0.00

Note: Th.: Theoretical Analysis; Si.: Simulation Analysis; Ex.: Experiments; N.A.: Represents the rotation direction of the rotor can not be determined.

2) Under conditions of using the 1D material properties to simulate the 3D piezoelectric stators, the gain of the maximum dimensionless vibration displacement or velocity of the first three modes of the mode shape 1 (MS1) piezoelectric stator is larger than other type (NS, MS2 and MS3) piezoelectric stators. Where the gain of the third vibration mode of the mode shape 1 (MS1) piezoelectric stator with respect to the normal shape (NS) piezoelectric stator is about 1.40, as shown in Table 4.

3) Under conditions of using the 2D material properties to simulate the 3D piezoelectric stators, the gain of the maximum dimensionless vibration displacement or velocity of the all mode shape 1 - 3 (MS1 - MS3) piezoelectric stators are larger than the normal shape (NS) piezoelectric stator. Where the gain of the orthogonal modes of the mode shape 1 - 3 (MS1 - 3) piezoelectric stators with respect to the normal shape (NS) piezoelectric stator are 1.30, 1.40 and 1.50, as shown in Table 4.

4) Under conditions of using the 3D material properties to simulate the 3D piezoelectric stators, the gain of the maximum dimensionless vibration displacement or velocity of the all mode shape 1 - 3 (MS1 - MS3) piezoelectric stators are larger than the normal shape (NS) piezoelectric stator on average. Where the gain of the orthogonal modes of the mode shape 1 - 3 (MS1 - 3) piezoelectric stators with respect to the normal shape (NS) piezoelectric stator are 1.20, 1.30 and 1.50, as shown in Table 4.

5) According to the results of experiments, shown as Figure 18 and Figure 19 and Table 4, we found the normal shape (NS) and mode shape 1 - 3 (MS1 - MS3) piezoelectric motors can be driven by the second vibration mode only, because the second vibration mode can

allow the stator to produce elliptical motion, and allow the rotor to generate orientation rotation. And we also found the gain of rotational speed of the mode shape 2 (MS2) piezoelectric motor is higher than other shapes piezoelectric motors (NS, MS1 and MS3) under driving condition of the second vibration mode. Where the gain of rotational speed of the second vibration mode of the mode shape 1 - 3 (MS1 - 3) piezoelectric stators with respect to the normal shape (NS) piezoelectric stator are 1.19, 1.31 and 1.29, as shown Table 4. In addition, the gain of rotational speed of the first vibration mode can't be determined. Although we observed that the rotational speed of rotor is very fast, but we also found that the direction of rotation of the rotor to be volatile. Therefore, the rotational speed of rotor of all type (NS and MS1 - MS3) piezoelectric stators is treated as zero under driving condition of the first vibration mode. Furthermore, the gain of rotational speed of all type (NS and MS1 - MS3) piezoelectric stators are zero under driving condition of the third vibration mode. Because its vibration energy is absorbed by the structure itself, so causing the rotor stagnant.

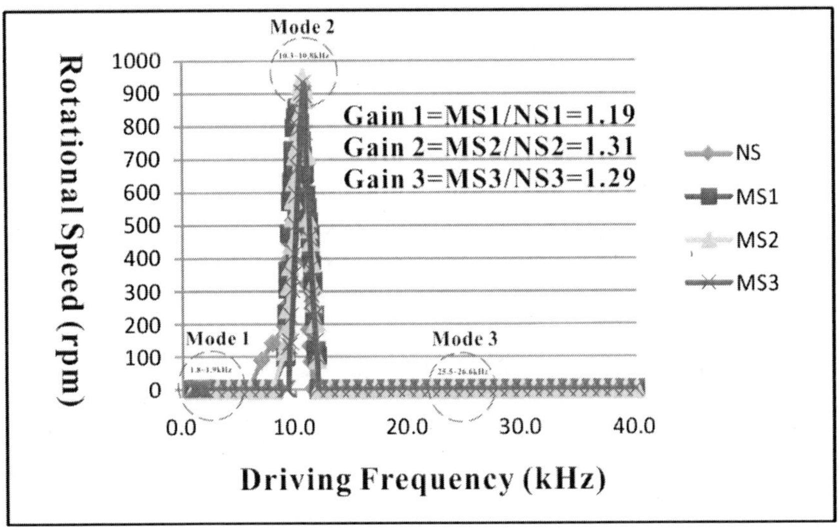

Figure 18. The gain and the rotational speed of the normal shape and mode shape piezoelectric motors under conditions of 180 V_{p-p} driving voltage, 1.0 - 40 kHz driving frequency and 13 gw net weight.

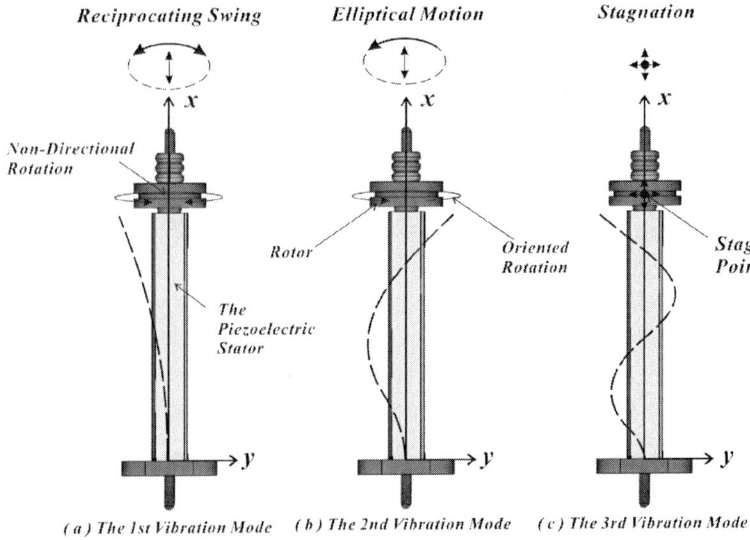

Figure 19. The real motion state of the normal shape and mode shape piezoelectric motors.

6) Shown as in Table 5, we can find the output power and efficiency of the mode shape 2 (MS2) piezoelectric motor is higher than other type (NS, MS1 and MS3) piezoelectric motors under conditions of 0° driving phase angle, 180 $V_{p\text{-}p}$ driving voltage, 13 gw net weight, maximum loading and different dynamic friction coefficient ($\mu_d = 0.22$ and $\mu_d = 0.27$) and the driving frequency of the second vibration mode. Where the maximum output power of the piezoelectric motors (NS, MS1, MS2 and MS3) is respectively 10.18 mW, 23.25 mW, 23.26 mW and 18.00 mW. And where the efficiency of the piezoelectric motors (NS, MS1, MS2 and MS3) is respectively 0.04%, 0.09%, 0.09% and 0.07%. In addition, the gain of the maximum output power or efficiency of the mode shape piezoelectric motors (MS1 - 3) is respectively 2.28, 2.28 and 1.77 with respect to the normal shape piezoelectric motor (NS) under conditions of 0° driving phase angle, 180 $V_{p\text{-}p}$ driving voltage, maximum loading, $\mu_d = 0.27$ different dynamic friction coefficient and the driving frequency of the second vibration mode.

7) Shown as in Table 6, we have found that the maximum torque and gain of the mode shape 2 (MS2) piezoelectric motor is higher than other type (NS, MS1 and MS3) piezoelectric motors under conditions of 180

V_{p-p} driving voltage, maximum loading, different dynamic friction coefficient ($\mu_d = 0.22$ and $\mu_d = 0.27$) and the driving frequency of the second vibration mode. Where the maximum torque of the piezoelectric motors (NS, MS1, MS2 and MS3) is respectively 0.59 mkgw-m, 0.81 mkgw-m, 0.91 mkgw-m and 0.70 mkgw-m. And where the gain of the maximum torque or loading of the mode shape piezoelectric motors (MS1 - 3) is respectively 1.36, 1.54 and 1.18 with respect to the normal shape piezoelectric motor (NS) under conditions of 0° driving phase angle, 180 V_{p-p} driving voltage, maximum loading, $\mu_d = 0.27$ different dynamic friction coefficient and the driving frequency of the second vibration mode.

8) Shown as in Figure 20, we found the noise level increase, represents the piezoelectric motors are entering into best vibration mode. One of the biggest noise level in the second vibration mode are about 107 - 110 dB.

9) Shown as in Figure 21, we can find the conversion efficiency of driving phase angle of the mode shape 2 (MS2) piezoelectric motor is higher than other type (NS, MS1 and MS3) piezoelectric motors. On the whole, the conversion efficiency of driving phase angle of the mode shape 1 - 3 (MS1 - MS3) piezoelectric motor is higher than the normal shape (NS) piezoelectric motor.

Table 5. The output power, efficiency and gain of the normal shape and mode shape piezoelectric stators under conditions of 0^0 driving phase angle, 180 V_{p-p} driving voltage, 13 gw net weight, maximum loading and the driving frequency of the second virbration mode.

	Pout of Net Weight (mW)		Pout of Maximum Loading (mW)		Efficiency (η)		Gain	
Items	$\mu_d = 0.22$	$\mu_d = 0.27$	$\mu_d = 0.22$	$\mu_d = 0.27$	$\mu_d = 0.22$	$\mu_d = 0.27$	$\mu_d = 0.22$	$\mu_d = 0.27$
NS	2.27	2.79	8.30	10.18	0.03%	0.04%	1.00	1.00
MS1	2.70	3.32	17.22	23.25	0.07%	0.09%	2.08	2.28
MS2	2.99	3.67	18.95	23.26	0.07%	0.09%	2.28	2.28
MS3	2.94	3.60	14.66	18.00	0.06%	0.07%	1.77	1.77

Note: Input Power_P_{In} = 26 Watts Under Condition of 180 V_{p-p} Driving Voltage. The Maximum Loading of the NS, MS1, MS2 and MS3 is 292 gw, 398 gw, 451 gw and 345 gw respectively. The Driving Frequency of the Second Virbration Mode of the NS, MS1, MS2 and MS3 is 10.3 kHz, 10.3 kHz, 10.7 kHz and 10.8 kHz respectively.

Table 6. The torque and gain of the normal shape and mode shape piezoelectric stators under conditions of 0^0 driving phase angle, 180 $V_{p\text{-}p}$, maximum loading and the driving frequency of the second virbration mode.

	Torque (mkgw-m)		Gain of Torque (mkgw-m/mkgw-m)	
ITEMS	$\mu_s = 0.22$	$\mu_d = 0.27$	$\mu_s = 0.22$	$\mu_d = 0.27$
NS	0.48	0.59	1.00	1.00
MS1	0.66	0.81	1.36	1.36
MS2	0.74	0.91	1.54	1.54
MS3	0.57	0.70	1.18	1.18

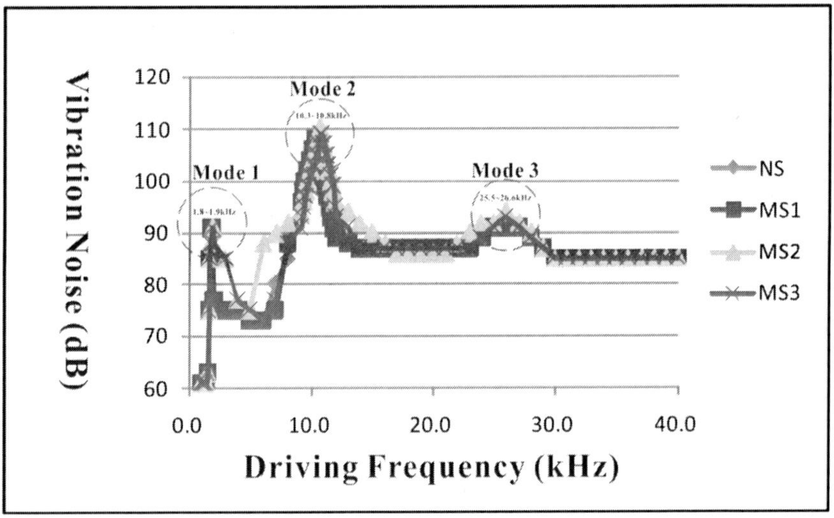

Figure 20. The vibration noise of the normal shape and mode shape piezoelectric motors under conditions of 180 $V_{p\text{-}p}$ driving voltage, 1.0 - 40 kHz driving frequency and 13 gw net weight.

10) Shown as in Figure 22, we can find the loading ability of the mode shape 2 (MS2) piezoelectric motor is higher than other type (NS, MS1 and MS3) piezoelectric motors. And loading ability of the mode shape1 - 3 (MS1 - MS3) piezoelectric motor is higher than the normal shape (NS) piezoelectric motor.

11) Shown as in Figure 23, we can find the rotational speed of the mode shape 2 (MS2) piezoelectric motor is higher than other type piezoelectric motors under conditions of different driving voltage (100 - 180 $V_{p\text{-}p}$) and 13gw net weight. And rotational speed of the mode shape piezoelectric motor is higher than the normal shape piezoelectric motor

on average. Where the r maximum rotational speed of the mode shape 2 (MS2) piezoelectric motor is about 946 rmp under conditions of 180 V_{p-p} driving voltage and 13gw net weight.

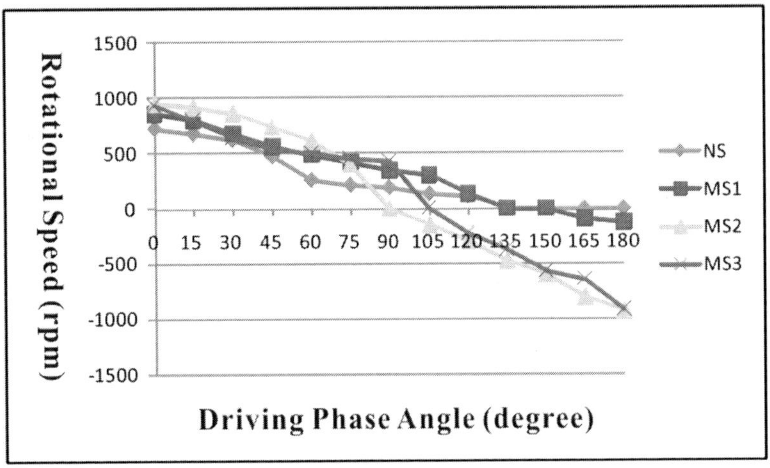

Figure 21. The rotational speed versus driving phase angle of the normal shape and mode shape piezoelectric motors under conditions of 180 V_{p-p} driving voltage, the second vibration mode and 13 gw net weight.

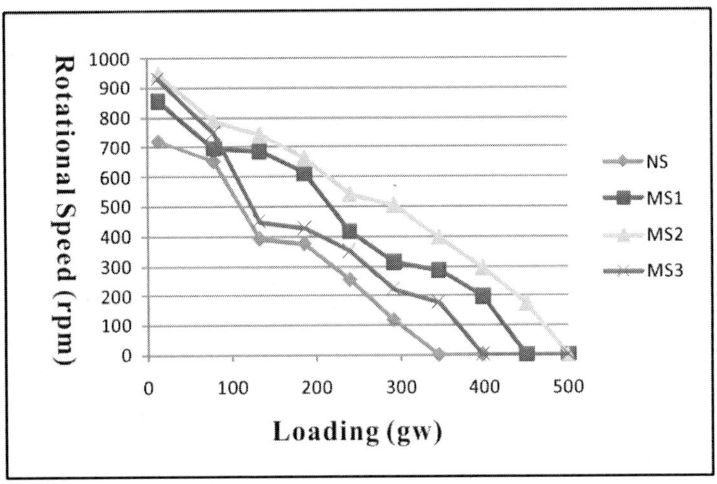

Figure 22. The rotational speed versus loading of the normal shape and mode shape piezoelectric motors under conditions of 180 V_{p-p} driving voltage and the second vibration mode.

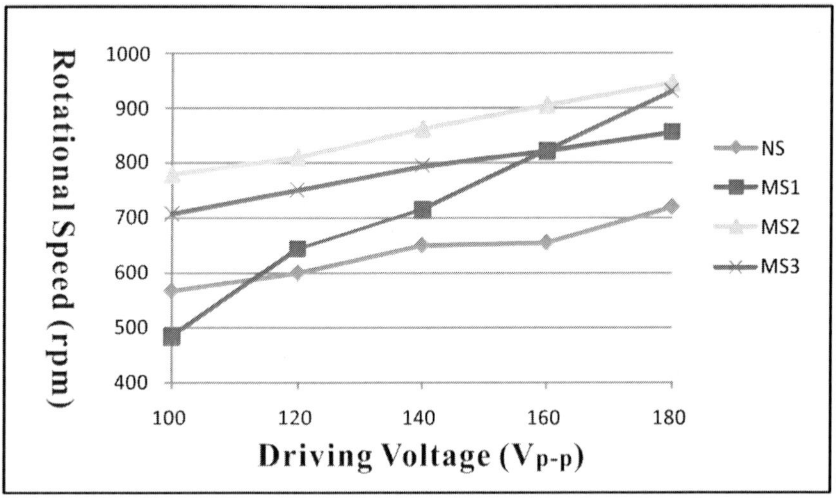

Figure 23. The rotational speed versus driving voltage of the normal shape and mode shape piezoelectric motors under conditions of 13 gw net weight and the second vibration mode.

7. CONCLUSION

According to the results of theoretical analysis, we have found that the gains of the mode shape piezoelectric stators are larger than the normal shape piezoelectric stator. And we also found that the mode shape piezoelectric stators can be effectively separated from the modal vibration. However, we have also found that the normal shape piezoelectric stator does not have this ability in theory. According to the results of simulation analysis, we have also found the gain of the mode shape piezoelectric stators can still appear in nonorthogonal vibration modes. Because the results of theoretical analysis are solutions in an ideal approximation, the results of simulation analysis are closer to reality. Therefore, the real results eventually must be confirmed through experiments. According to the experimental results, we have moreover found that the rotor can be rotated only by the second vibration mode. Here the gain of the rotational speed, loading ability and torque of the mode shape 2 (MS2) piezoelectric motor is higher than other shapes piezoelectric motors (NS, MS1 and MS3) under driving condition of the second vibration mode. And the gains of the rotational speed, loading ability, torque, driving phase angle conversion and efficiency of the mode shape (MS1 - 3) piezoelectric motors are higher than the normal shape piezoelectric motor (NS) under driving condition of the second vibration mode. In addition, we have also found that the first vibration mode allows the rotor to rotate very fast, but we can't determine the direction of rotation of the rotor. However, it can't make the rotation of the rotor orientation. Furthermore, we also found that the rotor cannot be rotated by the third vibration mode because its vibration energy is absorbed by the structure itself, so causing the rotor stagnation. At last, we have also found that the noise spectrum can help us quickly find the best vibration mode position of the normal shape and mode shape piezoelectric motors. When the noise level or decibel increases, the piezoelectric motors are entering into best vibration mode.

ACKNOWLEDGEMENTS

This study can be finished smoothly, I especially want to thank MOST of Taiwan of ROC sponsor on funding, (Project No.: MOST103-2221-E-230-007).

REFERENCES

1. Koc, B., Cagatay, S. and Uchino, K. (2002) A Piezoelectric Motor Using Two Orthogonal Bending Modes of a Hollow Cylinder. IEEE Transactions on Ultrasonics, Ferroelectrics and Frequency Control, 49, 495-500.
2. Cagatay, S., Koc, B. and Uchino, K. (2003) A 1.6-mm, Metal Tube Ultrasonic Motor. IEEE Transactions on Ultrasonics, Ferroelectrics and Frequency Control, 50, 782-786.
3. Kanda, T., Makino, A., Suzumori, K., Morita, T. and Kurosawa, M.K. (2004) A Cylindrical Micro Ultrasonic Motor Using a Micro-Machined Bulk Piezoelectric Transducer. IEEE Ultrasonics Symposium, 2, 1298-1301.
4. Kanda, T., Oomori, Y., Makino, A., Suzumori, K. and Kobayashi, A. (2005) Design and Testing of Rotors for a Cylindrical Micro-Machined Micro Ultrasonic Motor. IEEE Ultrasonics Symposium, 1, 301-304.

5. Zhu, H. and Zhao, C.S. (2006) Dynamic and Kinetic Analyses of the Stator of a Cylindrical Ultrasonic Motor. Proceedings of the 2006 IEEE International Conference on Robotics and Biomimetics, Kunming, 17-20 December 2006, 1346-1350.

6. Kobayashi, A. and Kanda, T. (2007) Driving Performance of a Cylindrical Micro Ultrasonic Motor. Proceedings of the 2007 IEEE/RSJ International Conference on Intelligent Robots and Systems, San Diego, 29 October-2 November 2007, 3809-3814.

7. Jou, J.M. (2009) A Study on the Metal Tube Type Ultrasonic Motor (MTTUSM). Proceedings of the European Frequency and Time Forum International Frequency Control Symposium, 609-612.

8. He, S.Y., Chiarot, P.R. and Park, S. (2011) A Single Vibration Mode Tubular Piezoelectric Ultrasonic Motor. IEEE Transactions on Ultrasonics, Ferroelectrics, and Frequency Control, 58, 1049-1061.

9. Cheng, T.-H., Guo, X.-D., Bao, G., Gao, H. and Xiao, C.F. (2012) Analysis and Development of Plate-Attached Cylindrical Rotary-Linear Ultrasonic Motor. Proceedings of the 2012 Symposium on Piezoelectricity, Acoustic Waves and Device Applications (SPAWDA), Shanghai, 23-25 November 2012, 270-273.

10. Guo, M.S., Hu, J.H., Zhu, H., Zhao, C.S. and Dong, S.X. (2013) Three-Degree-of-Freedom Ultrasonic Motor Using a 5-mm-Diameter Piezoelectric Ceramic Tube. Ultrasonics, Ferroelectrics and Frequency Control, IEEE Journals & Magazines, 60, 1446-1452.

11. Jou, J.-M. (2014) Theory and Simulation Analysis of the Mode Shape and Normal Shape Actuators and Sensors. Open Journal of Acoustics, 4, 184-203.

Chapter 12

Design and Application of Piezoelectric Stacks in Level Sensors

Andrzej Buchacz[1] and Andrzej Wróbel[1]

[1] *Silesian University of Technology, Poland*

1. INTRODUCTION

In recent years there is growing interest of materials, called smart materials. They have one or more properties that can be significantly changed. Smartness describes abilities of shape, size and state of aggregation changes. The main groups of smart materials are:

- piezoelectric plates,
- magneto-rheostatic materials,
- electro-rheostatic materials,
- shape memory alloys.

Those materials are widely used in technology and their numbers of applications still growing. Piezoelectric effect was discovered by French physicists Peter and Paul Curie in the 1880s. They described generation of electric charge on the surface with various shape during its deformation in different directions.

In their research, first of all, they focused on tourmaline crystal, salt and quartz. In 1881s Gabriel Lippman suggested the existence of the reverse piezoelectric phenomenon, which was confirmed experimentally by the

Curie brothers. As a solution of research such two, unique properties of piezoelectric materials were assigned:

- showing of simple piezoelectric effect, which rely on generating of voltage after deformation of material,
- reverse piezoelectric effect, which rely on changing of sizes (by around 4%) after applying a voltage to piezoelectric facing.

Designing of technological systems, which contains piezoelectric elements should not be framed only to mechanical system analysis, but should be taken under consideration also electrical part. The entity should be considered as complex system, which contains independent subsystem.

Problem with mechanical-physical systems synthesis, first of all electrical and mechanical ones, is well known and frequently published (Arczewski, 1988; Bellert 1981; Białas, 2012; Buchacz & Płaczek, 2012). In articles concerned theory and designing of filters not much space was devoted to mechanical systems with parameters distributed in continuous way. Determination tests of mechatronic systems characteristics, applications of graphs and structural numbers were carried out at Silesian Center repeatedly (Buchacz, 2004; Sękala & Świder, 2005; Wróbel, 2012). Those studies gave assumption to analysis of piezoelectric work. In many publications and papers, mechanical systems investigations on example of vibration beams and rods, were introduced. Moreover, rules of modellinig by non-classical method and attempts of analysis by using hypergraph skeletons (Buchacz & Świder, 2000), graphs with signal flow and matrix methods (Bishop et at.,1972) were shown in these works.

Nowadays, numerous piezoelectric advantages caused its multi-application in mechanics and in many replaced field of science (Shin et al.,2005; Ha, 2002). Many times beams configurations, with respect to different boundary conditions and during piezoelectric application in damping of vibrations, were analyzed. In the paper (Sherrit, 1999) capability of piezoelectric systems modelling using equivalent Manson models were presented. Analysis of longitudinal vibrations were made taking into consideration dielectric and piezoelement layer. Mason in (Mason, 1948) introduced one-dimensional, equivalent system parameters widely used in modelling systems both free, and loaded. The main disadvantage of such approach is the equivalent of the mechanical system by discreet model. In article (Shin et at.,2005) author presents 4-port equivalent system of piezoelectric plate, used to identification of system response on mechanical force. A matrix, size 5x5, input-output dependences, with different conditions of support, was also determined. Another type of piezoelectric transducer, which was based on Masons alternative systems of higher number of piezoelectric layer, were presented in many articles. Simulation was carried out in frequency domain, furthermore result was compared with values obtained by experimental method. Bellert in his volumes of chosen works (Bellert 1981) many times wrote about modelling of replacing systems, examined as 4-ports. In work (Bolkowski, 1986) author provide chain method of connection electric 4-ports. However, both: (Bolkowski, 1986) and (Bellert 1981) concerned primarily electric systems. In research work number N502 071 31/3719 attempts of active, mechanical systems, with damping in scope of graphs and structural numbers methods, were analyzed. In such, rich publications from field of vibration analysis, solution of piezoelectric plate itself with respect to dynamic characteristic was not undertaken, with the exceptions (Kacprzyk, 1995).

Previous presented solutions were conducted mainly in field of time and concerned single plate. Present paper is continuation of mentioned publications with stack of piezoelectric plates. This work is an author's idea of calculations of complex systems with many elements. The base of calculation is matrix method and application of aggregation of graphs to determination characteristic parameters of bimorphic systems, as well as to drawing its characteristics.

2. VIBRATING LEVEL SENSOR AS A PRACTICAL EXAMPLE OF THE APPLICATION OF PIEZOELECTRIC STACKS

An important characteristic of the designed and analyzed piezoelectric systems is the possibility of their practical application. This chapter presents options for further research related to the piezoelectric phenomenon of complex systems. Both a single piezoelectric plate, as well as complex systems, are often used in pressure, level, force and displacement sensors. As part of future research is proposed execution of laboratory stand for tests of piezoelectric plates used in vibration sensors. These sensors were used for level detection of loose materials in open or pressurized tanks. Output signal is a binary signal, transmitted to the automation systems via a relay. In Fig. 1 and Fig. 2 the level of vibration sensors manufactured by the "Nivomer" company from Gliwice were shown.

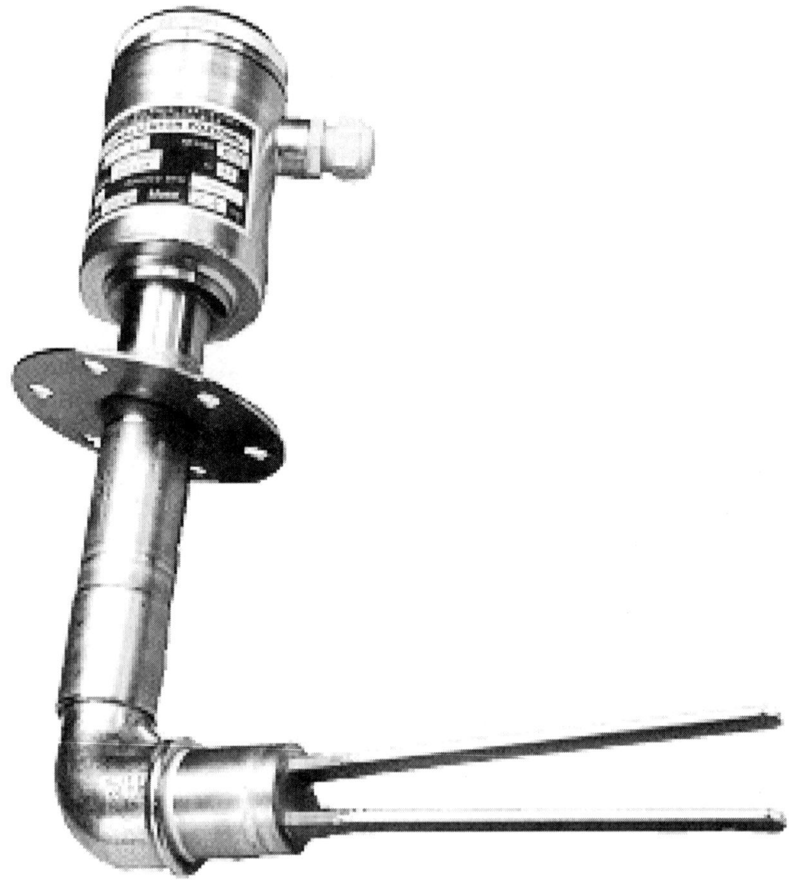

Figure 1. Angular vibrational sensor

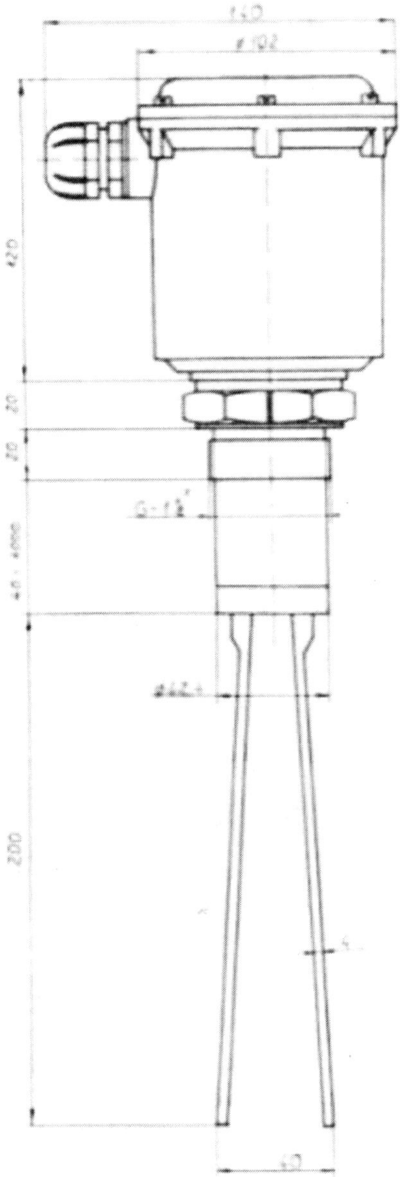

Figure 2. Approximate dimensions [documentation of the "Nivomer" company]

The application of stack tiles for intensification of the output signal. The sensors consist of two pairs of receiving plates and two or three supplying plates, connected in a bimorphic system (Fig. 3, Fig. 4). Variable voltage, which feeds the supply plates results in a change in their thickness proportionally to the value of applied voltage.

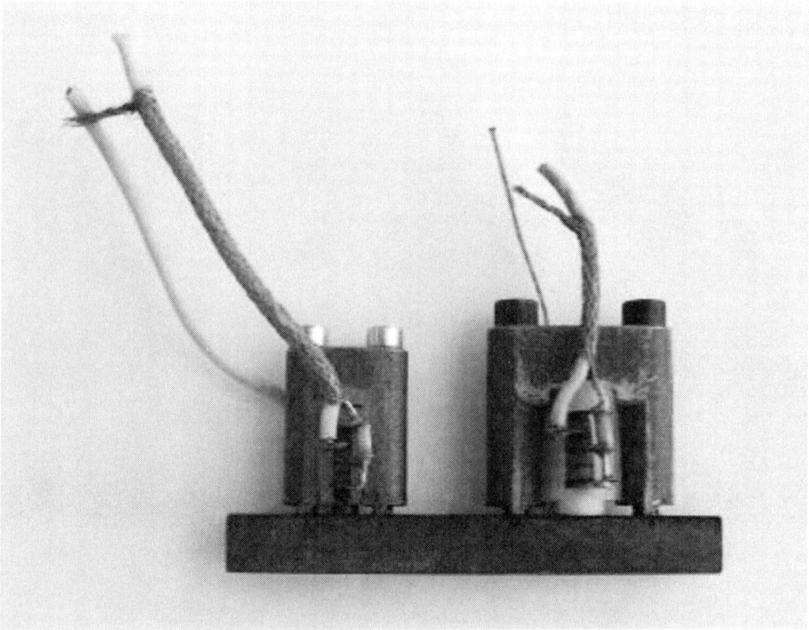

Figure 3. Stack of plates in the level sensor

Changes in the plate thickness causing mechanical vibrations of the element, so called "fork". When the "forks" are not covered by material, full deformation of supplying plates are transferred to the receiving ones. As a result of elongation of receiving plates, on its facing, there is a difference of potentials, proportional to the force. The value of this voltage is transformed by an electronic system (Fig. 5). In case of covered "forks", the receiving plates are no longer crushed and stretched. At the same time the potential is not generated on the facing of the plates.

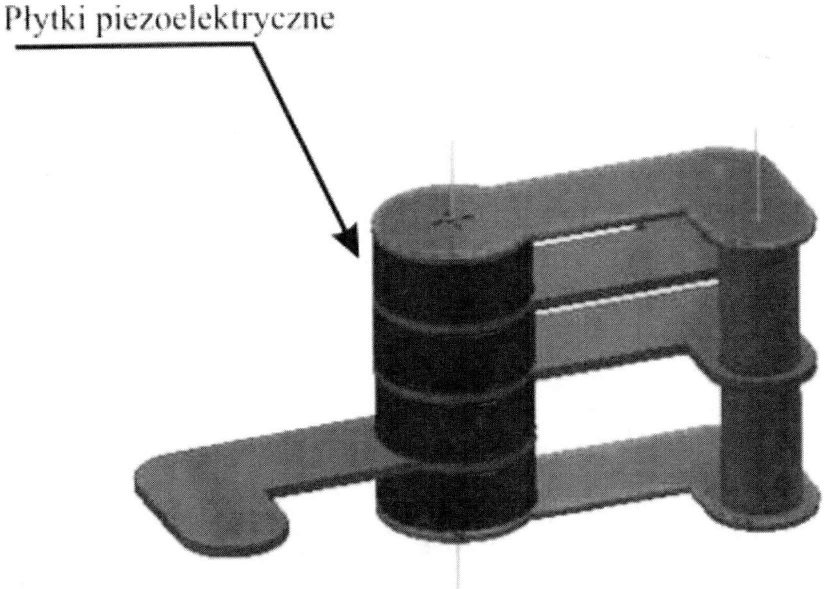

Figure 4. Computer model of the tiles stack

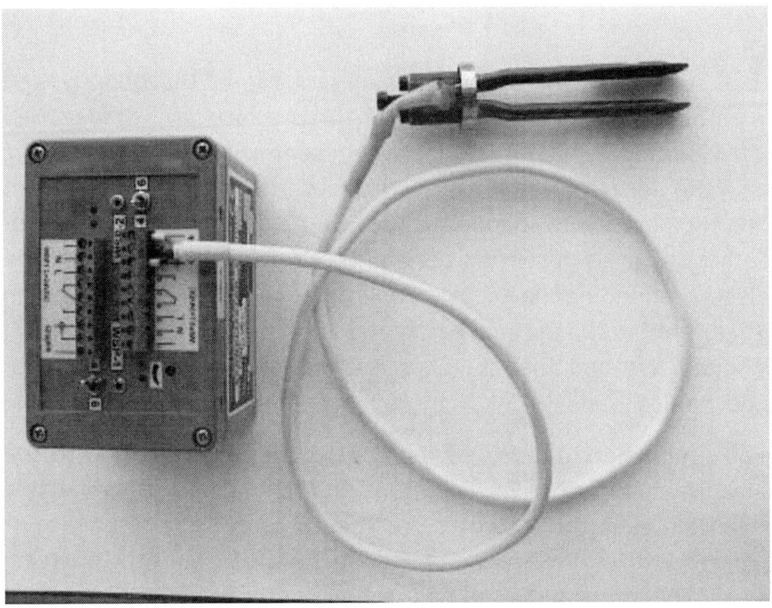

Figure 5. View of the sensor with control

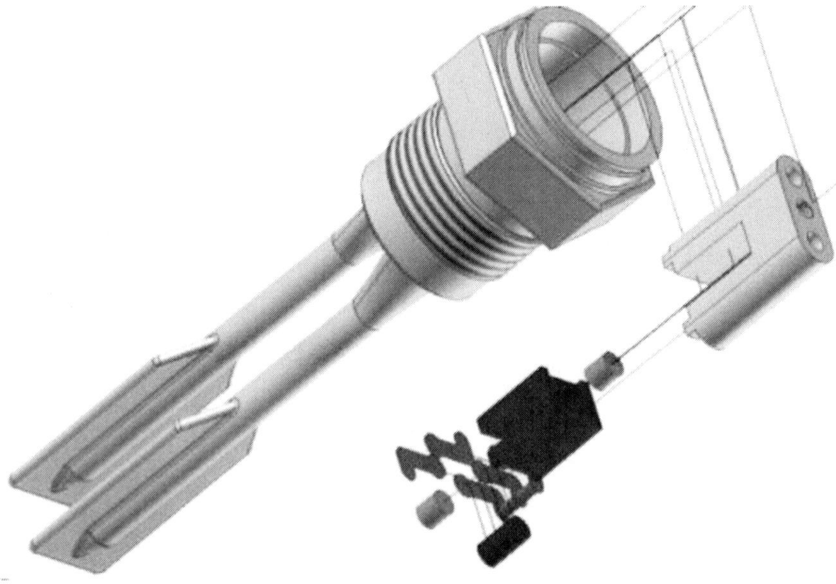

Figure 6. Construction of the level sensor by “Nivomer” company

Described sensor of "Nivomer" company is made of the body, ended with membrane which intensify a signal, to which are welded two identical vibrating rods (Fig. 6). In the presented sensor piezoelectric plates in the form of discs were used but there is a possibility of replacing the plates with shapes analyzed in previous chapters. Effect of plates stacks analysis in sensors designing The proposed analysis of the piezoelectric phenomenon of bimorphic systems allows at the design stage to determine the optimal parameters of piezoelectric plates. Well-chosen plate size and their number is crucial in the performance of other mechanical parts.

After the preliminary analysis of the construction of sensors with the company "Nivomer" from Gliwice found that by introducing a variable number of plates in the system is possible to choose the frequency of the generator and the maximum deflection of the fork carrying vibrations. The proposed methods and algorithms of work concern complex systems, can be used to design a stack of piezoelectric plates in the presented sensor level. In future work, it is proposed to conduct vibration test level sensors and a comparison of the algebraic method with the experimental method.

3. OBJECT MODEL UNDER EXAMINATION

Under consideration is vibrating piezoelectric plate with parameters distributed in a continuous way. The model has a section *A*, a thickness *d* and is made of a uniform material with a density *ρ*. Example of such system was shown in Fig. 7. Mechanical displacements are caused by the forces and voltage, while the current is generated by the difference of potentials on the plates of piezoelectric.

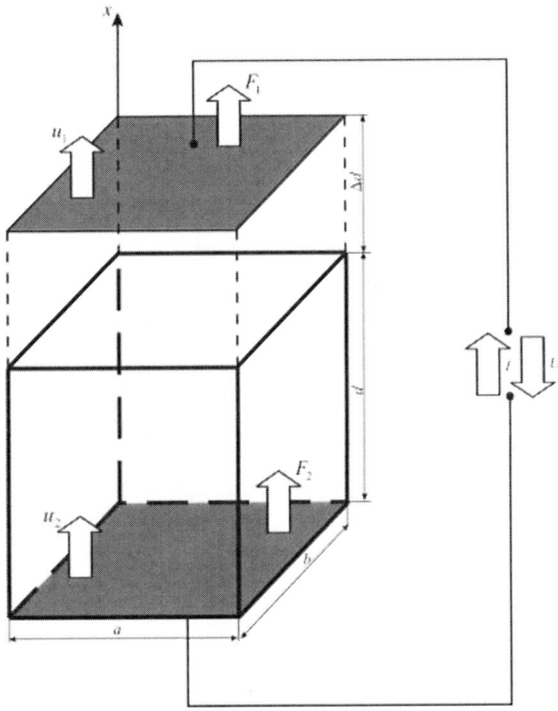

Figure 7. Continuous and limited piezoelectric model

In the analyzed example calculations are based on constitutive equations that include the assumed boundary conditions. In the study assumed that the test object is vibrating piezoelectric plate treated as a one-dimensional system. Piezoelectric plate constitutive equations are as follows (Soluch, 1980):

$$\begin{cases} \sigma = E\dfrac{\partial u}{\partial x} - \varepsilon E_p, \\ D = \varepsilon^S E_p + \varepsilon\dfrac{\partial u}{\partial x}, \end{cases} \tag{1}$$

where:

E- the modulus of longitudinal elasticity,

E_P - the intensity value of electric field,

ε - deformation,

ε^S - the electric permeability,

D - the electric induction.

The equation of motion of a given element is as follows:

$$\frac{\partial \sigma}{\partial x} = \rho \ddot{u}, \tag{2}$$

where:

σ - tension,

ρ - density of piezoelectric plate.

It was assumed that Poisson equation is:

$$\frac{\partial D}{\partial x} = 0. \tag{3}$$

Rearranging equation (1) due to E_p :

$$E_p = \frac{D}{\varepsilon^S} - \frac{\varepsilon}{\varepsilon^S}\frac{\partial u}{\partial x}. \tag{4}$$

Substituting expression (4) into equation (1), strain is given:

$$\sigma = c\frac{\partial u}{\partial x} - \frac{\varepsilon}{\varepsilon^S}D, \tag{5}$$

where:

$c = E + \varepsilon^2 / \varepsilon^S$ is the stiffened elastic constant.

From equation (3) result that *D=const*. Therefore, the equation of motion (2), that takes into account (5), is a one-dimensional wave equation:

$$c\frac{\partial^2 u}{\partial x^2} = \rho \ddot{u}, \tag{6}$$

or assuming that the volume wave equation in piezoelement is equal to:

$$V = \sqrt{\frac{c}{\rho}}, \tag{7}$$

equation of motion (6) was presented in the form:

$$\frac{\partial^2 u}{\partial x^2} - \frac{1}{V^2}\ddot{u}. \tag{8}$$

Assuming the expansion of the plate, mainly in the perpendicular plane to the axis, the following boundary conditions were defined:

$$\begin{cases} u = u_1, & when \quad x = x_1, \\ u = u_2, & when \quad x = x_2, \\ \sigma = -\sigma_1, & when \quad x = x_1, \\ \sigma = -\sigma_2, & when \quad x = x_2. \end{cases} \tag{9}$$

Replaced mechanical stress by force, using the formula *F=Aσ*, where *A* is the surface of piezoelectric plate and σ is the stress of piezoelectric plate. Determined forces in this case are:

$$\begin{cases} F = F_1, & when \quad x = x_1, \\ F = F_2, & when \quad x = x_2. \end{cases} \tag{10}$$

The solution of equation (8) is harmonical displacement:

$$u = a_1 e^{j(\omega t - kx)} + a_1 e^{j(\omega t + kx)}, \tag{11}$$

or assuming, that

$$A_1 = a_1 e^{j\omega t}, A_2 = a_2 e^{j\omega t}, A_2 = a_2 e^{j\omega t}. \tag{12}$$

written as:

$$u = A_1 e^{-jkx} + A_2 e^{jkx}. \tag{13}$$

Taking into account the boundary conditions included in the model (9) was obtained:

$$\begin{cases} u_1 = A_1 e^{-jkx_1} + A_2 e^{jx_1}, \\ u_2 = A_1 e^{-jkx_2} + A_2 e^{jx_2}. \end{cases} \tag{14}$$

In order to determine the coefficients A_1 and A_2, presented in equation (12) the determinants method was used. The determinant of the set of equations (14) takes the form:

$$W = \begin{vmatrix} e^{-jkx_1} & e^{jkx_1} \\ e^{-jkx_2} & e^{jkx_2} \end{vmatrix}, \tag{15}$$

$$|W| = e^{-jkx_1} e^{jkx_2} - e^{jkx_1} e^{-jkx_2}, \tag{16}$$

$$|W| = e^{-jkx_1 + jkx_2} - e^{-jkx_2 + jkx_1}. \tag{17}$$

Using the dependences between the trigonometric and exponential functions:

$$\sin x = \frac{1}{2j}\left(e^{jx} - e^{-jx}\right), \tag{18}$$

$$\cos x = \frac{1}{2}\left(e^{jx} + e^{-jx}\right), \tag{19}$$

equation (17) was written as:

$$|W| = 2j \sin(k(x_2 - x_1)), \tag{20}$$

and assuming that the thickness of the plate $d=x_2-x_1$, the equation was determined:

$$|W| = 2j\sin kd. \tag{21}$$

Determinant W_{A1} results from equation (2.14) and is:

$$W_{A_1} = \begin{vmatrix} u_1 & e^{jkx_1} \\ u_2 & e^{jkx_2} \end{vmatrix}, \tag{22}$$

so:

$$|W_{A_1}| = u_1 e^{jkx_2} - u_2 e^{jkx_1}. \tag{23}$$

Using the record:

$$A_1 = \frac{|W_A|}{|W|}, \tag{24}$$

and substituting the expression (21), (23) into equation (24), A_1 was obtained:

$$A_1 = \frac{u_1 e^{jkx_2} - u_2 e^{jkx_1}}{2j\sin kd}. \quad 25)$$

Furthermore, the determinant W_{A2} was calculated:

$$W_{A_2} = \begin{vmatrix} e^{-jkx_1} & u_1 \\ e^{-jkx_2} & u_2 \end{vmatrix}, \tag{26}$$

so:

$$\left| W_{A_2} \right| = u_2 e^{-jkx_1} - u_1 e^{-jkx_2}, \tag{27}$$

Analogously to (25) A_2 was determined:

$$A_2 = \frac{u_2 e^{-jkx_1} - u_1 e^{-jkx_2}}{2j \sin kd}. \tag{28}$$

Substituting (11) into equation (5) was written:

$$\sigma = c(-jkA_1 e^{-jkx} + jkA_2 e^{jkx}) - \frac{\varepsilon}{\varepsilon^S} D. \tag{29}$$

Using (10), the force was defined F_1:

$$-F_1 = -Ack(jA_1 e^{-jkx_1} - jA_2 e^{jkx_1}) - \frac{\varepsilon}{\varepsilon^S} D, \tag{30}$$

substituting determined A_1, A_2 to the equation (30), was obtained:

$$-F_1 = -Ack\left[j\left(\frac{u_1 e^{jkx_2} - u_2 e^{jkx_1}}{2j \sin kd} \right) e^{-jkx_1} - j\left(\frac{u_2 e^{-jkx_1} - u_1 e^{-jkx_2}}{2j \sin kd} \right) e^{jku_1} \right] - \frac{\varepsilon}{\varepsilon^S} D. \tag{31}$$

Carrying out the multiplication of expressions in brackets (31):

$$-F_1 = -Ack\left[\left(\frac{u_1 e^{jk(x_2-x_1)} - u_2 e^{jk(x_1-x_1)}}{2\sin kd}\right) - \left(\frac{u_2 e^{-jk(x_1-jx_1)} - u_1 e^{-jk(x_2-x_1)}}{2\sin kd}\right)\right] - \frac{\varepsilon}{\varepsilon^S} D, \tag{32}$$

and excluding the displacement u_1 and u_2 before the bracket:

$$-F_1 = -Ack\left[\frac{u_1\left(e^{jk(x_2-x_1)} + e^{-jk(x_2-x_1)}\right) - u_2\left(e^{jk(x_1-x_1)} + e^{-jk(x_1-x_1)}\right)}{2\sin kd}\right] - \frac{\varepsilon}{\varepsilon^S} D \tag{33}$$

And taking into account the dependence:

$$\cos kd = \frac{\left(e^{jk(x_2-x_1)} + e^{-jk(x_2-x_1)}\right)}{2}, \tag{34}$$

was obtained:

$$-F_1 = -Ack\left[\frac{u_1 \cos kd}{\sin kd} - \frac{u_2}{\sin kd}\right] - \frac{\varepsilon}{\varepsilon^S} D. \tag{35}$$

Finally, introducing trigonometric functions, equation (35) was written as:

$$F_1 = Ack\left[\frac{u_1}{\tan kd} - \frac{u_2}{\sin kd}\right] + \frac{\varepsilon}{\varepsilon^S} D, \tag{36}$$

In (36) the ralationship of force F_1, acting on the beginning of the system (Fig.2), on the displacements

u_1 and u_2 and its influence on the piezoelectric effect $\frac{\varepsilon}{\varepsilon^S} D$ were shown. Force F_2, determined from the equation (30) is equal to:

$$-F_2 = -Ack\left(jA_1 e^{-jkx_2} - jA_2 e^{jkx_2}\right) - \frac{\varepsilon}{\varepsilon^S} D, \tag{37}$$

and substituting the determinants A_1 (25),A_2 (28) to the equation (30), was obtained:

$$-F_2 = -Ack\left[j\left(\frac{u_1 e^{jkx_2} - u_2 e^{jkx_1}}{2j\sin kd} \right) e^{-jkx_2} - j\left(\frac{u_2 e^{-jkx_1} - u_1 e^{-jkx_2}}{2j\sin kd} \right) e^{jku_2} \right] - \frac{\varepsilon}{\varepsilon^S} D. \tag{38}$$

Carrying out the multiplication of expressions in brackets:

$$-F_2 = -Ack\left[\left(\frac{u_1 e^{jk(x_2 - x_2)} - u_2 e^{-jk(x_2 - x_1)}}{2\sin kd} \right) - \left(\frac{u_2 e^{jk(x_2 - jx_1)} - u_1 e^{-jk(x_2 - x_2)}}{2\sin kd} \right) \right] - \frac{\varepsilon}{\varepsilon^S} D, \tag{39}$$

and excluding displacements u_1 I u_2 before brackets:

$$-F_2 = -Ack\left[\frac{u_1\left(e^{jk(x_2 - x_2)} + e^{-jk(x_2 - x_2)} \right) - u_2\left(e^{-jk(x_2 - x_1)} + e^{jk(x_2 - x_1)} \right)}{2\sin kd} \right] - \frac{\varepsilon}{\varepsilon^S} D, \tag{40}$$

and using dependences (18, 19) written:

$$-F_2 = -Ack\left[\frac{u_1}{\sin kd} - \frac{u_2 \cos kd}{\sin kd} \right] - \frac{\varepsilon}{\varepsilon^S} D. \tag{41}$$

Finally, taking into account the trigonometric functions,

$$F_2 = Ack\left[\frac{u_1}{\sin kd} - \frac{u_2}{\tan kd} \right] + \frac{\varepsilon}{\varepsilon^S} D \tag{42}$$

was calculated.

Dependences of the forces and displacements acting on the system under consideration, taking into account the piezoelectric effect, are taking the following form:

$$\begin{cases} F_1 = \rho VA\left(\dfrac{u_1}{\tan kd} - \dfrac{u_2}{\sin kd}\right) + \dfrac{\varepsilon}{\varepsilon^S}D, \\ F_2 = \rho VA\left(\dfrac{u_1}{\sin kd} - \dfrac{u_2}{\tan kd}\right) + \dfrac{\varepsilon}{\varepsilon^S}D. \end{cases} \quad (43)$$

Analyzing the effects occurring in the piezoelectric plate, also the electrical parameters such as voltage on the plates of piezoelectric and current value were took into account. Voltage U is therefore expressed as a function of electric field:

$$U = \int_{u1}^{u2} E_p du, \quad (44)$$

where:

U- value of generated voltage on the linings of piezoelectric,

Ep - electric field intensity.

Integrating (44):

$$U = \frac{Dd}{\varepsilon^s} - \frac{\varepsilon}{\varepsilon^s}(u_2 - u_1), \quad (45)$$

where:

D- electric induction module, defined as:

$$D = \frac{i}{\omega A}, \quad (46)$$

finally the voltage in a function of current was shown as:

$$U = \frac{h}{\omega}\left(u_2 - u_1\right) + \frac{1}{\omega C_0} i, \tag{47}$$

where:

$$h = \frac{\varepsilon}{\varepsilon^S}. \tag{48}$$

Capacitance, depends directly on the dimensions of the plates, aswell as physicochemical properties, written in the form:

$$C_0 = \frac{\varepsilon^S A}{d}. \tag{49}$$

Introducing (46) into equations (36), (42), (46), dependences of replacement set of piezoelectric plate were received, plate characterized by three equations:

$$F_1 = Ack\left[\frac{u_1}{\tan kd} - \frac{u_2}{\sin kd}\right] + \frac{\varepsilon}{\varepsilon^S}\frac{i}{\omega A}, \tag{50}$$

$$F_2 = Ack\left[\frac{u_1}{\tan kd} - \frac{u_2}{\sin kd}\right] + \frac{\varepsilon}{\varepsilon^S}\frac{i}{\omega A}, \tag{51}$$

$$U = \frac{h}{\omega}\left(u_2 - u_1\right) + \frac{1}{\omega C_0} i\ . \tag{52}$$

Assuming signs :

$$k = \frac{\omega}{V}, \tag{53}$$

and

$$Z = \rho V A, \tag{54}$$

obtained the relations between the electrical and mechanical values of piezoelectric plates:

$$F_1 = Z\left[\frac{u_1}{\tan kd} - \frac{u_2}{\sin kd}\right] + \frac{hi}{\omega},$$ (55)

$$F_2 = Z\left[\frac{u_1}{\tan kd} - \frac{u_2}{\sin kd}\right] + \frac{hi}{\omega},$$ (56)

$$U = \frac{h}{\omega}(u_2 - u_1) + \frac{1}{\omega C_0} i$$ (57)

which are also written in a matrix form:

$$\begin{bmatrix} F_1 \\ U \\ F_2 \end{bmatrix} = \begin{bmatrix} \frac{Z}{\tan kd} & \frac{h}{\omega} & -\frac{Z}{\sin kd} \\ \frac{h}{\omega} & \frac{1}{\omega C_0} & -\frac{h}{\omega} \\ \frac{Z}{\sin kd} & \frac{h}{\omega} & -\frac{Z}{\tan kd} \end{bmatrix} \begin{bmatrix} u_1 \\ i \\ u_2 \end{bmatrix}.$$ 58)

4. MAPPING MATRIX INTO GRAPH

The values of susceptibility, admittance and characteristics were determined from the formula:

$$Y = \frac{1}{\det Z}\left((Z^M)^D\right)^T,$$ (59)

The matrix (58) was written as respond of system for operating extortion. The individual elements of matrix, presented as a flexibility, admittance and characteristics were recorded as follows: determined dependences of force from displacement are given in Table 1.

Table 1. Dependences of force from displacement

	F_1	U	F_2
u_1	$\dfrac{\left(\dfrac{h^2}{\omega^2}\right)-\left(\dfrac{\rho VA}{\omega C_0 \tan kd}\right)}{\det Z}$	-	$\dfrac{-\left(\dfrac{h^2}{\omega^2}\right)+\left(\dfrac{1}{\omega C_0}\dfrac{\rho VA}{\sin kd}\right)}{\det Z}$
i	-	-	-
u_2	$\dfrac{\left(\dfrac{h^2}{\omega^2}\right)-\left(\dfrac{1}{\omega C_0}\dfrac{\rho VA}{\sin kd}\right)}{\det Z}$	-	$\dfrac{-\left(\dfrac{h^2}{\omega^2}\right)+\left(\dfrac{\rho VA}{\omega C_0 \tan kd}\right)}{\det Z}$

The dependences between mechanical and electrical parameters were shown in Table 2.

Table 2. Dependences between mechanical and electrical parameters

	F_1	U	F_2
u_1	-	$\dfrac{\dfrac{h\rho VA}{\omega}\left(\dfrac{1}{\tan kd}-\dfrac{1}{\sin kd}\right)}{\det Z}$	-
i	$\dfrac{\dfrac{h\rho VA}{\omega}\left(\dfrac{1}{\tan kd}-\dfrac{1}{\sin kd}\right)}{\det Z}$	-	$\dfrac{\dfrac{h\rho VA}{\omega}\left(\dfrac{1}{\tan kd}-\dfrac{1}{\sin kd}\right)}{\det Z}$
u_2	-	$\dfrac{-\dfrac{h\rho VA}{\omega}\left(\dfrac{1}{\tan kd}+\dfrac{1}{\sin kd}\right)}{\det Z}$	-

In table 3 electrical dependences: voltage and current, so called admittance of piezoelectric system, were listed.

Table 3. Electrical dependences: voltage and current

	F_1	U	F_2
u_1	-	-	-
i	-	$\dfrac{-\left(\dfrac{\rho VA}{\tan kd}\right)^2+\left(\dfrac{\rho VA}{\sin kd}\right)^2}{\det Z}$	-
u_2	-	-	-

In order to determine graph, representing modeled system of piezoelectric plates, as the symbols in the matrix were used. The elements of matrix (58) assigned to the edges of graph are presented as:

$$Y_{11}=(x_1,F_1)=\frac{1}{\det Z}\left\{-\frac{\rho VA}{\omega C_0 tgkd}-\frac{h^2}{\omega^2}\right\} \tag{60}$$

$$Y_{12}=(x_1,U)=\frac{1}{\det Z}\left\{-\frac{\rho VA}{\sin kd}\frac{h}{\omega}-\frac{h}{\omega}\frac{\rho VA}{\sin kd}\right\} \tag{61}$$

$$Y_{13}=(x_1,F_2)=\frac{1}{\det Z}\left\{\frac{h^2}{\omega^2}-\frac{1}{\omega C_0}\frac{\rho VA}{\sin kd}\right\} \tag{62}$$

$$Y_{21} = (i, F_1) = \frac{1}{\det Z}\left\{\frac{h}{\omega}\frac{\rho VA}{tgkd} + \frac{\rho VA}{\sin kd}\frac{h}{\omega}\right\} \tag{63}$$

$$Y_{22} = (i, U) = \frac{1}{\det Z}\left\{-\left(\frac{\rho VA}{tgkd}\right)^2 + \left(\frac{\rho VA}{\sin kd}\right)^2\right\} \tag{64}$$

$$Y_{23} = (i, F_2) = \frac{1}{\det Z}\left\{-\frac{\rho VA}{\omega C_0 tgkd} - \frac{h^2}{\omega^2}\right\} \tag{65}$$

$$Y_{31} = (x_2, F_1) = \frac{1}{\det Z}\left\{\frac{h^2}{\omega^2} + \frac{\rho VA}{\omega C_0 \sin kd}\right\} \tag{66}$$

$$Y_{32} = (x_2, U) = \frac{1}{\det Z}\left\{\frac{h}{\omega}\frac{\rho VA}{tgkd} - \frac{\rho VA}{\sin kd}\frac{h}{\omega}\right\} \tag{67}$$

$$Y_{33} = (x_2, F_2) = \frac{1}{\det Z}\left\{\frac{h^2}{\omega^2}\frac{\rho VA}{tgkd} - \frac{\rho VA}{\omega C_0 tgkd}\right\} \tag{68}$$

The graphical representation of mapping is shown as:

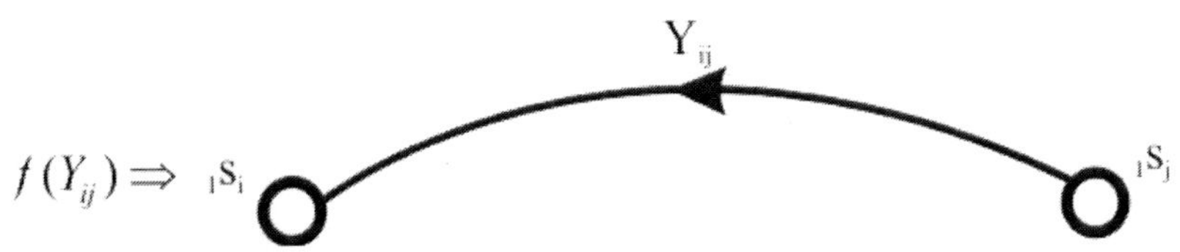

Figure 8. Mapping Y_{ij}

The symbol Y_{ij} is the mechanical flexibility, electrical admittance or characteristics of the system. In mapping of the parameters into the graph, mark

Y_{ij} means the relationship between the vertex of graph, directed from the apex i to apex j, with the symbol $i=j$, then the following relationship were true:

$$Y_{11}=Y_{10},$$

$$Y_{22}=Y_{20}, \tag{69}$$

$$Y_{33}=Y_{30}.$$

Dependences according to the index $j=0$ maps a connection of the vertex with the base vertex. Following this systematic, assignment by an edge of following relations was made:

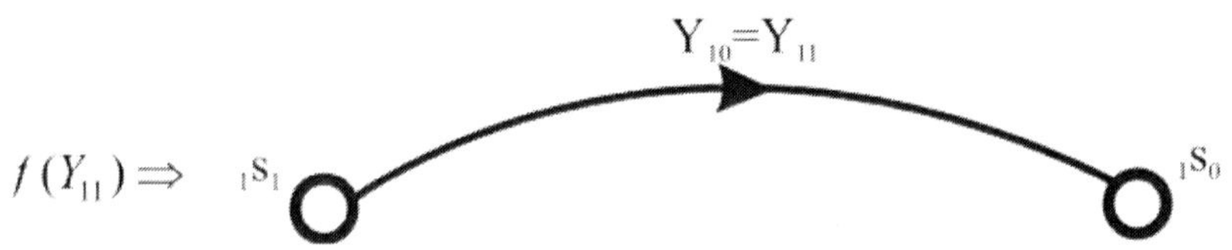

Figure 9. Mapping Y_{10}

where $f(Y_{11})$ is the mechanical flexibility;

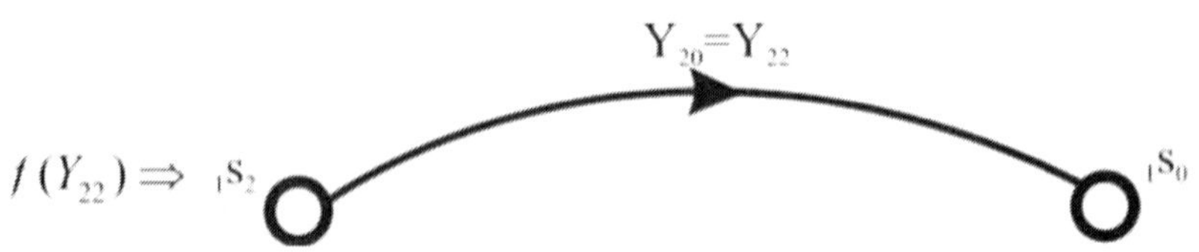

Figure 10. Mapping Y_{20}

where $f(Y_{22})$ is admittance of electrical system;

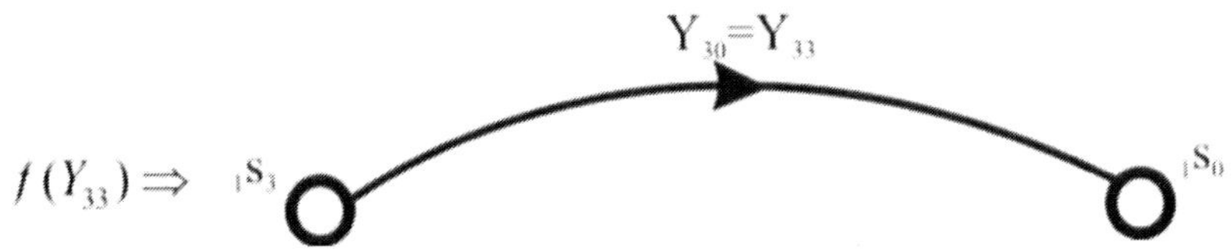

Figure 11. Mapping Y_{30}

where $f(Y_{33})$ is mechanical flexibility;

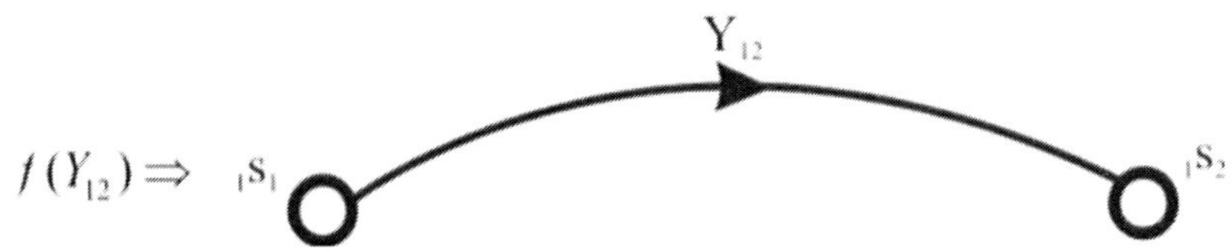

Figure 12. Mapping Y_{12}

where $f(Y_{12})$ is system characteristic;

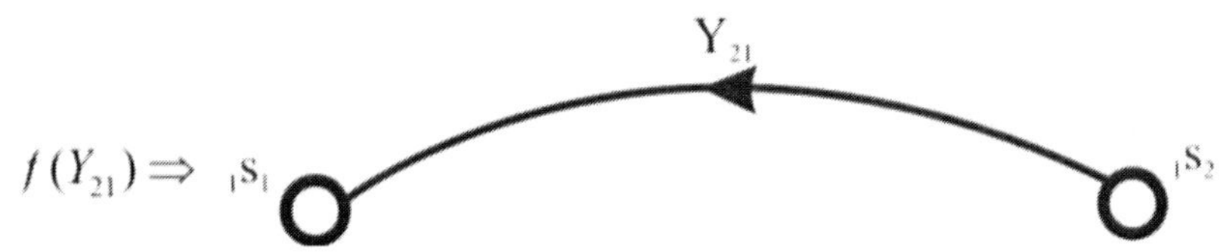

Figure 13. Mapping Y_{21}

where $f(Y_{21})$ is system characteristic;

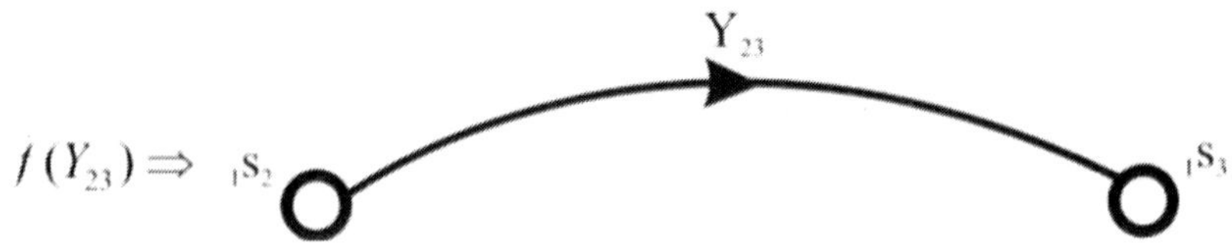

Figure 14. Mapping Y_{23}

where $f(Y_{23})$ is system characteristic;

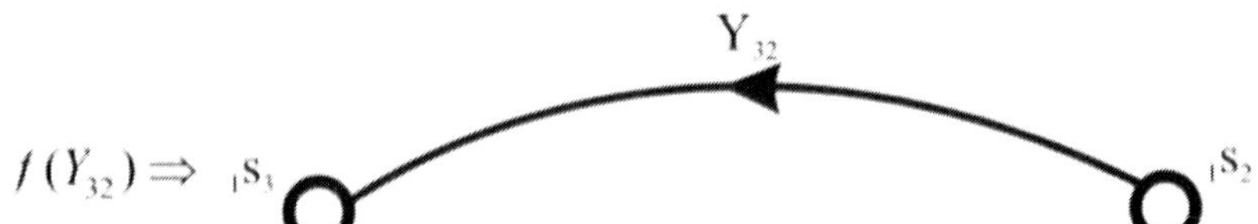

Figure 15. Mapping Y_{32}

where $f(Y_{32})$ is mechanical flexibility;

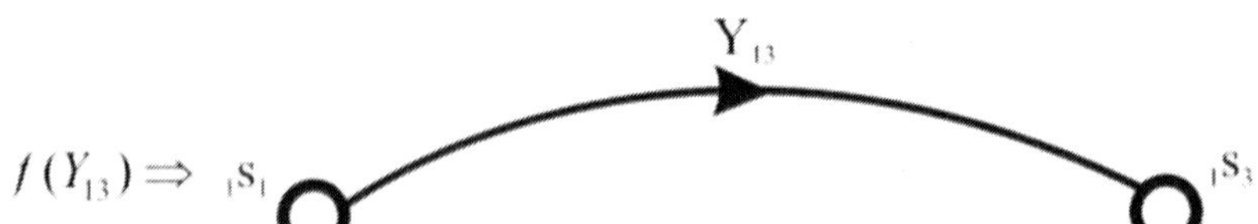

Figure 16. Mapping Y_{13}

where $f(Y_{13})$ is mechanical flexibility;

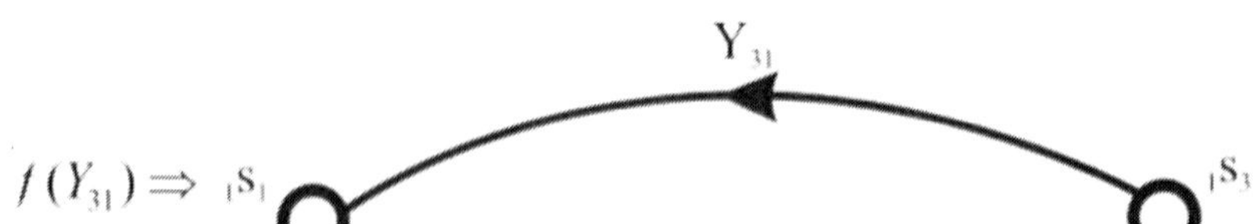

Figure 17. Mapping Y_{31}

where $f(Y_{31})$ is mechanical flexibility.

$$ {}_2X = \{Y_{11}, Y_{22}, Y_{33}, Y_{12}, Y_{21}, Y_{31}, Y_{13}, Y_{23}, Y_{32}\} \quad (70) $$

A set of drawings of the relation (fig. 9.) – (fig.17.), represents 4-vertex graph, were created and presented in Fig. 18.

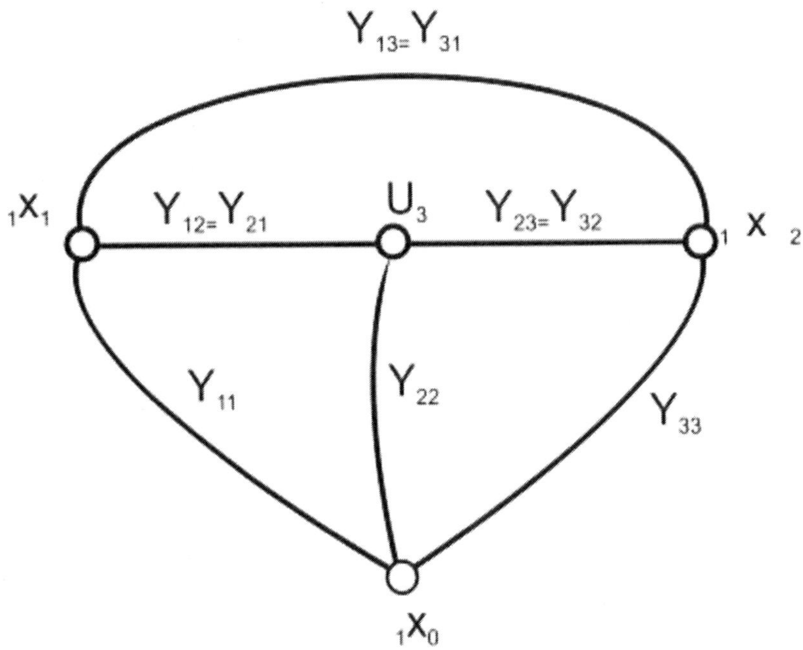

Figure 18. Geometric representation of mapping in the graph

In the rest of the work earlier created 4-vertex graph was replaced by structural number method to the 3-vertex graph.

5. CONSTRUCTION OF THE REPLACEMENT GRAPH

Furthermore, the use of an extended 4-vertex graph may prove to complicated calculations. In such case, a modelling of system using the replaced graph was performed. In order to maintain clearness of mapping, characteristics determined in paragraph 4 are indicated by Arabic numerals in parentheses, in accordance with (Bellert, 1981). As a consequence of introduction of the replaced graph, a graph presented in Fig. 19 was obtained. It is the basis for further network analysis methods.

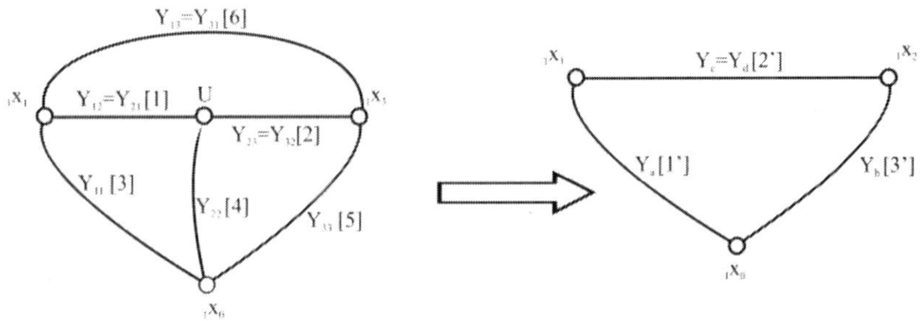

Figure 19. Construction of the replacement graph

As a result of insertion of replaced graph, replaced flexibility of the system was calculated by structural number method:

$$Y_b = Y_{3'} = \frac{\det\limits_{Z}(A_{1'} \cap A_{2'})}{\det\limits_{Z} A_{1'2'}} = \frac{Y_4(Y_2 + Y_5) + Y_5(Y_1 + Y_3)}{Y_1 + Y_2 + Y_4}. \quad (71)$$

$$Y_a = Y_{1'} = \frac{\det\limits_{Z}(A_{2'} \cap A_{3'})}{\det\limits_{Z} A_{2'3'}} = \frac{Y_1(Y_3 + Y_4) + Y_3(Y_2 + Y_4)}{Y_1 + Y_2 + Y_4}. \quad (72)$$

6. CHAIN EQUATION OF SIMPLE AND COMPLEX SYSTEM PLATE

On the figure 14 a piezoelectric plate with parameters distributed in the continuous way, the left and right end is free, was presented. The model of a single plate was marked by (i). Currently considered a model system is reduced system in the previous graph from 4-vertex to 3-vertex graph, as shown inFig. 20.

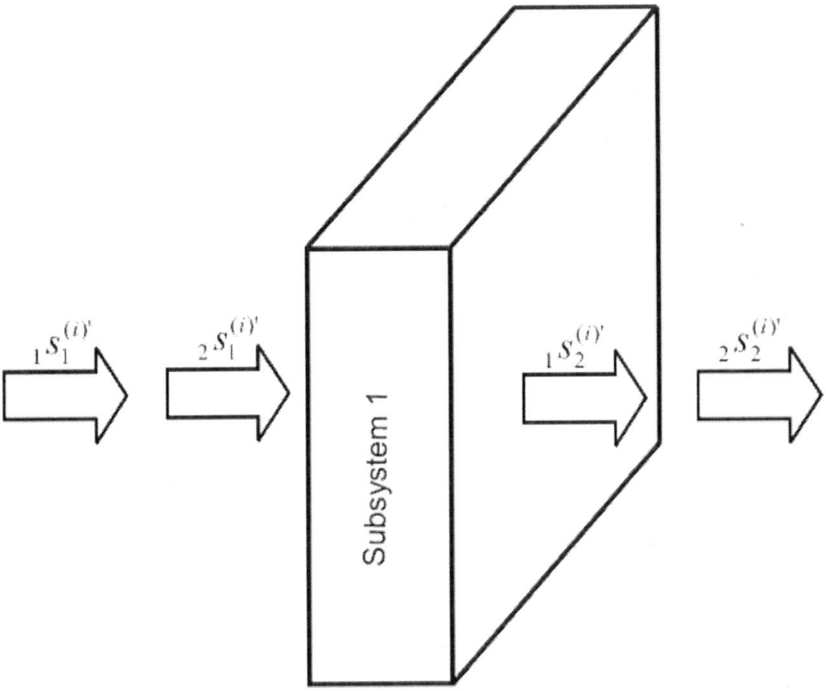

Figure 20. Model of single piezoelectric plate after reduction

Longitudinal vibrations of piezoelectric plate were considered, in the literature described also as thickness. The parameters specifying the system, in accordance with the previously accepted assumptions, were the sizes of input $_1S_1$, $_2S_1$ and output $_1S_2$, $_2S_2$ values, which were presented as:

$$ {}_1S^{(i)} = Y\,{}_2S^{(i)} \tag{73}$$

where:

Y is a value characterized input-output dependences.

The relations between displacements of plate, and the forces acting on them, written in matrix form:

$$\begin{bmatrix} {}_1s_1^{(i)} \\ {}_1s_2^{(i)} \end{bmatrix} = \begin{bmatrix} Y_a^{(i)} & Y_c^{(i)} \\ Y_d^{(i)} & Y_b^{(i)} \end{bmatrix} \begin{bmatrix} {}_2s_2^{(i)} \\ {}_2s_2^{(i)} \end{bmatrix} \tag{74}$$

Transforming the matrix (26) to the chain form expects to receive in the form of matrices:

$$\begin{bmatrix} {}_2s_1^{(i)} \\ {}_1s_1^{(i)} \end{bmatrix} = \begin{bmatrix} A_{11}^{(i)} & A_{12}^{(i)} \\ A_{21}^{(i)} & A_{22}^{(i)} \end{bmatrix} \begin{bmatrix} {}_2s_2^{(i)} \\ {}_1s_2^{(i)} \end{bmatrix} \tag{75}$$

where:

$$\begin{cases} A_{11}^{(i)} = \dfrac{Y_b^{(i)}}{Y_c^{(i)}}, \\[2ex] A_{12}^{(i)} = \dfrac{1}{Y_c^{(i)}}, \\[2ex] A_{21}^{(i)} = \dfrac{Y_c^{(i)} Y_d^{(i)} - Y_a^{(i)} Y_b^{(i)}}{Y_d^{(i)}}, \\[2ex] A_{22}^{(i)} = \dfrac{Y_a^{(i)}}{Y_d^{(i)}}. \end{cases} \tag{76}$$

In Fig. 21 the free system, consisting of two plates was presented. Superscript indicates the subsequent number of subsystem.

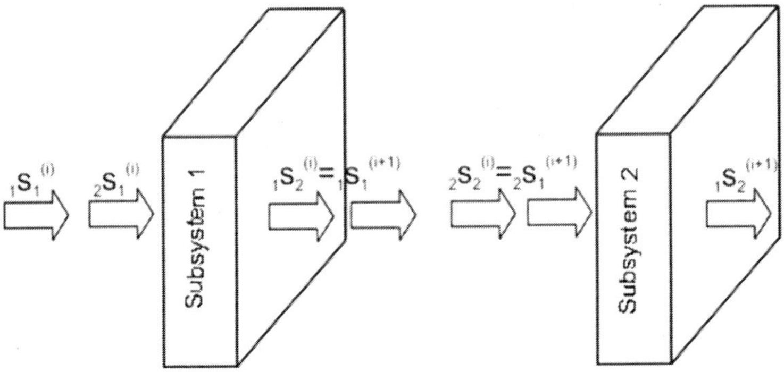

Figure 21. Diagram of a connection between the two cells and the relation between them

$$B = \begin{bmatrix} {}_2s_1^{(i)} \\ {}_1s_1^{(i)} \end{bmatrix} = \begin{bmatrix} A_{11}^{(i)} & A_{12}^{(i)} \\ A_{21}^{(i)} & A_{22}^{(i)} \end{bmatrix} \begin{bmatrix} A_{11}^{(i+1)} & A_{12}^{(i+1)} \\ A_{21}^{(i+1)} & A_{22}^{(i+1)} \end{bmatrix} \begin{bmatrix} {}_2s_2^{(i+1)} \\ {}_1s_2^{(i+1)} \end{bmatrix} \tag{77}$$

Finally, chain equation was written in general form:

$$A^{(k)} = A^{(i)} A^{(i+1)}. \tag{78}$$

After the operations carried out according to (79) it was found, that the chain matrix with cascade structure is the ratio of chain matrix of individual cells of the complex system. Obtained transition matrix is presented as:

$$B = \begin{bmatrix} {}_2s_1^{(i)} \\ {}_1s_1^{(i)} \end{bmatrix} = \begin{bmatrix} A_{11}^{(i+1)} A_{11}^{(i)} + A_{12}^{(i)} A_{21}^{(i+1)} & A_{11}^{(i)} A_{12}^{(i+1)} + A_{12}^{(i)} A_{22}^{(i+1)} \\ A_{21}^{(i)} A_{11}^{(i+1)} + A_{22}^{(i)} A_{21}^{(i+1)} & A_{21}^{(i)} A_{12}^{(i)} + A_{22}^{(i)} A_{22}^{(i+1)} \end{bmatrix} \begin{bmatrix} {}_2s_2^{(i+1)} \\ {}_1s_2^{(i+1)} \end{bmatrix} \tag{79}$$

Calculated coefficients (80) were substituted and the final form of the transition matrix was received:

$$B = \begin{bmatrix} B_{11}^{(k)} & B_{12}^{(k)} \\ B_{21}^{(k)} & B_{22}^{(k)} \end{bmatrix} \tag{80}$$

$$\begin{cases} B_{11}^{(k)} = \dfrac{Y_b^{(i)} Y_b^{(i+1)}}{Y_d^{(i)} Y_d^{(i+1)}} + \dfrac{-Y_a^{(i+1)} Y_b^{(i+1)} + Y_c^{(i+1)} Y_d^{(i+1)}}{Y_d^{(i)} Y_d^{(i+1)}}, \\ B_{12}^{(k)} = \dfrac{Y_b^{(i)}}{Y_d^{(i)} Y_d^{(i+1)}} + \dfrac{Y_a^{(i+1)}}{Y_c^{(i)} Y_d^{(i+1)}}, \\ B_{21}^{(k)} = \dfrac{(-Y_a^{(i)} Y_b^{(i)} + Y_c^{(i)} Y_d^{(i)}) Y_b^{(i+1)}}{Y_d^{(i)} Y_d^{(i+1)}} + \dfrac{Y_a^{(i)}(-Y_a^{(i+1)} Y_b^{(i+1)} + Y_c^{(i+1)} Y_d^{(i+1)})}{Y_d^{(i)} Y_d^{(i+1)}}, \\ B_{22}^{(k)} = \dfrac{-Y_a^{(i)} Y_b^{(i)} + Y_c^{(i)} Y_d^{(i)}}{Y_d^{(i)} Y_d^{(i+1)}} + \dfrac{Y_a^{(i)} Y_a^{(i+1)}}{Y_d^{(i)} Y_d^{(i+1)}}, \end{cases} \tag{81}$$

In order to obtain the flexibility of the complex system, calculated coefficients of chain equation (79), was transformed to the basic form:

$$\begin{cases} Y_a^{(k)} = \dfrac{B_{22}{}^{(k)}}{B_{12}{}^{(k)}}, \\ Y_c^{(k)} = -\dfrac{B_{11}{}^{(k)} B_{22}{}^{(k)}}{B_{12}{}^{(k)}} - B_{21}{}^{(k)}, \\ Y_d^{(k)} = \dfrac{1}{B_{12}{}^{(k)}}, \\ Y_b^{(k)} = \dfrac{B_{11}{}^{(k)}}{B_{12}{}^{(k)}}. \end{cases} \tag{82}$$

Equation (83) is the components of the complex characteristics of the matrix taking into account obtained chain parameters of complex system.

7. CHARTS OF SIMPLE AND BIMORPH SYSTEM

In this paragraph, graphical charts of characteristics of piezoelectric plates were shown. The parameters adopted for graphs plotting was presented in table 4.

Table 4. The parameters adopted for graphs plotting

No.	Symbol	Value	Unit
1	ρ	7.5	$[\frac{g}{cm^3}]$
2	$E = c_{33}$	150	$[GPa]$
3	A	3.1	$[cm^2]$
4	d	1	$[mm]$

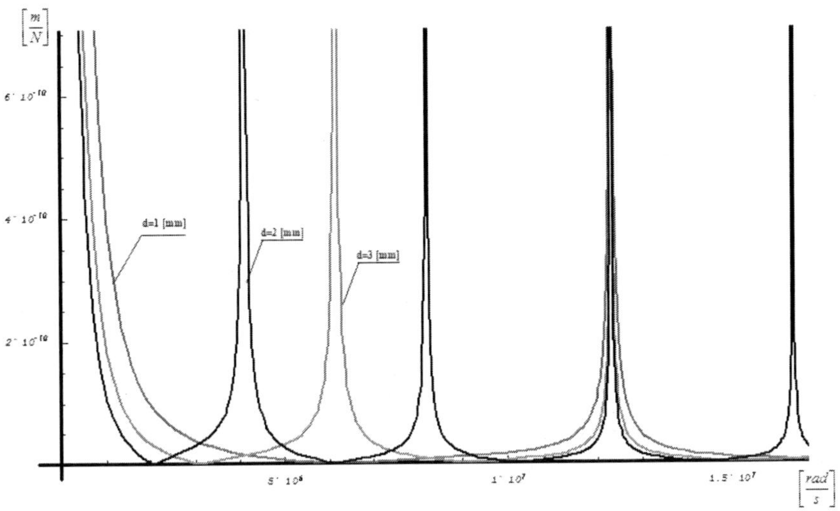

Figure 22. Characteristics of a single piezoelectric plate in the frequency domain, depending on the thickness of the plate

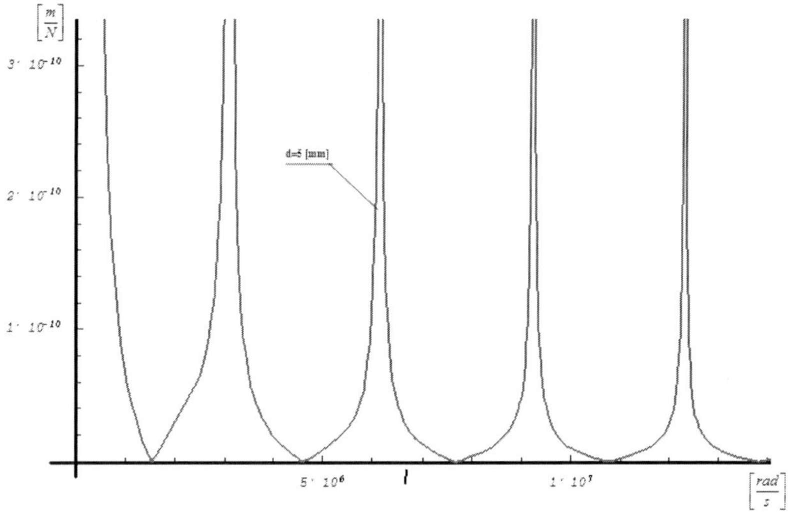

Figure 23. Characteristics of the combined plates of thickness 2and 3[mm] in frequency domain

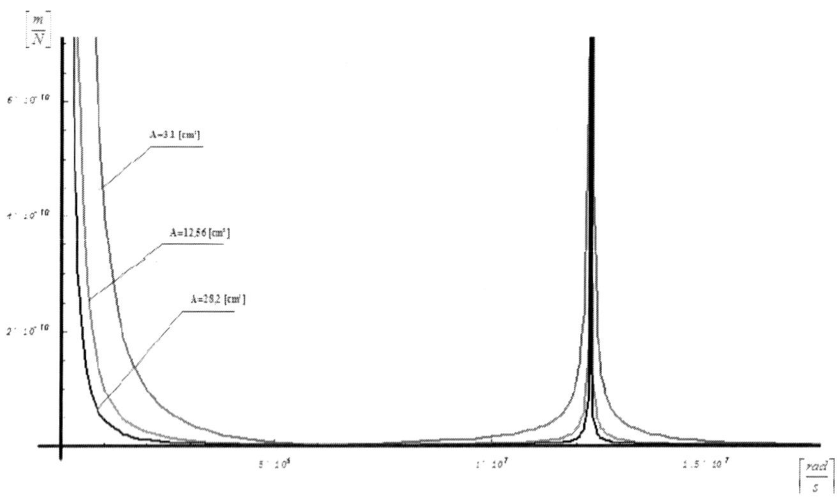

Figure 24. Characteristics of a single piezoelectric plate in the frequency domain, depending on the plate surface area

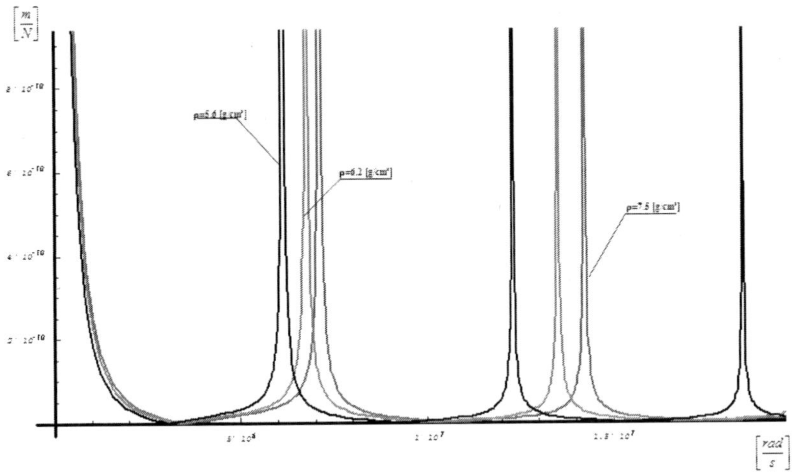

Figure 25. Characteristics of a single piezoelectric plate in the frequency domain, depending on the piezoelectric density

8. CONCLUSIONS

The chapter concerns the analysis of simple and complex piezoelectric systems, in order to determine the impact of piezoelectric plates parameters on the characteristics of the system. For a long time in the machine building are used subassemblies, whose operation is based on the piezoelectric phenomenon. In a researches of machine elements, on their surface piezoelectric sensors are glued, whereas to monitor the state plates are used transducers made from piezoelectric foil. Piezoelectric are often used in machine building also as assemblies, subassemblies or executive elements. Implementation of the piezoelectric system, which acts as a sensor or actuator is based on the selection of geometric dimensions of the plate, and their basic material parameters. In systems composed of several layers is also important piezoelectric plate number. Moreover, there are new problems at the design stage for the designers working in the field of machine building, concerning the application of both: single and stack plates. This matter is extremely important in terms of practical applications. For this reason it is necessary to conduct research whose main objective is to understand the phenomena associated with vibrations of complex piezoelectric systems.

Work is a continuation and development of decades researches at the Gliwice Center, consist in making the analysis of both the classical methods and non-classical. Take advantage of non-classical methods is a more general proposes from modeling in classical meaning. Resolves simple and complex systems irrespective of the type and number of elements included in the test system.

Applied method of structural numbers method was presented and used previously in modeling mechanical systems.

REFERENCES

1. K Arczewski, 1988Structural Methods of the Complex Mechanical Systems Analysis, WPW, Warsaw
2. S Behrens, A. J Fleming, S. O. R Moheimani, 2003*Abroadband controller for shunt piezoelectric damping of structural vibration*. Smart Materials and Structures, 121828
3. S Bellert, 1981*Chosen works,* PWN, 8-30100-247-6
4. K Bialas, 2012*Mechanical and electrical elements in reduction of vibrations,* Journal of vibroengineering, 1411231281392-8716
5. R. E. D Bishop, G. M. L Gladwell, S Michaelson, 1972*Matrix analysis of vibration.* WNT, Warsaw
6. S Bolkowski, 1986Theoretical electrical engineering. WNT, Warsaw
7. A Buchacz, 2004*Hypergrphs and their subgraphs in modelling and investigation of robots.* Journal of materials processing technology. 157-158Complete, Elsevier, 3744
8. A Buchacz, M Placzek, 2012*The analysis of a composite beam with piezoelectric actuator based on the approximate method,* Journal of vibroengineering, 1411111161392-8716
9. A Buchacz, J Swider, 2000Skeletons hypergraph in modeling, examination and position robot's manipulator and subassembly of machines. Silesian University of Technology Press, Gliwice.
10. S. K Ha, 2002Analysis of a piezoelectric multimorph in extensional and flexular motions. Journal of Sound and Vibration, 253, 3, 10011014

11. R Kacprzyk, E Motyl, J. B Gajewski, A Pasternak, 1995*Piezoelectric properties of nouniform electrets*, Journal of Electrostatics 35, 161166

12. W. P Mason, 1948Electromechanical Transducers and Wale Filters. Van Nostrand

13. A Sekala, J Swider, 2005*Hybrid Graphs in Modelling and Analysis of Discrete-Continuous Mechanical Systems.* Journal of Materials Processing Technology, 164-165Complete Elsevier 14361443

14. S Sherrit, S. P Leary, B. P Dolgin, 1999Comparison of the Mason and KLM equivalent circuits for piezoelectric resonators in thickness mode, Ultrasonics Symposium.

15. H Shin, H Ahn, D. Y Han, 2005*Modeling and analysis of multilayer piezoelectric transformer*, Materials chemistry and physics 92616620

16. W Soluch, 1980*The introduction to piezoelectronics*, WKiŁ, 8-32060-041-3

17. A Wróbel, 2012*Model of piezoelectric including material damping,* Proceedings of 16th International Conference ModTech 2012, 2069-673610611064

Index